Introduction to Environmental Data Science

Introduction to Environmental Data Science focuses on data science methods in the R language applied to environmental research, with sections on exploratory data analysis in R including data abstraction, transformation, and visualization; spatial data analysis in vector and raster models; statistics & modelling ranging from exploratory to modelling, considering confirmatory statistics and extending to machine learning models; time series analysis, focusing especially on carbon and micrometeorological flux; and communication. *Introduction to Environmental Data Science.* It is an ideal textbook to teach undergraduate to graduate level students in environmental science, environmental studies, geography, earth science, and biology, but can also serve as a reference for environmental professionals working in consulting, NGOs, and government agencies at the local, state, federal, and international levels.

Features

- Gives thorough consideration of the needs for environmental research in both spatial and temporal domains.
- Features examples of applications involving field-collected data ranging from individual observations to data logging.
- Includes examples also of applications involving government and NGO sources, ranging from satellite imagery to environmental data collected by regulators such as EPA.
- Contains class-tested exercises in all chapters other than case studies. Solutions manual available for instructors.
- All examples and exercises make use of a GitHub package for functions and especially data.

Introduction to Environmental Data Science

Jerry D. Davis

CRC Press is an imprint of the
Taylor & Francis Group, an **informa** business
A CHAPMAN & HALL BOOK

Designed cover image: By Anna Studwell and Jerry D. Davis

First edition published 2023
by CRC Press
6000 Broken Sound Parkway NW, Suite 300, Boca Raton, FL 33487-2742

and by CRC Press
4 Park Square, Milton Park, Abingdon, Oxon, OX14 4RN

CRC Press is an imprint of Taylor & Francis Group, LLC

ISBN: 978-1-032-32218-6 (hbk)
ISBN: 978-1-032-33034-1 (pbk)
ISBN: 978-1-003-31782-1 (ebk)

DOI: 10.1201/9781003317821

Typeset in LM Roman
by KnowledgeWorks Global Ltd.

Publisher's note: This book has been prepared from camera-ready copy provided by the authors.

"Dandelion fluff – Ephemeral stalk sheds seeds to the universe" by Anna Studwell

Contents

Author/editor biographies

Jerry Douglas Davis is a Professor of Geography & Environment (https://geog.sfsu.edu/) and the Director of the Institute for Geographic Information Science (https://gis.sfsu.edu/) at San Francisco State University, and borrows heavily from his and his students' field-based environmental research for examples in the book.

List of Figures

1

Background, Goals and Data

1.1 Environmental Data Science

Data science is *an interdisciplinary field that uses scientific methods, processes, algorithms and systems to extract knowledge and insights from noisy, structured and unstructured data* (Wikipedia). A data science approach is especially suitable for applications involving large and complex data sets, and environmental data is a prime example, with rapidly growing collections from automated sensors in space and time domains.

Environmental data science is data science applied to environmental science research. In general, data science can be seen as being the intersection of math and statistics, computer science/IT, and some research domain, and in this case it's environmental (Figure 1.1).

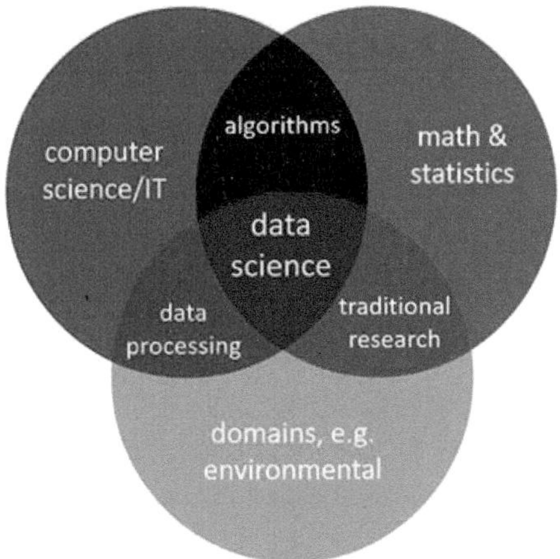

FIGURE 1.1 Environmental data science

1.2 Environmental Data and Methods

The methods needed for environmental research can include many things since environmental *data* can include many things, including environmental measurements in space and time domains.

- data analysis and transformation methods
  - importing and other methods to create data frames
  - reorganization and creation of fields
  - filtering observations
  - data joins
  - reorganizing data, including pivots
- visualization
  - graphics
  - maps
  - imagery
- spatial analysis
  - vector and raster spatial analysis
    * spatial joins
    * distance analysis
    * overlay analysis
    * terrain modeling
  - spatial statistics
  - image analysis
- statistical summaries, tests and models
  - statistical summaries and visualization
  - stratified/grouped summaries
  - confirmatory statistical tests
  - physical, statistical and machine learning models
  - classification models
- temporal data and time series
  - analyzing and visualizing long-term environmental data
  - analyzing and visualizing high-frequency data from loggers

1.3 Goals

While the methodological reach of data science is very great, and the spectrum of environmental data is as well, our goal is to lay the foundation and provide useful introductory methods in the areas outlined above, but as a "live" book be able to extend into more advanced methods and provide a growing suite of research examples with associated data sets. We'll briefly explore some data mining methods that can be applied to so-called "big data" challenges, but our focus is on **exploratory data analysis** in general, applied to environmental data in *space and time domains*. For clarity in understanding the methods and products, much of our data will be in fact be quite *small*, derived from field-based environmental measurements where we can best understand how the data were collected, but these methods extend to much larger data sets. It will primarily be in the areas of time-series and imagery, where automated data capture and machine learning are employed, when we'll dip our toes into big data.

1.3.1 Some definitions:

Machine Learning: *building a model using training data in order to make predictions without being explicitly programmed to do so.* Related to artificial intelligence methods. Used in:

- image and imagery classification, including computer vision methods
- statistical modeling
- data mining

Data Mining: *discovering patterns in large data sets*

- databases collected by government agencies
- imagery data from satellite, aerial (including drone) sensors
- time-series data from long-term data records or high-frequency data loggers
- methods may involve machine learning, artificial intelligence and computer vision

Big Data: *data having a size or complexity too big to be processed effectively by traditional software*

- data with many cases or dimensions (including imagery)
- many applications in environmental science due to the great expansion of automated environmental data capture in space and time domains
- big data challenges exist across the spectrum of the environmental research process, from data capture, storage, sharing, visualization, querying

Exploratory Data Analysis: *procedures for analyzing data, techniques for interpreting the results of such procedures, ways of structuring data to make its analysis easier*

- summarizing
- restructuring
- visualization

1.4 Exploratory Data Analysis

Just as *exploration* is a part of what *National Geographic* has long covered, it's an important part of geographic and environmental science research. **Exploratory data analysis** is exploration applied to data, and has grown as an alternative approach to traditional statistical analysis. This basic approach perhaps dates back to the work of Thomas Bayes in the eighteenth century, but Tukey (1962) may have best articulated the basic goals of this approach in defining the "data analysis" methods he was promoting: "Procedures for analyzing data, techniques for interpreting the results of such procedures, ways of planning the gathering of data to make its analysis easier, more precise or more accurate, and all the machinery and results of (mathematical) statistics which apply to analyzing data." Some years later Tukey (1977) followed up with *Exploratory Data Analysis*.

Exploratory data analysis (EDA) is an approach to analyzing data via summaries and graphics. The key word is *exploratory*, and while one might view this in contrast to *confirmatory* statistics, in fact they are highly complementary. The objectives of EDA include (a) suggesting hypotheses; (b) assessing assumptions on which inferences will be based; (c) selecting appropriate statistical tools; and (d) guiding further data collection. This philosophy led to the development of S at Bell Labs (led by John Chambers, 1976), then to R.

1.5 Software and Data

First, we're going to use the R language, designed for statistical computing and graphics. It's not the only way to do data analysis – Python is another important data science language – but R with its statistical foundation is an important language for academic research, especially in the environmental sciences.

```
## [1] "This book was produced in RStudio using R version 4.2.1 (2022-06-23 ucrt)"
```

For a start, you'll need to have R and RStudio installed, then you'll need to install various packages to support specific chapters and sections.

- In **Introduction to R** (Chapter 2), we will mostly use the base installation of R, with a few packages to provide data and enhanced table displays:
  - `igisci`
  - `palmerpenguins`
  - `DT`
  - `knitr`
- In **Abstraction** (Chapter 3) and **Transformation** (Chapter 5), we'll start making a lot of use of *tidyverse* 3.1 packages such as:
  - `ggplot2`
  - `dplyr`
  - `stringr`
  - `tidyr`
  - `lubridate`
- In **Visualization** (Chapter 4), we'll mostly use ggplot2, but also some specialized visualization packages such as:
  - `GGally`
- In **Spatial** (starting with Chapter 6), we'll add some spatial data, analysis and mapping packages:
  - `sf`
  - `terra`
  - `tmap`
  - `leaflet`
- In **Statistics and Modeling** (starting with Chapter 10), no additional packages are needed, as we can rely on base R's rich statistical methods and ggplot2's visualization.

- In **Time Series** (Chapter 13), we'll find a few other packages handy:
  - `xts` (Extensible Time Series)
  - `forecast` (for a few useful functions like a moving average)

And there will certainly be other packages we'll explore along the way, so you'll want to install them when you first need them, which will typically be when you first see a `library()` call in the code, or possibly when a function is prefaced with the package name, something like `dplyr::select()`, or maybe when R raises an error that it can't find a function you've called or that the package isn't installed. One of the earliest we'll need is the suite of packages in the "tidyverse" (Wickham and Grolemund (2016)), which includes some of the ones listed above: `ggplot2`, `dplyr`, `stringr`, and `tidyr`. You can install these individually, or all at once with:

```
`install.packages("tidyverse")`
```

This is usually done from the console in RStudio and not included in an R script or markdown document, since you don't want to be installing the package over and over again. You can also respond to a prompt from RStudio when it detects a package called in a script you open that you don't have installed.

From time to time, you'll want to update your installed packages, and that usually happens when something doesn't work and maybe the dependencies of one package on another gets broken with a change in a package. Fortunately, in the R world, especially at the main repository at CRAN, there's a lot of effort put into making sure packages work together, so usually there are no surprises if you're using the most current versions. *Note that there can be exceptions to this, and occasionally new package versions will create problems with other packages due to inter-package dependencies and the introduction of functions with names that duplicate other packages. The packages installed for this book were current as of that version of R, but new package versions may occasionally introduce errors.*

Once a package like `dplyr` is installed, you can access all of its functions and data by adding a library call, like ...

```
library(dplyr)
```

... which you *will* want to include in your code, or to provide access to multiple libraries in the tidyverse, you can use `library(tidyverse)`. Alternatively, if you're only using maybe one function out of an installed package, you can call that function with the `::` separator, like `dplyr::select()`. This method has another advantage in avoiding problems with duplicate names – and for instance we'll generally call `dplyr::select()` this way.

1.5.1 Data

We'll be using data from various sources, including data on CRAN like the code packages above which you install the same way – so use `install.packages("palmerpenguins")`.

We've also created a repository on GitHub that includes data we've developed in the Institute for Geographic Information Science (iGISc) at SFSU, and you'll need to install that package a slightly different way.

GitHub packages require a bit more work on the user's part since we need to first install `remotes`[1], then use that to install the GitHub data package:

```
install.packages("remotes")
remotes::install_github("iGISc/igisci")
```

Then you can access it just like other built-in data by including:

```
library(igisci)
```

To see what's in it, you'll see the various datasets listed in:

```
data(package="igisci")
```

For instance, Figure 1.2 is a map of California counties using the CA_counties `sf` feature data. We'll be looking at the `sf` (Simple Features) package later in the Spatial section of the book, but seeing `library(sf)`, this is one place where you'd need to have installed another package, with `install.packages("sf")`.

```
library(tidyverse); library(igisci); library(sf)
ggplot(data=CA_counties) + geom_sf()
```

The package datasets can be used directly as `sf` data or data frames. And similarly to functions, you can access the (previously installed) data set by prefacing with `igisci::` this way, without having to load the library. This might be useful in a one-off operation:

```
mean(igisci::sierraFeb$LATITUDE)
```

```
## [1] 38.3192
```

Raw data such as `.csv` files can also be read from the `extdata` folder that is installed on your computer when you install the package, using code such as:

```
csvPath <- system.file("extdata","TRI/TRI_1987_BaySites.csv", package="igisci")
TRI87 <- read_csv(csvPath)
```

[1]Note: you can also use `devtools` instead of `remotes` if you have that installed. They do the same thing; `remotes` is a subset of `devtools`. If you see a message about Rtools, you can ignore it since that is only needed for building tools from C++ and things like that.

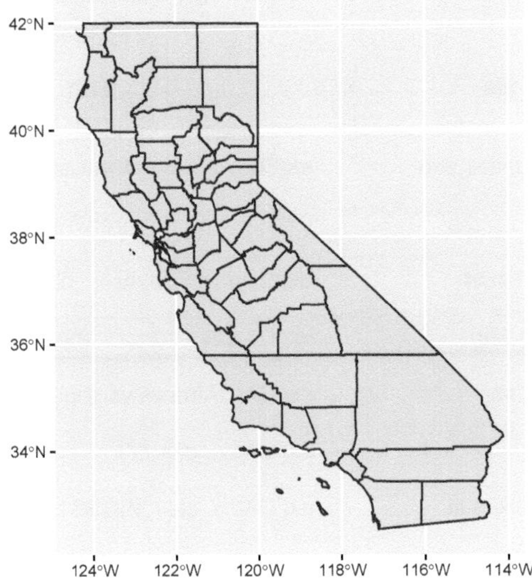

FIGURE 1.2 California counties simple features data in igisci package

or something similar for shapefiles, such as:

```
shpPath <- system.file("extdata","marbles/trails.shp", package="igisci")
trails <- st_read(shpPath)
```

And we'll find that including most of the above arcanity in a function will help. We'll look at functions later, but here's a function that we'll use a lot for setting up reading data from the extdata folder:

```
ex <- function(dta){system.file("extdata",dta,package="igisci")}
```

And this `ex()`function is needed so often that it's installed in the `igisci` package, so if you have `library(igisci)` in effect, you can just use it like this:

```
trails <- st_read(ex("marbles/trails.shp"))
```

But how do we see what's in the `extdata` folder? We can't use the `data()` function, so we would have to dig for the folder where the igisci package gets installed, which is buried pretty deeply in your user profile. So I wrote another function `exfiles()` that creates a data frame showing all of the files and the paths to use. In RStudio you could access it with `View(exfiles())` or we could use a datatable (you'll need to have installed "DT"). You can use the path using the `ex()` function with any function that needs it to read data, like `read.csv(ex('CA/CA_ClimateNormals.csv'))`, or just enter that `ex()` call in the console like `ex('CA/CA_ClimateNormals.csv')` to display where on your computer the installed data reside.

```
DT::datatable(exfiles(), options=list(scrollX=T), rownames=F)
```

Show 10 entries Search:

dir	file	path	type
BayArea	BayAreaCounties.shp	ex('BayArea/BayAreaCounties.shp')	shapefile
BayArea	BayAreaTracts.shp	ex('BayArea/BayAreaTracts.shp')	shapefile
BayArea	BayArea_hillsh.tif	ex('BayArea/BayArea_hillsh.tif')	TIFF
CA	CA_counties.shp	ex('CA/CA_counties.shp')	shapefile
CA	CAfreeways.shp	ex('CA/CAfreeways.shp')	shapefile
CA	ca_elev.tif	ex('CA/ca_elev.tif')	TIFF
CA	ca_elev_WGS84.tif	ex('CA/ca_elev_WGS84.tif')	TIFF
CA	ca_hillsh_WGS84.tif	ex('CA/ca_hillsh_WGS84.tif')	TIFF
CA	CA_ClimateNormals.csv	ex('CA/CA_ClimateNormals.csv')	CSV
CA	CA_MdInc.csv	ex('CA/CA_MdInc.csv')	CSV

Showing 1 to 10 of 94 entries Previous 1 2 3 4 5 … 10 Next

1.6 Acknowledgements

This book was immensely aided by extensive testing by students in San Francisco State's GEOG 604/704 *Environmental Data Science* class, including specific methodological contributions from some of the students and a contributed data wrangling exercise by one from the first offering (Josh von Nonn) in Chapter 5. Thanks to Andrew Oliphant, Chair of the Department of Geography and Environment, for supporting the class (as long as I included time series) and then came through with some great data sets from eddy covariance flux towers as well as guest lectures. Many thanks to Adam Davis, California Energy Commission, for suggestions on R spatial methods and package development, among other things in the R world. Thanks to Anna Studwell, recent Associate Director of the IGISc, for ideas on statistical modeling of birds and marine environments, and the nice water-color for the front cover. And a lot of thanks goes to Nancy Wilkinson, who put up with my obsessing on R coding puzzles at all hours and pretended to be impressed with what you can do with R Markdown.

Cover art "Dandelion fluff – Ephemeral stalk sheds seeds to the universe" by Anna Studwell.

Part I

Exploratory Data Analysis

2

Introduction to R

This section lays the foundation for exploratory data analysis using the R language and packages especially within the tidyverse. This foundation progresses through:

- Introduction : An introduction to the R language
- Abstraction : Exploration of data via reorganization using `dplyr` and other packages in the tidyverse (Chapter 3)
- Visualization : Adding visual tools to enhance our data exploration (Chapter 4)
- Transformation : Reorganizing our data with pivots and data joins (Chapter 5)

In this chapter we'll introduce the R language, using RStudio to explore its basic data types, structures, functions and programming methods in base R. We're assuming you're either new to R or need a refresher. Later chapters will add packages that extend what you can do with base R for data abstraction, transformation, and visualization, then explore the spatial world, statistical models, and time series applied to environmental research.

The following code illustrates a few of the methods we'll explore in this chapter:

```
temp <- c(10.7, 9.7, 7.7, 9.2, 7.3)
elev <- c(52, 394, 510, 564, 725)
lat <- c(39.52, 38.91, 37.97, 38.70, 39.09)
elevft <- round(elev / 0.3048)
deg <- as.integer(lat)
min <- as.integer((lat-deg) * 60)
sec <- round((lat-deg-min/60)*3600)
sierradata <- cbind(temp, elev, elevft, lat, deg, min, sec)
mydata <- as.data.frame(sierradata)
mydata
```

```
##   temp elev elevft   lat deg min sec
## 1 10.7   52    171 39.52  39  31  12
## 2  9.7  394   1293 38.91  38  54  36
## 3  7.7  510   1673 37.97  37  58  12
## 4  9.2  564   1850 38.70  38  42   0
## 5  7.3  725   2379 39.09  39   5  24
```

RStudio

If you're new to RStudio, or would like to learn more about using it, there are plenty of resources you can access to learn more about using it. As with many of the major packages we'll explore, there's even a cheat sheet: https://www.rstudio.com/resources/cheatsheets/.

Have a look at this cheat sheet while you have RStudio running, and use it to learn about some of its different components:

- The **Console**, where you'll enter short lines of code, install packages, and get help on functions. Messages created from running code will also be displayed here. There are other tabs in this area (e.g. Terminal, R Markdown) we may explore a bit, but mostly we'll use the console.
- The **Source Editor**, where you'll write full R scripts and R Markdown documents. You should get used to writing complete scripts and R Markdown documents as we go through the book.
- Various **Tab Panes** such as the **Environment** pane, where you can explore what scalars and more complex objects contain.
- The **Plots** pane in the lower right for static plots (graphs and maps that aren't interactive), which also lets you see a listing of **Files**, or **View** interactive maps and maps.

2.1 Data Objects

As with all programming languages, R works with *data* and since it's an object-oriented language, these are *data objects*. Data objects can range from the most basic type – the *scalar* which holds one value, like a number or text – to everything from an array of values to spatial data for mapping or a time series of data.

2.1.1 Scalars and assignment

We'll be looking at a variety of types of data objects, but scalars are the most basic type, holding individual values, so we'll start with it. Every computer language, like in math, stores values by assigning them constants or results of expressions. These are often called "variables," but we'll be using that name to refer to a column of data stored in a data frame, which we'll look at later in this chapter. R uses a lot of objects, and not all are data objects; we'll also create functions 2.8.1, a type of object that does something (runs the function code you've defined for it) with what you provide it.

To create a scalar (or other data objects), we'll use the most common type of statement, the *assignment statement*, that takes an *expression* and assigns it to a new data object that we'll name. The *class* of that data object is determined by the class of the expression provided, and that expression might be something as simple as a *constant* like a number or a character string of text. Here's an example of a very basic assignment statement that assigns the value of a constant `5` to a new scalar `x`:

```
x <- 5
```

Note that this uses the assignment operator `<-` that is standard for R. You can also use `=` as most languages do (and I sometimes do), but we'll use `=` for other types of assignments.

All object names must start with a letter, have no spaces, and must not use any names that are built into the R language or used in package libraries, such as reserved words like `for` or function names like `log`. Object names are case-sensitive (which you'll probably discover at some point by typing in something wrong and getting an error).

```
x <- 5
y <- 8
Longitude <- -122.4
Latitude <- 37.8
my_name <- "Inigo Montoya"
```

To check the value of a data object, you can just enter the name in the console, or even in a script or code chunk.

```
x
```

```
## [1] 5
```

```
y
```

```
## [1] 8
```

```
Longitude
```

```
## [1] -122.4
```

```
Latitude
```

```
## [1] 37.8
```

```
my_name
```

```
## [1] "Inigo Montoya"
```

This is counter to the way printing out values commonly works in other programming languages, and you will need to know how this method works as well because you will want to use your code to develop tools that accomplish things, and there are also limitations to what you can see by just naming objects.

To see the values of objects in programming mode, you can also use the `print()` function (but we rarely do); or to concatenate character string output, use `paste()` or `paste0`.

```
print(x)
paste0("My name is ", my_name, ". You killed my father. Prepare to die.")
```

Numbers concatenated with character strings are converted to characters.

```
paste0(paste("The Ultimate Answer to Life", "The Universe",
             "and Everything is ... ", sep=", "),42,"!")
```

```
paste("The location is latitude", Latitude, "longitude", Longitude)
```

```
## [1] "The location is latitude 37.8 longitude -122.4"
```

Review the code above and what it produces. Without looking it up, what's the difference between `paste()` and `paste0()`?

> We'll use `paste0()` a lot in this book to deal with long file paths which create problems for the printed/pdf version of this book, basically extending into the margins. Breaking the path into multiple strings and then combining them with `paste0()` is one way to handle them. For instance, in the Imagery and Classification Models chapter, the Sentinel2 imagery is provided in a very long file path. So here's how we use `paste0()` to recombine after breaking up the path, and we then take it one more step and build out the full path to the 20 m imagery subset.

```
imgFolder <- paste0("S2A_MSIL2A_20210628T184921_N0300_R113_T10TGK_20210628T230915.",
                    "SAFE/GRANULE/L2A_T10TGK_A031427_20210628T185628")
img20mFolder <- paste0("~/sentinel2/",imgFolder,"/IMG_DATA/R20m")
```

2.2 Functions

Just as in regular mathematics, R makes a lot of use of *functions* that accept an input and create an output:

```
log10(100)
log(exp(5))
cos(pi)
sin(90 * pi/180)
```

But functions can be much more than numerical ones, and R functions can return a lot of different data objects. You'll find that most of your work will involve functions, from those in base R to a wide variety in packages you'll be adding. You will likely have already used the `install.packages()` and `library()` functions that add in an array of other functions.

Later in this chapter, we'll also learn how to *write our own functions*, a capability that is easy to accomplish and also gives you a sense of what developing your own package might be like.

Arithmetic operators There are, of course, all the normal arithmetic operators (that are actually functions) like plus `+` and minus `-` or the key-stroke approximations of multiply `*` and divide `/` operators. You're probably familiar with these approximations from using equations in Excel if not in some other programming language you may have learned. These operators look a bit different from how they'd look when creating a nicely formatted equation.

For example, $\frac{NIR-R}{NIR+R}$ instead has to look like `(NIR-R)/(NIR+R)`.

Similarly `*` *must* be used to multiply; there's no implied multiplication that we expect in a math equation like $x(2+y)$, which would need to be written `x*(2+y)`.

In contrast to those four well-known operators, the symbol used to exponentiate – raise to a power – varies among programming languages. R uses either `**` or `^` so the the Pythagorean theorem $c^2 = a^2 + b^2$ might be written `c**2 = a**2 + b**2` or `c^2 = a^2 + b^2` except for the fact that it wouldn't make sense as a statement to R. Why?

And how would you write an R statement that assigns the variable `c` an expression derived from the Pythagorean theorem? (And don't use any new functions from a Google search – from deep math memory, how do you do $\sqrt{x}$ using an exponent?)

It's time to talk more about expressions and statements.

2.3 Expressions and Statements

The concepts of expressions and statements are very important to understand in any programming language.

An **expression** in R (or any programming language) has a *value* just like an object has a value. An expression will commonly combine data objects and functions to be *evaluated* to derive the value of the expression. Here are some examples of expressions:

```
5
x
x*2
sin(x)
(a^2 + b^2)^0.5
(-b+sqrt(b**2-4*a*c))/2*a
paste("My name is", aname)
```

Note that some of those expressions used previously assigned objects – `x, a, b, c, aname`.

An expression can be entered in the console to display its current value, and this is commonly done in R for objects of many types and complexity.

```
cos(pi)
```

```
## [1] -1
```

```
Nile
```

```
## Time Series:
## Start = 1871
## End = 1970
## Frequency = 1
##   [1] 1120 1160  963 1210 1160 1160  813 1230 1370 1140  995  935 1110  994 1020
##  [16]  960 1180  799  958 1140 1100 1210 1150 1250 1260 1220 1030 1100  774  840
##  [31]  874  694  940  833  701  916  692 1020 1050  969  831  726  456  824  702
##  [46] 1120 1100  832  764  821  768  845  864  862  698  845  744  796 1040  759
##  [61]  781  865  845  944  984  897  822 1010  771  676  649  846  812  742  801
##  [76] 1040  860  874  848  890  744  749  838 1050  918  986  797  923  975  815
##  [91] 1020  906  901 1170  912  746  919  718  714  740
```

Whoa, what was that? We entered the expression `Nile` and got a bunch of stuff! `Nile` is a type of data object called a time series that we'll be looking at much later, and since it's in the built-in data in base R, just entering its name will display it. And since time series are also *vectors* which are like entire columns, rows, or variables of data, we can *vectorize* it (apply mathematical operations and functions element-wise) in an expression:

```
Nile * 2
```

```
## Time Series:
## Start = 1871
## End = 1970
## Frequency = 1
##   [1] 2240 2320 1926 2420 2320 2320 1626 2460 2740 2280 1990 1870 2220 1988 2040
##  [16] 1920 2360 1598 1916 2280 2200 2420 2300 2500 2520 2440 2060 2200 1548 1680
##  [31] 1748 1388 1880 1666 1402 1832 1384 2040 2100 1938 1662 1452  912 1648 1404
##  [46] 2240 2200 1664 1528 1642 1536 1690 1728 1724 1396 1690 1488 1592 2080 1518
##  [61] 1562 1730 1690 1888 1968 1794 1644 2020 1542 1352 1298 1692 1624 1484 1602
##  [76] 2080 1720 1748 1696 1780 1488 1498 1676 2100 1836 1972 1594 1846 1950 1630
##  [91] 2040 1812 1802 2340 1824 1492 1838 1436 1428 1480
```

More on that later, but we'll start using vectors here and there. Back to expressions and statements:

A **statement** in R *does something*. It represents a directive we're assigning to the computer, or maybe the environment we're running on the computer (like RStudio, which then runs

R). A simple `print()` *statement* seems a lot like what we just did when we entered an expression in the console, but recognize that it *does something*:

```
print("Hello, World")
```

```
## [1] "Hello, World"
```

Which is the same as just typing `"Hello, World"`, but either way we write it, it *does something.*

Statements in R are usually put on one line, but you can use a semicolon to have multiple statements on one line, if desired:

```
x <- 5; print(x); print(x**2); x; x^0.5
```

```
## [1] 5
```

```
## [1] 25
```

```
## [1] 5
```

```
## [1] 2.236068
```

What's the print function for? It appears that you don't really need a print function, since you can just enter an object you want to print in a statement, so the `print()` is implied. And indeed we'll rarely use it, though there are some situations where it'll be needed, for instance in a structure like a loop. It also has a couple of parameters you can use like setting the number of significant digits:

```
print(x^0.5, digits=3)
```

```
## [1] 2.24
```

Many (perhaps most) statements don't actually display anything. For instance:

```
x <- 5
```

doesn't display anything, but it does assign the constant `5` to the object `x`, so it simply *does something.* It's an **assignment statement**, easily the most common type of statement that we'll use in R, and uses that special assignment operator `<-` . Most languages just use `=` which the designers of R didn't want to use, to avoid confusing it with the equal sign meaning "is equal to".

An assignment statement assigns an expression to a object. If that object already exists, it is reused with the new value. For instance it's completely legit (and commonly done in coding) to update the object in an assignment statement. This is very common when using a counter scalar:

```
i = i + 1
```

You're simply updating the index object with the next value. This also illustrates why it's *not* an equation: `i=i+1` doesn't work as an equation (unless `i` is actually ∞ but that's just really weird.)

And `c**2 = a**2 + b**2` doesn't make sense as an R statement because `c**2` isn't an object to be created. The `**` part is interpreted as *raise to a power*. What is to the left of the assignment operator `=` *must* be an object to be assigned the value of the expression.

2.4 Data Classes

Scalars, constants, vectors, and other data objects in R have data classes. Common types are numeric and character, but we'll also see some special types like Date.

```
x <- 5
class(x)
```

```
## [1] "numeric"
```

```
class(4.5)
```

```
## [1] "numeric"
```

```
class("Fred")
```

```
## [1] "character"
```

```
class(as.Date("2021-11-08"))
```

```
## [1] "Date"
```

2.4.1 Integers

By default, R creates double-precision floating-point numeric data objects. To create integer objects:

- append an L to a constant, e.g. `5L` is an integer 5
- convert with `as.integer`

We're going to be looking at various `as.` functions in R, more on that later, but we should look at `as.integer()` now. Most other languages use `int()` for this, and what it does is convert *any number* into an integer, *truncating* it to an integer, not rounding it.

```
as.integer(5)
```

```
## [1] 5
```

```
as.integer(4.5)
```

```
## [1] 4
```

To round a number, there's a `round()` function or you can instead use `as.integer` adding 0.5:

```
x <- 4.8
y <- 4.2
as.integer(x + 0.5)
```

```
## [1] 5
```

```
round(x)
```

```
## [1] 5
```

```
as.integer(y + 0.5)
```

```
## [1] 4
```

```
round(y)
```

```
## [1] 4
```

Integer division is really the first kind of division you learned about in elementary school, and is the kind of division that each step in long division employs, where you first get the highest integer you can get ...

```
5 %/% 2
```

```
## [1] 2
```

... but then there's a remainder from division, which we can call the modulus. To see the modulus we use `%%` instead of `%/%`:

```
5 %% 2
```

```
## [1] 1
```

That modulus is handy for *periodic* data (like angles of a circle, hours of the day, days of the year), where if we use the length of that period (like 360°) as the divisor, the remainder will always be the value's position in the repeated period. We'll use a vector created by the `seq` function, and then apply a modulus operation:

```
ang = seq(90,540,90)
ang
```

```
## [1]  90 180 270 360 450 540
```

```
ang %% 360
```

```
## [1]  90 180 270   0  90 180
```

Surprisingly, the values returned by integer division or the remainder are not stored as integers. R seems to prefer floating point...

2.5 Rectangular Data

A common data format used in most types of research is **rectangular** data such as in a spreadsheet, with rows and columns, where rows might be **observations** and columns might be **variables** (Figure 2.1). We'll read this type of data in from spreadsheets or even more commonly from comma-separated-variable (CSV) files, though some of these package data sets are already available directly as data frames.

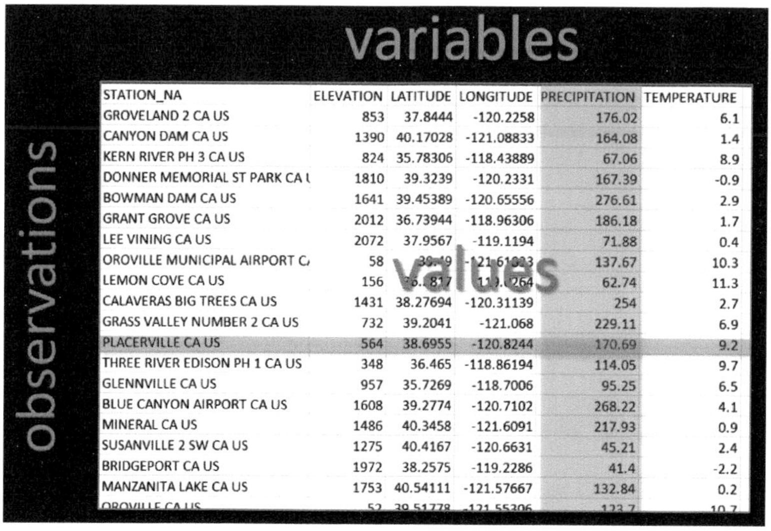

STATION_NA	ELEVATION	LATITUDE	LONGITUDE	PRECIPITATION	TEMPERATURE
GROVELAND 2 CA US	853	37.8444	-120.2258	176.02	6.1
CANYON DAM CA US	1390	40.17028	-121.08833	164.08	1.4
KERN RIVER PH 3 CA US	824	35.78306	-118.43889	67.06	8.9
DONNER MEMORIAL ST PARK CA U	1810	39.3239	-120.2331	167.39	-0.9
BOWMAN DAM CA US	1641	39.45389	-120.65556	276.61	2.9
GRANT GROVE CA US	2012	36.73944	-118.96306	186.18	1.7
LEE VINING CA US	2072	37.9567	-119.1194	71.88	0.4
OROVILLE MUNICIPAL AIRPORT C	58	[illegible]	[illegible]	137.67	10.3
LEMON COVE CA US	156	[illegible]	[illegible]	62.74	11.3
CALAVERAS BIG TREES CA US	1431	38.27694	-120.31139	254	2.7
GRASS VALLEY NUMBER 2 CA US	732	39.2041	-121.068	229.11	6.9
PLACERVILLE CA US	564	38.6955	-120.8244	170.69	9.2
THREE RIVER EDISON PH 1 CA US	348	36.465	-118.86194	114.05	9.7
GLENNVILLE CA US	957	35.7269	-118.7006	95.25	6.5
BLUE CANYON AIRPORT CA US	1608	39.2774	-120.7102	268.22	4.1
MINERAL CA US	1486	40.3458	-121.6091	217.93	0.9
SUSANVILLE 2 SW CA US	1275	40.4167	-120.6631	45.21	2.4
BRIDGEPORT CA US	1972	38.2575	-119.2286	41.4	-2.2
MANZANITA LAKE CA US	1753	40.54111	-121.57667	132.84	0.2
OROVILLE CA US	52	39.51778	-121.55306	123.7	10.7

FIGURE 2.1 Variables, observations, and values in rectangular data

```
library(igisci)
sierraFeb
```

```
## # A tibble: 82 x 7
##    STATION_NAME                     COUNTY ELEVA~1 LATIT~2 LONGI~3 PRECI~4 TEMPE~5
##    <chr>                            <chr>    <dbl>   <dbl>   <dbl>   <dbl>   <dbl>
##  1 GROVELAND 2, CA US               Tuolu~    853.    37.8   -120.   176.      6.1
##  2 CANYON DAM, CA US                Plumas   1390.    40.2   -121.   164.      1.4
##  3 KERN RIVER PH 3, CA US           Kern      824.    35.8   -118.    67.1     8.9
##  4 DONNER MEMORIAL ST PARK, CA US Nevada   1810.    39.3   -120.   167.     -0.9
##  5 BOWMAN DAM, CA US                Nevada   1641.    39.5   -121.   277.      2.9
##  6 BRUSH CREEK RANGER STATION, C~ Butte    1085.    39.7   -121.   296.     NA
##  7 GRANT GROVE, CA US               Tulare   2012.    36.7   -119.   186.      1.7
##  8 LEE VINING, CA US                Mono     2072.    38.0   -119.    71.9     0.4
##  9 OROVILLE MUNICIPAL AIRPORT, C~ Butte      57.9   39.5   -122.   138.     10.3
## 10 LEMON COVE, CA US                Tulare    156.    36.4   -119.    62.7    11.3
## # ... with 72 more rows, and abbreviated variable names 1: ELEVATION,
## #   2: LATITUDE, 3: LONGITUDE, 4: PRECIPITATION, 5: TEMPERATURE
```

2.6 Data Structures in R

We've already started using the most common data structures – scalars and vectors – but haven't really talked about vectors yet, so we'll start there.

2.6.1 Vectors

A vector is an ordered collection of numbers, strings, vectors, data frames, etc. What we mostly refer to as simply as vectors are formally called **atomic vectors**, which require that they be *homogeneous* sets of whatever type we're referring to, such as a vector of numbers, a vector of strings, or a vector of dates/times.

You can create a simple vector with the `c()` function:

```
lats <- c(37.5,47.4,29.4,33.4)
lats
```

```
## [1] 37.5 47.4 29.4 33.4
```

```
states <- c("VA", "WA", "TX", "AZ")
states
```

```
## [1] "VA" "WA" "TX" "AZ"
```

```
zips <- c(23173, 98801, 78006, 85001)
zips
```

```
## [1] 23173 98801 78006 85001
```

The class of a vector is the type of data it holds

```
temp <- c(10.7, 9.7, 7.7, 9.2, 7.3, 6.7)
class(temp)
```

```
## [1] "numeric"
```

Let's also introduce the handy `str()` function, which in one step gives you a view of the class of an item and its content – so its structure. We'll often use it in this book when we want to tell the reader what a data object contains, instead of listing a vector and its class separately, so instead of ...

```
temp
```

```
## [1] 10.7  9.7  7.7  9.2  7.3  6.7
```

```
class(temp)
```

```
## [1] "numeric"
```

... we'll just use `str()`:

```
str(temp)
```

```
##  num [1:6] 10.7 9.7 7.7 9.2 7.3 6.7
```

Vectors can only have one data class, and if mixed with character types, numeric elements will become character:

```
mixed <- c(1, "fred", 7)
str(mixed)
```

```
##  chr [1:3] "1" "fred" "7"
```

```
mixed[3]   # gets a subset, example of coercion
```

```
## [1] "7"
```

2.6.1.1 NA

Data science requires dealing with missing data by storing some sort of null value, called various things:

- null
- nodata
- `NA` "not available" or "not applicable"

```
as.numeric(c("1","Fred","5")) # note NA introduced by coercion
```

```
## [1]  1 NA  5
```

Note that `NA` doesn't really have a data class. The above example created a numeric vector with the one it couldn't figure out being assigned `NA`. *Remember that vectors (and matrices and arrays) have to be all the same data class.* A character vector can also include `NA`s. Both of the following are valid vectors, with the second item being `NA`:

```
c(5,NA,7)
c("alpha",NA,"delta")
```

```
## [1]  5 NA  7
## [1] "alpha" NA      "delta"
```

> Note that we typed `NA` without quotations. It's kind of like a special constant, like the `TRUE` and `FALSE` logical values, neither of which uses quotations.

We often want to ignore `NA` in statistical summaries. Where normally the summary statistic can only return `NA`...

```
mean(as.numeric(c("1", "Fred", "5")))
```

```
## [1] NA
```

... with `na.rm=T` you can still get the result for all actual data:

```
mean(as.numeric(c("1", "Fred", "5")), na.rm=T)
```

```
## [1] 3
```

Don't confuse with `nan` ("not a number"), which is used for things like imaginary numbers (explore the help for more on this), as you can see here:

```
is.nan(NA)
```

```
## [1] FALSE
```

```
is.na(as.numeric(''))
```

```
## [1] TRUE
```

```
is.nan(as.numeric(''))
```

```
## [1] FALSE
```

```
i <- sqrt(-1)
is.na(i) # interestingly nan is also na
```

```
## [1] TRUE
```

```
is.nan(i)
```

```
## [1] TRUE
```

2.6.1.2 Creating a vector from a sequence

We often need sequences of values, and there are a few ways of creating them. The following three examples are equivalent:

```
seq(1,10)
1:10
c(1,2,3,4,5,6,7,8,9,10)
```

The `seq()` function has special uses like using a step parameter:

```
seq(2,10,2)
```

```
## [1]  2  4  6  8 10
```

2.6.1.3 Vectorization and vector arithmetic

Arithmetic on vectors operates element-wise, a process called *vectorization*.

```
elev <- c(52,394,510,564,725,848,1042,1225,1486,1775,1899,2551)
elevft <- elev / 0.3048
elevft
```

```
##  [1]  170.6037 1292.6509 1673.2283 1850.3937 2378.6089 2782.1522 3418.6352
##  [8] 4019.0289 4875.3281 5823.4908 6230.3150 8369.4226
```

Another example, with two vectors:

```
temp03 <- c(13.1,11.4,9.4,10.9,8.9,8.4,6.7,7.6,2.8,1.6,1.2,-2.1)
temp02 <- c(10.7,9.7,7.7,9.2,7.3,6.7,4.0,5.0,0.9,-1.1,-0.8,-4.4)
tempdiff <- temp03 - temp02
tempdiff
```

```
##  [1] 2.4 1.7 1.7 1.7 1.6 1.7 2.7 2.6 1.9 2.7 2.0 2.3
```

2.6.1.4 Plotting vectors

Vectors of Feb temperature, elevation, and latitude at stations in the Sierra:

```
temp <- c(10.7,  9.7,   7.7,   9.2,   7.3,   6.7,   4.0,   5.0,   0.9, -1.1, -0.8,-4.4)
elev <-    c(52,  394,   510,   564,   725,   848,  1042,  1225,  1486,  1775,  1899,  2551)
lat <- c(39.52,38.91,37.97,38.70,39.09,39.25,39.94,37.75,40.35,39.33,39.17,38.21)
```

Plot individually by index vs a scatterplot

We'll use the `plot()` function to visualize what's in a vector. The `plot()` function will create an output based upon its best guess of what you're wanting to see, and will depend on the nature of the data you provide it. We'll be looking at a lot of ways to visualize data soon, but it's often useful to just see what `plot()` gives you. In this case, it just makes a bivariate plot where the x dimension is the sequential index of the vector from 1 through the length of the vector, and the values are in the y dimension. For comparison is a scatterplot with `elevation` on the x axis (Figure 2.2).

```
plot(temp)
plot(elev,temp)
```

2.6.1.5 Named indices

Vectors themselves have names (like `elev`, `temp`, and `lat` above), but individual indices can also be named.

```
fips <- c(16, 30, 56)
str(fips)
```

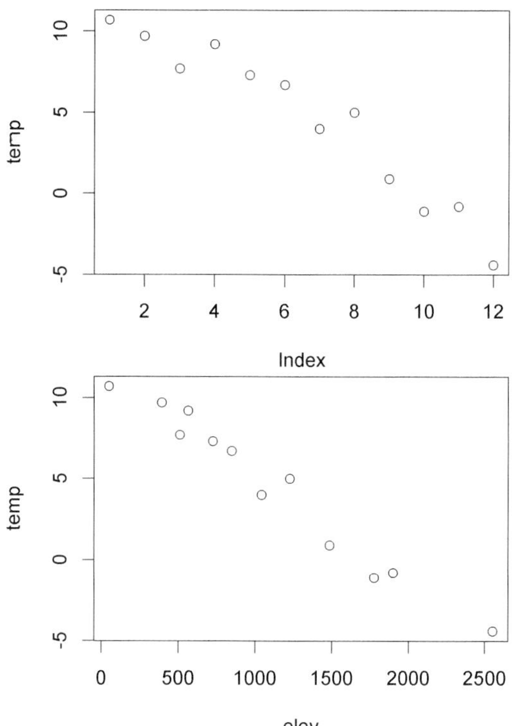

FIGURE 2.2 Temperature plotted by index (left) and elevation (right)

```
##  num [1:3] 16 30 56
```

```
fips <- c(idaho = 16, montana = 30, wyoming = 56)
str(fips)
```

```
##  Named num [1:3] 16 30 56
##  - attr(*, "names")= chr [1:3] "idaho" "montana" "wyoming"
```

The reason we might do this is so you can refer to observations by name instead of index, maybe to filter observations based on criteria where the name will be useful. The following are equivalent:

```
fips[2]
```

```
## montana
##      30
```

```
fips["montana"]
```

```
## montana
##      30
```

The `names()` function can be used to display a character vector of names, or assign names from a character vector:

```
names(fips) # returns a character vector of names
```

```
## [1] "idaho"   "montana" "wyoming"
```

```
names(fips) <- c("Idaho","Montana","Wyoming")
names(fips)
```

```
## [1] "Idaho"   "Montana" "Wyoming"
```

2.6.2 Lists

Lists can be heterogeneous, with multiple class types. Lists are actually used a lot in R, and are created by many operations, but they can be confusing to get used to especially when it's unclear what we'll be using them for. We'll avoid them for a while, and look into specific examples as we need them.

2.6.3 Matrices

Vectors are commonly used as a column in a matrix (or as we'll see, a data frame), like a variable

```
temp <- c(10.7,  9.7,  7.7,  9.2,  7.3,  6.7,  4.0,  5.0,  0.9, -1.1, -0.8,-4.4)
elev <-   c(52,  394,  510,  564,  725,  848, 1042, 1225, 1486, 1775, 1899, 2551)
lat <- c(39.52,38.91,37.97,38.70,39.09,39.25,39.94,37.75,40.35,39.33,39.17,38.21)
```

Building a matrix from vectors as columns

```
sierradata <- cbind(temp, elev, lat)
class(sierradata)
```

```
## [1] "matrix" "array"
```

```
str(sierradata)
```

```
##  num [1:12, 1:3] 10.7 9.7 7.7 9.2 7.3 6.7 4 5 0.9 -1.1 ...
##  - attr(*, "dimnames")=List of 2
##   ..$ : NULL
##   ..$ : chr [1:3] "temp" "elev" "lat"
```

```
sierradata
```

```
##        temp elev   lat
##  [1,] 10.7   52 39.52
##  [2,]  9.7  394 38.91
##  [3,]  7.7  510 37.97
##  [4,]  9.2  564 38.70
##  [5,]  7.3  725 39.09
##  [6,]  6.7  848 39.25
##  [7,]  4.0 1042 39.94
##  [8,]  5.0 1225 37.75
##  [9,]  0.9 1486 40.35
## [10,] -1.1 1775 39.33
## [11,] -0.8 1899 39.17
## [12,] -4.4 2551 38.21
```

2.6.3.1 Dimensions for arrays and matrices

Note: a matrix is just a 2D array. Arrays have 1, 3, or more dimensions.

```
dim(sierradata)
```

```
## [1] 12  3
```

It's also important to remember that a matrix or an array is a vector with dimensions, and we can change those dimensions in various ways as long as they work for the length of the vector.

```
a <- 1:12
dim(a) <- c(3, 4)    # matrix
class(a)
```

```
## [1] "matrix" "array"
```

```
dim(a) <- c(2,3,2)  # 3D array
class(a)
```

```
## [1] "array"
```

```
dim(a) <- 12         # 1D array
class(a)
```

```
## [1] "array"
```

```
b <- matrix(1:12, ncol=1)  # 1 column matrix is allowed
```

We just saw that we can change the dimensions of an existing matrix or array. But what if the matrix has names for its columns? I wasn't sure so following my basic philosophy of *empirical programming* I just tried it:

```
dim(sierradata) <- c(3,12)
sierradata
```

```
##      [,1] [,2] [,3] [,4] [,5] [,6] [,7] [,8]  [,9] [,10] [,11] [,12]
## [1,] 10.7  9.2  4.0 -1.1   52  564 1042 1775 39.52 38.70 39.94 39.33
## [2,]  9.7  7.3  5.0 -0.8  394  725 1225 1899 38.91 39.09 37.75 39.17
## [3,]  7.7  6.7  0.9 -4.4  510  848 1486 2551 37.97 39.25 40.35 38.21
```

So the answer is that it gets rid of the column names, and we can also see that redimensioning changes a lot more about how the data appears (though `dim(sierradata) <- c(12,3)` will return it to its original structure, but without column names). It's actually a little odd that matrices can have column names, because that really just makes them seem like data frames, so let's look at those next. Let's consider a situation where we want to create a rectangular data set from some data for a set of states:

```
abb <- c("CO","WY","UT")
area <- c(269837, 253600, 84899)
pop <- c(5758736, 578759, 3205958)
```

We can use `cbind` to create a matrix out of them, just like we did with the `sierradata` above

```
cbind(abb,area,pop)
```

```
##      abb  area     pop
## [1,] "CO" "269837" "5758736"
## [2,] "WY" "253600" "578759"
## [3,] "UT" "84899"  "3205958"
```

But notice what it did – area and pop were converted to character type. This reminds us that *matrices are still atomic vectors – all of the same class.* So to comply with this, the numbers were converted to character strings, since you can't convert character strings to numbers.

This isn't very satisfactory as a data object, so we'll need to use a data frame, which is *not* a vector, though its individual column variables are vectors.

2.6.4 Data frames

A data frame is a database with variables in columns and rows of observations. They're kind of like a spreadsheet with rules (like the first row is field names) or a matrix that can have variables of unique types. Data frames will be very important for data analysis and GIS.

Before we get started, we're going to use the `palmerpenguins` data set, so you need to install it if you haven't yet, and then load the library with:

```
library(palmerpenguins)
```

I'd encourage you to learn more about this dataset at https://allisonhorst.github.io/palmerpenguins/articles/intro.html(Horst, Hill, and Gorman (2020)) (Figures 2.3 and 2.4). It will be useful for a variety of demonstrations using numerical morphometric variables as well as a couple of categorical factors (species and island).

FIGURE 2.3 The three penguin species in palmerpenguins. Photos by KB Gorman. Used with permission

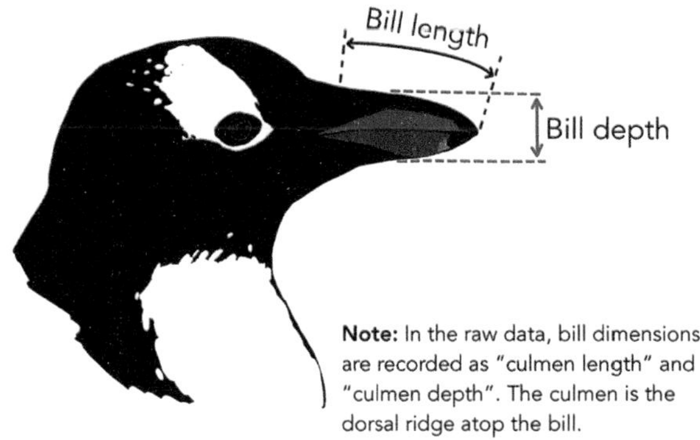

FIGURE 2.4 Diagram of penguin head with indication of bill length and bill depth (from Horst, Hill, and Gorman (2020), used with permission)

We'll use a couple of alternative table display methods, first a simple one...

```
penguins
```

```
## # A tibble: 344 x 8
##    species island    bill_length_mm bill_depth_mm flipper_~1 body_~2 sex    year
##    <fct>   <fct>              <dbl>         <dbl>      <int>   <int> <fct> <int>
##  1 Adelie  Torgersen           39.1          18.7        181    3750 male   2007
##  2 Adelie  Torgersen           39.5          17.4        186    3800 fema~  2007
##  3 Adelie  Torgersen           40.3          18          195    3250 fema~  2007
##  4 Adelie  Torgersen           NA            NA           NA      NA <NA>   2007
##  5 Adelie  Torgersen           36.7          19.3        193    3450 fema~  2007
##  6 Adelie  Torgersen           39.3          20.6        190    3650 male   2007
##  7 Adelie  Torgersen           38.9          17.8        181    3625 fema~  2007
##  8 Adelie  Torgersen           39.2          19.6        195    4675 male   2007
##  9 Adelie  Torgersen           34.1          18.1        193    3475 <NA>   2007
## 10 Adelie  Torgersen           42            20.2        190    4250 <NA>   2007
## # ... with 334 more rows, and abbreviated variable names 1: flipper_length_mm,
## #   2: body_mass_g
```

... then a nicer table display using `DT::datatable`, with a bit of improvement using an option.

```
DT::datatable(penguins, options=list(scrollX=T))
```

Show 10 entries Search:

	species	island	bill_length_mm	bill_depth_mm	flipper_length_mm	body_mass_g	sex	year
1	Adelie	Torgersen	39.1	18.7	181	3750	male	2007
2	Adelie	Torgersen	39.5	17.4	186	3800	female	2007
3	Adelie	Torgersen	40.3	18	195	3250	female	2007
4	Adelie	Torgersen						2007
5	Adelie	Torgersen	36.7	19.3	193	3450	female	2007
6	Adelie	Torgersen	39.3	20.6	190	3650	male	2007
7	Adelie	Torgersen	38.9	17.8	181	3625	female	2007
8	Adelie	Torgersen	39.2	19.6	195	4675	male	2007
9	Adelie	Torgersen	34.1	18.1	193	3475		2007
10	Adelie	Torgersen	42	20.2	190	4250		2007

Showing 1 to 10 of 344 entries Previous 1 2 3 4 5 … 35 Next

2.6.4.1 Creating a data frame out of a matrix

There are many functions that start with `as.` that convert things to a desired type. We'll use `as.data.frame()` to create a data frame out of a matrix, the same `sierradata` we created earlier, but we'll build it again so it'll have variable names, and use yet another table display method from the **`knitr`** package (which also has a lot of options you might want to explore), which works well for both the html and pdf versions of this book, and creates numbered table headings, so I'll use it a lot (Table 2.1).

TABLE 2.1 Temperatures (Feb), elevations, and latitudes of 12 Sierra stations

temp	elev	lat
10.7	52	39.52
9.7	394	38.91
7.7	510	37.97
9.2	564	38.70
7.3	725	39.09
6.7	848	39.25
4.0	1042	39.94
5.0	1225	37.75
0.9	1486	40.35
-1.1	1775	39.33
-0.8	1899	39.17
-4.4	2551	38.21

```
temp <- c(10.7,  9.7,  7.7,  9.2,  7.3,  6.7,  4.0,  5.0,  0.9, -1.1, -0.8,-4.4)
elev <-   c(52,  394,  510,  564,  725,  848, 1042, 1225, 1486, 1775, 1899, 2551)
lat <- c(39.52,38.91,37.97,38.70,39.09,39.25,39.94,37.75,40.35,39.33,39.17,38.21)
sierradata <- cbind(temp, elev, lat)
mydata <- as.data.frame(sierradata)
knitr::kable(mydata,
  caption = 'Temperatures (Feb), elevations, and latitudes of 12 Sierra stations')
```

Then to plot the two variables that are now part of the data frame, we'll need to make vectors out of them again using the `$` accessor (Figure 2.5).

```
plot(mydata$elev, mydata$temp)
```

2.6.4.2 Read a data frame from a CSV

We'll be looking at this more in the next chapter, but a common need is to read data from a spreadsheet stored in the CSV format. Normally, you'd have that stored with your project and can just specify the file name, but we'll access CSVs from the `igisci` package. Since you have this installed, it will already be on your computer, but not in your project folder. The path to it can be derived using the `system.file()` function.

Reading a csv in `readr` (part of the tidyverse that we'll be looking at in the next chapter) is done with `read_csv()`. We'll use the `DT::datatable` for this because it lets you interactively scroll across the many variables, and this table is included as Figure 2.6.

```
library(readr)
csvPath <- system.file("extdata","TRI/TRI_2017_CA.csv", package="igisci")
TRI2017 <- read_csv(csvPath)
```

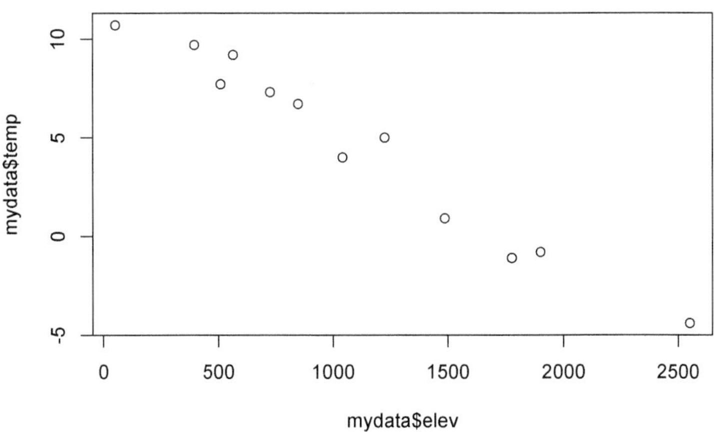

FIGURE 2.5 Temperature and elevation scatter plot

```
DT::datatable(TRI2017, options=list(scrollX=T))
```

Show 10 entries Search:

	YEAR	TRI_FACILITY_ID	FRS_ID	FACILITY_NAME	STREET_ADDRESS
1	2017	92101SLRTR2200P	110000478741	SOLAR TURBINES INC HARBOR DRIVE	2200 PACIFIC HWY MZ T-2
2	2017	93501NKLRC1667P	110046524249	TRICAL MOJAVE	1667 PURDY AVE
3	2017	92227HLLYS395WE	110000479090	SPRECKELS SUGAR CO INC	395 W KEYSTONE RD
4	2017	91730WSTRN8875I	110000477626	WESTERN METAL DECORATING CO	8875 INDUSTRIAL AVE
5	2017	92123SLRTR4200R	110000478974	SOLAR TURBINES INC KEARNY MESA	4200 RUFFIN RD MZ T-2
6	2017	91767SLPKN47EBN	110030909257	SILPAK INC	470 E. BONITA AVE.

Showing 1 to 10 of 3,685 entries Previous 1 2 3 4 5 … 369 Next

FIGURE 2.6 TRI dataframe – DT datatable output

Note that we could have used the built-in `read.csv` function, but as you'll see later, there are advantages of `readr::read_csv` so we should get in the habit of using that instead.

2.6.4.3 Reorganizing data frames

There are quite a few ways to reorganize your data in R, and we'll learn other methods in the next chapter where we start using the tidyverse, which makes data abstraction and transformation much easier. For now, we'll work with a simpler CSV I've prepared:

```
csvPath <- system.file("extdata","TRI/TRI_1987_BaySites.csv", package="igisci")
TRI87 <- read_csv(csvPath)
TRI87
```

```
## # A tibble: 335 x 9
##    TRI_FACILITY_ID count FACILI~1 COUNTY air_r~2 fugit~3 stack~4 LATIT~5 LONGI~6
##    <chr>           <dbl> <chr>    <chr>    <dbl>   <dbl>   <dbl>   <dbl>   <dbl>
##  1 91002FRMND585BR     2 FORUM I~ SAN M~    1423    1423       0    37.5   -122.
##  2 92052ZPMNF2970C     1 ZEP MFG~ SANTA~     337     337       0    37.4   -122.
##  3 93117TLDYN3165P     2 TELEDYN~ SANTA~   12600   12600       0    37.4   -122.
##  4 94002GTWSG477HA     2 MORGAN ~ SAN M~   18700   18700       0    37.5   -122.
##  5 94002SMPRD120SE     2 SEM PRO~ SAN M~    1500     500    1000    37.5   -122.
##  6 94025HBLNN151CO     2 HEUBLEI~ SAN M~     500       0     500    37.5   -122.
##  7 94025RYCHM300CO    10 TE CONN~ SAN M~  144871   47562   97309    37.5   -122.
##  8 94025SNFRD9900B     1 SANFORD~ SAN M~    9675    9675       0    37.5   -122.
##  9 94026BYPCK3575H     2 BAY PAC~ SAN M~   80000   32000   48000    37.5   -122.
## 10 94026CDRSY3475E     2 CDR SYS~ SAN M~  126800       0  126800    37.5   -122.
## # ... with 325 more rows, and abbreviated variable names 1: FACILITY_NAME,
## #   2: air_releases, 3: fugitive_air, 4: stack_air, 5: LATITUDE, 6: LONGITUDE
```

Sort, Index, and Max/Min

One simple task is to sort data (numerically or by alphabetic order), such as a variable extracted as a vector.

```
head(sort(TRI87$air_releases))
```

```
## [1]  2  5  5  7  9 10
```

... or create an index vector of the order of our vector/variable...

```
index <- order(TRI87$air_releases)
```

... where the index vector is just used to store the order of the `TRI87$air_releases` vector/variable; then we can use that index to display facilities in order of their air releases.

```
head(TRI87$FACILITY_NAME[index])
```

```
## [1] "AIR PRODUCTS MANUFACTURING CORP"
## [2] "UNITED FIBERS"
## [3] "CLOROX MANUFACTURING CO"
```

```
## [4] "ICI AMERICAS INC WESTERN RESEARCH CENTER"
## [5] "UNION CARBIDE CORP"
## [6] "SCOTTS-SIERRA HORTICULTURAL PRODS CO INC"
```

This is similar to filtering for a subset. We can also pull out individual values using functions like `which.max` to find the desired index value:

```
i_max <- which.max(TRI87$air_releases)
TRI87$FACILITY_NAME[i_max]    # was NUMMI at the time
```

```
## [1] "TESLA INC"
```

2.6.5 Factors

Factors are vectors with predefined values, normally used for categorical data, and as R is a statistical language are frequently used to stratify data, such as defining groups for analysis of variance among those groups. They are built on an *integer* vector, and *levels* are the set of predefined values, which are commonly character data.

```
nut <- factor(c("almond", "walnut", "pecan", "almond"))
str(nut)    # note that levels will be in alphabetical order
```

```
##  Factor w/ 3 levels "almond","pecan",..: 1 3 2 1
```

```
typeof(nut)
```

```
## [1] "integer"
```

As always, there are multiple ways of doing things. Here's an equivalent conversion that illustrates their relation to integers:

```
nutint <- c(1, 2, 3, 2) # equivalent conversion
nut <- factor(nutint, labels = c("almond", "pecan", "walnut"))
str(nut)
```

```
##  Factor w/ 3 levels "almond","pecan",..: 1 2 3 2
```

2.6.5.1 Categorical data and factors

While character data might be seen as categorical (e.g. `"urban"`, `"agricultural"`, `"forest"` land covers), to be used as categorical variables they must be made into factors. So we have something to work with, we'll generate some random memberships in one of three vegetation moisture categories using the `sample()` function:

```
veg_moisture_categories <- c("xeric", "mesic", "hydric")
veg_moisture_char <- sample(veg_moisture_categories, 42, replace = TRUE)
veg_moisture_fact <- factor(veg_moisture_char, levels = veg_moisture_categories)
veg_moisture_char
```

```
##  [1] "hydric" "xeric"  "xeric"  "mesic"  "xeric"  "xeric"  "mesic"  "mesic"
##  [9] "mesic"  "hydric" "mesic"  "mesic"  "mesic"  "mesic"  "hydric" "mesic"
## [17] "hydric" "mesic"  "hydric" "xeric"  "hydric" "mesic"  "xeric"  "mesic"
## [25] "mesic"  "xeric"  "xeric"  "xeric"  "mesic"  "hydric" "xeric"  "hydric"
## [33] "mesic"  "xeric"  "mesic"  "mesic"  "xeric"  "mesic"  "xeric"  "hydric"
## [41] "xeric"  "xeric"
```

```
veg_moisture_fact
```

```
##  [1] hydric xeric  xeric  mesic  xeric  xeric  mesic  mesic  mesic  hydric
## [11] mesic  mesic  mesic  mesic  hydric mesic  hydric mesic  hydric xeric
## [21] hydric mesic  xeric  mesic  mesic  xeric  xeric  xeric  mesic  hydric
## [31] xeric  hydric mesic  xeric  mesic  mesic  xeric  mesic  xeric  hydric
## [41] xeric  xeric
## Levels: xeric mesic hydric
```

To make a categorical variable a factor:

```
nut <- c("almond", "walnut", "pecan", "almond")
farm <- c("organic", "conventional", "organic", "organic")
ag <- as.data.frame(cbind(nut, farm))
ag$nut <- factor(ag$nut)
ag$nut
```

```
## [1] almond walnut pecan  almond
## Levels: almond pecan walnut
```

Factor example

```
library(igisci)
sierraFeb$COUNTY <- factor(sierraFeb$COUNTY)
str(sierraFeb$COUNTY)
```

```
##  Factor w/ 21 levels "Amador","Butte",..: 20 14 7 12 12 2 19 11 2 19 ...
```

2.7 Accessors and Subsetting

The use of *accessors* in R can be confusing, but they're very important to understand. An accessor is "a method for accessing data in an object usually an attribute of that object" (Brown (n.d.)), so a method for subsetting, and for R these are `[]`, `[[]]`, and `$`, but it can be confusing to know when you might use which one. There are good reasons to have these three types for code clarity, however you can also use `[]` with a bit of clumsiness for all purposes.

We've already been using these in this chapter and will continue to use them throughout the book. Let's look at the various accessors and subsetting options:

2.7.1 `[]` Subsetting

You use this to get a subset of any R object, whether it be a vector, list, or data frame. For a vector, the subset might be defined by indices ...

```
lats <- c(37.5,47.4,29.4,33.4)
str(lats[4])
```

```
##  num 33.4
```

```
str(lats[2:3])
```

```
##  num [1:2] 47.4 29.4
```

```
str(letters[24:26])
```

```
##  chr [1:3] "x" "y" "z"
```

... or a Boolean with `TRUE` or `FALSE` values representing which to keep or not.

```
str(lats[c(TRUE,FALSE,TRUE,TRUE)])
```

```
##  num [1:3] 37.5 29.4 33.4
```

An initially surprising but very handy way of creating one of these is to *reference the vector itself* in a relational expression. The following returns the same as the above, with the purpose of selecting all lats that are less than 40:

```
str(lats[lats<40])
```

```
##  num [1:3] 37.5 29.4 33.4
```

Getting one element from a data frame will return a data frame, in this case a data frame with just one variable. For columns especially, you might expect the result to be a vector, but it's a data frame with just one variable. Note that you can use either the column number or the name of the variable.

```
str(cars)
```

```
## 'data.frame':	50 obs. of  2 variables:
##  $ speed: num  4 4 7 7 8 9 10 10 10 11 ...
##  $ dist : num  2 10 4 22 16 10 18 26 34 17 ...
```

```
str(cars[1])
```

```
## 'data.frame':	50 obs. of  1 variable:
##  $ speed: num  4 4 7 7 8 9 10 10 10 11 ...
```

```
str(cars["speed"])
```

```
## 'data.frame':	50 obs. of  1 variable:
##  $ speed: num  4 4 7 7 8 9 10 10 10 11 ...
```

Similarly, subsetting to get a single *observation* (a row of values) with the `[n,]` method returns a very small data frame, or it can be used to create a subset of a range or selection of observations:

```
cars[1,]
```

```
##   speed dist
## 1     4    2
```

```
cars[3:5,]
```

```
##   speed dist
## 3     7    4
## 4     7   22
## 5     8   16
```

```
cars[cars["speed"]<10,]
```

```
##   speed dist
## 1     4    2
## 2     4   10
## 3     7    4
## 4     7   22
## 5     8   16
## 6     9   10
```

Getting a data frame this way is very useful because many functions you'll want to use require a data frame as input.

But sometimes you just want a vector. If you select an individual variable using the `[,n]` method, you'll get one. You'll want to assign it to a meaningful name like the original variable name. Note that you can either use the variable name or the variable column number:

```
dist <- cars[,"dist"] # or cars[,2]
dist
```

```
##  [1]   2  10   4  22  16  10  18  26  34  17  28  14  20  24  28  26  34  34  46
## [20]  26  36  60  80  20  26  54  32  40  32  40  50  42  56  76  84  36  46  68
## [39]  32  48  52  56  64  66  54  70  92  93 120  85
```

2.7.2 `[[]]` The mysterious double bracket

Double brackets extract just one element, so just one value from a vector or one vector from a data frame. You're going one step further into the structure.

```
str(cars[,"speed"])
```

```
##  num [1:50] 4 4 7 7 8 9 10 10 10 11 ...
```

```
str(cars[["speed"]])
```

```
##  num [1:50] 4 4 7 7 8 9 10 10 10 11 ...
```

Note that the str result is telling you that these are both simply vectors, not something like a data frame. Though uncommon, you can also use the index of variables instead of its name. Similar to the examples above, the variable can be specified either with the index (or column number) or the variable name. Since "speed" is the first variable in the cars data frame, these return the same thing:

```
str(cars[[1]])
```

```
##  num [1:50] 4 4 7 7 8 9 10 10 10 11 ...
```

```
str(cars[["speed"]])
```

```
##  num [1:50] 4 4 7 7 8 9 10 10 10 11 ...
```

2.7.3 $ Accessing a vector from a data frame

The $ accessor is really just a shortcut, but any shortcut reduces code and thus increases clarity, so it's a good idea and this accessor is commonly used. Their only limitation is that you can't use the integer indices, which would allow you to loop through a numerical sequence.

These accessor operations do the same thing:

```
cars$speed
cars[,"speed"]
cars[["speed"]]
```

```
##  num [1:50] 4 4 7 7 8 9 10 10 10 11 ...
```

Again, to contrast, the following accessor operations do the same thing – create a data frame:

```
sp1 <- cars[1] # gets the first variable, "speed"
sp2 <- cars["speed"]
str(sp1)
```

```
## 'data.frame':    50 obs. of  1 variable:
##  $ speed: num  4 4 7 7 8 9 10 10 10 11 ...
```

```
identical(sp1,sp2)
```

```
## [1] TRUE
```

2.8 Programming scripts in RStudio

Given the exploratory nature of the R language, we sometimes forget that it provides significant capabilities as a programming language where we can solve more complex problems by coding procedures and using logic to control the process and handle a range of possible scenarios.

Programming languages are used for a wide range of purposes, from developing operating systems built from low-level code to high-level *scripting* used to run existing functions in libraries. R and Python are commonly used for scripting, and you may be familiar with using arcpy to script ArcGIS geoprocessing tools. But whether low- or high-level, some common operational structures are used in all computer programming languages:

- functions (defining your own)
- conditional operations: *if* a condition is true, *do this*

- loops

Common to all of these is an input in parentheses followed by what to do in braces:

- *obj* `<- function(`*input*`){`*process ending in an expression result*`}`
- `if(`*condition*`){`*process*`}`
- `if(`*condition*`){`*process*`} else {`*process*`}`
- `for(`*objectinlist*`){`*process*`}`

The `{ }` braces are only needed if the process goes past the first line, so for instance the following examples don't need them (and actually some of the spaces aren't necessary either, but I left them in for clarity):

```
a <- 2
f <- function(a) a*2
print(f(4))
if(a==3) print('yes') else print('no')
for(i in 1:5) print(i)
```

2.8.1 `function` : creating your own

`myFunction <- function(input){`*Do this and return the resulting expression*`}`

The various packages that we're installing, all those that aren't purely data (like `igisci`), are built primarily of functions and perhaps most of those functions are written in R. Many of these simply make existing R functions work better or at least differently, often for a particular data science task commonly needed in a discipline or application area.

In geospatial environmental research for instance, we are often dealing with direction, for instance the movement of marine mammals who might be influenced by ship traffic. An agent-based model simulation of marine mammal movement might have the animal respond by shifting to the left or right, so we might want a `turnLeft()` or `turnRight()` function. Given the nature of circular data however, the code might be sufficiently complex to warrant writing a function that will make our main code easier to read:

```
turnright <- function(ang){(ang + 90) %% 360}
```

Then in our code later on, after we've prepared some data ...

```
id <- c("5A", "12D", "3B")
direction <- c(260, 270, 300)
whale <- dplyr::bind_cols(id = id,direction = direction) # better than cbind
whale
```

```
## # A tibble: 3 x 2
##   id    direction
##   <chr>     <dbl>
```

```
## 1 5A          260
## 2 12D         270
## 3 3B          300
```

... we can call this function:

```
whale$direction <- turnright(whale$direction)
whale
```

```
## # A tibble: 3 x 2
##   id    direction
##   <chr>     <dbl>
## 1 5A          350
## 2 12D           0
## 3 3B           30
```

Another function I found useful for our external data in `igisci` is to simplify the code needed to access the external data. I found I had to keep looking up the syntax for that task that we use a lot. It also makes the code difficult to read. Adding this function to the top of your code helps for both:

```
ex <- function(fnam){system.file("extdata",fnam,package="igisci")}
```

Then our code that accesses data is greatly simplified, with `read.csv` calls looking a lot like reading data stored in our project folder. For example, where if we had `fishdata.csv` stored locally in our project folder we might read it with ...

```
read.csv("fishdata.csv")
```

... reading from the data package's extdata folder looks pretty similar:

```
read.csv(ex("fishdata.csv"))
```

> This simple function was so useful that it's now included in the igisci package, so you can just call it with `ex()` if you have `library(igisci)` in your code.

2.8.2 `if` : conditional operations

`if (condition)` {*Do this*} `else` {*Do this other thing*}

Probably not used as much in R for most users, as it's mostly useful for building a new tool (or function) that needs to run as a complete operation. (But I found it essential in writing this book to create some outputs differently for the output to html than for the output to LaTeX for the pdf/book version.)

Common in coding is the need to be able to handle a variety of inputs and *avoid errors*. Here's an admittedly trivial example of some short code that avoids an error:

```
getRatio <- function(x,y){
  if (y!=0){x/y} else {1e1000}
}

getRatio(5,0)
```

```
## [1] Inf
```

```
getRatio(2,5)
```

```
## [1] 0.4
```

Note that the `else {` needs to follow the close of the `if (...) {...}` section on the same line.

2.8.3 `for` loops

`for(`*counter*`in`*list*`){*Do something*}`‘

Loops are very common in computer languages (in FORTRAN they were called `DO` loops) as they allow us to iterate through a list. The list could be just a series of numbers, and if starting from 1 might be called a *counter*, but could also be a list of objects like data frames or spatial features. The following example of a counter is trivial but illustrates a simple loop that prints a series of results, starting with the counter `i` itself:

```
for(i in 1:10) print(paste(i, 1/i))
```

```
## [1] "1 1"
## [1] "2 0.5"
## [1] "3 0.333333333333333"
## [1] "4 0.25"
## [1] "5 0.2"
## [1] "6 0.166666666666667"
## [1] "7 0.142857142857143"
## [1] "8 0.125"
```

```
## [1] "9 0.111111111111111"
## [1] "10 0.1"
```

Note the use of `print` above – this is one example where you have to use it.

2.8.3.1 A `for` loop with an internal `if..else`

Here's a more complex loop that builds river data for a map and profile that we'll look at again in the visualization chapter. It also includes a conditional operation with `if..else`, embedded within the for loop (note the { } enclosures for each structure.)

```
x <- c(1000, 1100, 1300, 1500, 1600, 1800, 1900)
y <- c(500, 780, 820, 950, 1250, 1320, 1600)
elev <- c(0, 1, 2, 5, 25, 75, 150)
```

First we'll create a crude map from the xy coordinates using the base plot system (Figure 2.7):

```
plot(x,y,asp=1,add=TRUE)
lines(x,y)
```

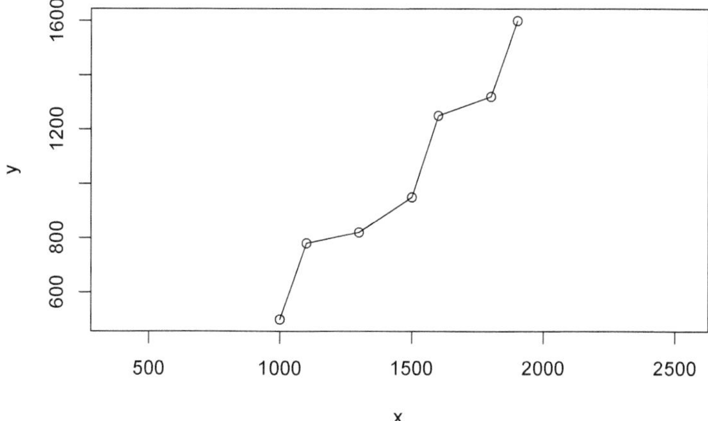

FIGURE 2.7 Crude river map using x y coordinates

Options for the `plot` function in base R: In the visualization chapter, we'll be making use of many options in `ggplot2`, but for now we're using the `plot` function of base R, which has quite a few options that we'll introduce from time to time as needed. We'll continue to use `plot` either for a quick output of our results, or where various packages like `terra` have extended its functionality for its data types. In the above code, we used two options in plot: `asp=1` for setting the x and y scales to be the same (like on a map) with an aspect ratio of 1; and `add=TRUE` to allow the lines to be added to the same plot, using the same coordinate space.

Now we'll use a loop to create a longitudinal graph by determining a cumulative longitudinal distance along the path of the river. You could look at it as straightening out the curves.

First we'll need empty vectors to populate with the longitudinal profile data:

```
d <- double()      # creates an empty numeric vector
longd <- double()  # ("double" means double-precision floating point)
s <- double()
```

Then the `for` loop that goes through all of the points, but uses an `if` statement to assign an initial distance and longitudinal distance of zero, since the longitudinal profile data is just started and we don't have a distance yet.

```
for(i in 1:length(x)){
  if(i==1){longd[i] <- 0; d[i] <- 0} else {
    d[i] <- sqrt((x[i]-x[i-1])^2 + (y[i]-y[i-1])^2)
    longd[i] <- longd[i-1] + d[i]
    s[i-1] <- (elev[i]-elev[i-1])/d[i]
    }
  }
```

What well-known theorem is used to derive each distance? Can you find $c^2 = a^2 + b^2$ or $c = \sqrt{a^2 + b^2}$ in the above code? Consider how Cartesian (x and y) coordinates work and where you might find a right triangle and a hypotenuse in such a coordinate system. Hint: what if $a = \Delta x$ and $b = \Delta y$? Draw it out on a piece of paper to help.

There is no known slope for the last point (since we have no next point), so the last slope is assigned NA.

```
s[length(x)] <- NA
```

Then we'll create a data frame out of the vectors we just built, and then display the longitudinal profile (Figure 2.8).

```
riverData <- as.data.frame(cbind(x=x,y=y,elev=elev,d=d,longd=longd,s=s))
riverData
```

```
##      x    y elev        d     longd           s
## 1 1000  500    0   0.0000    0.0000 0.003363364
## 2 1100  780    1 297.3214  297.3214 0.004902903
## 3 1300  820    2 203.9608  501.2822 0.012576654
## 4 1500  950    5 238.5372  739.8194 0.063245553
## 5 1600 1250   25 316.2278 1056.0471 0.235964589
## 6 1800 1320   75 211.8962 1267.9433 0.252252298
## 7 1900 1600  150 297.3214 1565.2647          NA
```

```
plot(riverData$longd, riverData$elev, add=T)
lines(riverData$longd, riverData$elev)
```

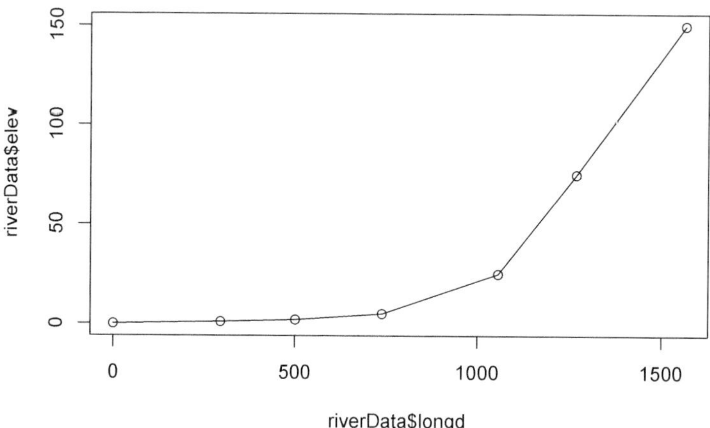

FIGURE 2.8 Longitudinal profile built from cumulative distances and elevation

2.8.4 Subsetting with logic

We'll use the 2022 USDOE fuel efficiency data to list all of the car lines with at least 50 miles to the gallon. Note the creation of a Boolean logical expression with an admittedly long variable `... >= 50`; this will return `TRUE` or `FALSE`, so `i` will be a logical (Boolean) vector which can be used to subset a couple of variables which we'll then paste together (using vectorization of course) to create a character string vector.

```
library(readxl)
fuelEff22 <- read_excel(ex("USDOE_FuelEfficiency2022.xlsx"))
i <- fuelEff22$`Hwy FE (Guide) - Conventional Fuel` >= 50
str(i)
```

```
##  logi [1:394] FALSE FALSE FALSE FALSE FALSE FALSE ...
```

```
efficientCars <- paste(fuelEff22$Division[i],fuelEff22$Carline[i])
str(efficientCars)
```

```
##  chr [1:9] "TOYOTA COROLLA HYBRID" "HYUNDAI MOTOR COMPANY Elantra Hybrid" ...
```

```
efficientCars
```

```
## [1] "TOYOTA COROLLA HYBRID"
## [2] "HYUNDAI MOTOR COMPANY Elantra Hybrid"
## [3] "HYUNDAI MOTOR COMPANY Elantra Hybrid Blue"
## [4] "TOYOTA PRIUS"
## [5] "TOYOTA PRIUS Eco"
## [6] "HYUNDAI MOTOR COMPANY Ioniq"
## [7] "HYUNDAI MOTOR COMPANY Ioniq Blue"
## [8] "HYUNDAI MOTOR COMPANY Sonata Hybrid"
## [9] "HYUNDAI MOTOR COMPANY Sonata Hybrid Blue"
```

The **which** function allows us to do something similar, returning the indices of the variable or vector for which a condition `TRI87$air_releases > 1e6` is `TRUE`. Then we can use that index to do the subset.

```
library(readr)
csvPath = system.file("extdata","TRI/TRI_1987_BaySites.csv", package="igisci")
TRI87 <- read_csv(csvPath)
i <- which(TRI87$air_releases > 1e6)
str(i)
```

```
##  int [1:4] 85 96 127 328
```

```
TRI87$FACILITY_NAME[i]
```

```
## [1] "VALERO REFINING CO-CALI FORNIA BENICIA REFINERY"
```

```
## [2] "TESLA INC"
## [3] "TESORO REFINING & MARKETING CO LLC"
## [4] "HGST INC"
```

The **%in%** operator is very useful when we have very specific items to select.

```
library(readr)
csvPath = system.file("extdata","TRI/TRI_1987_BaySites.csv", package="igisci")
TRI87 <- read_csv(csvPath)
i <- TRI87$COUNTY %in% c("NAPA","SONOMA")
TRI87$FACILITY_NAME[i]
```

```
## [1] "SAWYER OF NAPA"
## [2] "BERINGER VINEYARDS"
## [3] "CAL-WOOD DOOR INC"
## [4] "SOLA OPTICAL USA INC"
## [5] "KEYSIGHT TECHNOLOGIES INC"
## [6] "SANTA ROSA STAINLESS STEEL"
## [7] "OPTICAL COATING LABORATORY INC"
## [8] "MGM BRAKES"
## [9] "SEBASTIANI VINEYARDS INC, SONOMA CASK CELLARS"
```

2.8.5 Apply functions

There are many apply functions in R, and they often obviate the need for looping. For instance:

- `apply` derives values at margins of rows and columns, e.g. to sum across rows or down columns.

```
# matrix apply – the same would apply to data frames
matrix12 <- 1:12
dim(matrix12) <- c(3,4)
rowsums <- apply(matrix12, 1, sum)
colsums <- apply(matrix12, 2, sum)
sum(rowsums)
```

```
## [1] 78
```

```
sum(colsums)
```

```
## [1] 78
```

```
zero <- sum(rowsums) - sum(colsums)
matrix12
```

```
##      [,1] [,2] [,3] [,4]
```

```
## [1,]    1    4    7   10
## [2,]    2    5    8   11
## [3,]    3    6    9   12
```

Apply functions satisfy one of the needs that spreadsheets are used for. Consider how often you use sum, mean, or similar functions in Excel.

sapply

sapply applies functions to either:

- all elements of a vector – unary functions only

```
sapply(1:12, sqrt)
```

```
##  [1] 1.000000 1.414214 1.732051 2.000000 2.236068 2.449490 2.645751 2.828427
##  [9] 3.000000 3.162278 3.316625 3.464102
```

- or all variables of a data frame (not a matrix), where it works much like a column-based apply (since variables are columns) but more easily interpreted without the need of specifying columns with 2:

```
sapply(cars,mean)  # same as apply(cars,2,mean)
```

```
## speed  dist
## 15.40 42.98
```

```
temp02 <- c(10.7,9.7,7.7,9.2,7.3,6.7,4.0,5.0,0.9,-1.1,-0.8,-4.4)
temp03 <- c(13.1,11.4,9.4,10.9,8.9,8.4,6.7,7.6,2.8,1.6,1.2,-2.1)
sapply(as.data.frame(cbind(temp02,temp03)),mean) # has to be a data frame
```

```
##   temp02   temp03
## 4.575000 6.658333
```

While various `apply` functions are in base R, the purrr package takes these further. See https://www.rstudio.com/resources/cheatsheets/ for more information on this and other packages in the RStudio/tidyverse world.

2.9 RStudio projects

So far, you've been using RStudio, and it organizes your code and data into a project folder. You should familiarize yourself with where things are being saved and where you can find things. Start by seeing your *working directory* with :

```
getwd()
```

When you create a new RStudio project with `File/New Project...`, it will set the working directory to the *project folder*, where you create the project. (You can change the working directory with `setwd()` but I don't recommend it.) The project folder is useful for keeping things organized and allowing you to use relative paths to your data and allow everything to be moved somewhere else and still work. The project file has the extension `.Rproj` and it will reside in the project folder. If you've saved any scripts (`.R`)or R Markdown (`.Rmd`) documents, they'll also reside in the project folder; and if you've saved any data, or if you want to read any data without providing the full path or using the `extdata` access method, those data files (e.g. `.csv`) will be in that project folder. You can see your scripts, R Markdown documents, and data files using the Files tab in the default lower right pane of RStudio.

RStudio projects are going to be the way we'll want to work for the rest of this book, so you'll often want to create new ones for particular data sets so things don't get messy. And you may want to create data folders within your project folder, as we have in the `igisci` extdata folder, to keep things organized. Since we're using our `igisci` package, this is less of an issue since at least input data aren't stored in the project folder. However you're going to be creating data, so you'll want to manage your data in individual projects. You may want to start a new project for each data set, using File/New Project, and try to keep things organized (things can get messy fast!)

In this book, we'll be making a lot of use of data provided for you from various data packages such as built-in data, `palmerpenguins` (Horst, Hill, and Gorman 2020), or `igisci`, but they correspond to specific research projects, such as Sierra Climate to which several data frames and spatial data apply. For this chapter, you can probably just use one project, but later you'll find it useful to create separate projects for each data set – *such as a* **`sierra`** *project* and return to it every time it applies.

In that project, you'll build a series of scripts, many of which you'll re-use to develop new methods. When you're working on your own project with your own data files, you should store these in a **`data`** folder inside the project folder. With the project folder as the default working directory, you can use *relative paths*, and everything will work even if the project folder is moved. So, for instance, you can specify **"`data/mydata.csv`"** *as the path* to a csv of that name. You can still access package data, including extdata folders and files, but your processed and saved or imported data will reside with your project.

An *absolute path* to somewhere on your computer in contrast won't work for anyone else trying to run your code; absolute paths should only be used for servers that other users have access to and URLs on the web.

2.9.1 R Markdown

An alternative to writing scripts is writing *R Markdown* documents, which includes both formatted text (such as you're seeing in this book, like *italics* created using asterisks) and code chunks. R lends itself to running code in chunks, as opposed to creating complete tools that run all of the way through. This book is built from R Markdown documents organized in a bookdown structure, and most of the figures are created from R code chunks. There are also many good resources on writing R Markdown documents, including the very thorough *R Markdown: The Definitive Guide* (Xie, Allaire, and Grolemund 2019).

2.10 Exercises: Introduction to R

Exercise 2.1. Assign scalars for your name, city, state and zip code, and use `paste()` to combine them, and assign them to the object `me`. What is the class of `me`?

Exercise 2.2. Knowing that trigonometric functions require angles (including azimuth directions) to be provided in radians, and that degrees can be converted into radians by dividing by 180 and multiplying that by pi, derive the sine of 30 degrees with an R expression. (Base R knows what pi is, so you can just use `pi`.)

Exercise 2.3. If two sides of a right triangle on a map can be represented as dX and dY and the direct line path between them c, and the coordinates of two points on a map might be given as $(x1, y1)$ and $(x2, y2)$, with $dX = x2 - x1$ and $dY = y2 - y1$, use the Pythagorean theorem to derive the distance between them and assign that expression to c. Start by assigning the input scalars.

Exercise 2.4. You can create a vector uniform random numbers from 0 to 1 using `runif(n=30)` where n=30 says to make 30 of them. Use the `round()` function to round *each* of the values (it vectorizes them), and provide what you created and explain what happened.

Exercise 2.5. Create two vectors `x` and `y` of 10 numbers each with the c() function, then assigning to x and y. Then plot(x,y), and provide the three lines of code you used to do the assignment and plot.

Exercise 2.6. Change your code from the previous question so that one value is NA (entered simply as `NA`, no quotation marks), and derive the mean value for x. Then add `,na.rm=T` to the parameters for `mean()`. Also do this for y. Describe your results and explain what happens.

Exercise 2.7. Create two sequences, `a` and `b`, with `a` all odd numbers from 1 to 99, `b` all even numbers from 2 to 100. Then derive c through vector division of `b/a`. Plot a and c together as a scatterplot.

Exercise 2.8. Referring to the **Matrices** section, create the same `sierradata` matrix using the same data vectors repeated here ...

... then convert it to a data frame (using the same `sierradata` object name), and *from that data frame* plot temperature (`temp`) against latitude (`lat`).

Exercise 2.9. From that `sierradata` data frame, derive colmeans using the `mean` parameter on the columns `2` for `apply()`.

Exercise 2.10. Do the same thing with the sierra data frame with `sapply()`.

3

Data Abstraction

Abstracting data from large data sets (or even small ones) is critical to data science. The most common first step to visualization is abstracting the data in a form that allows for the visualization goal in mind. If you've ever worked with data in spreadsheets, you commonly will be faced with some kind of data manipulation to create meaningful graphs, unless that spreadsheet is specifically designed for it, but then doing something else with the data is going to require some work.

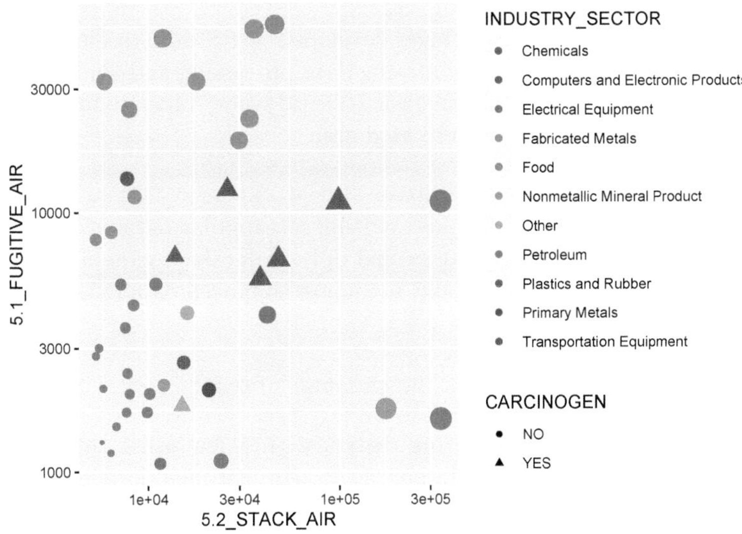

FIGURE 3.1 Visualization of some abstracted data from the EPA Toxic Release Inventory

Figure 3.1 started with abstracting some data from EPA's Toxic Release Inventory (TRI) program, which holds data reported from a large number of facilities that must report either "stack" or "fugitive" air. Some of the abstraction had already happened when I used the EPA website to download data for particular years and only in California. But there's more we need to do, and we'll want to use some `dplyr` functions to help with it.

At this point, we've learned the basics of working with the R language. From here we'll want to explore how to analyze data, statistically, spatially, and temporally. One part of this is abstracting information from existing data sets by selecting variables and observations and summarizing their statistics.

In the previous chapter, we learned some abstraction methods in base R, such as selecting parts of data frames and applying some functions across the data frame. There's a lot we can do with these methods, and we'll continue to use them, but they can employ some fairly arcane language. There are many packages that extend R's functionality, but some of the most important for data science can be found in the various packages of "The Tidyverse" (Wickham and Grolemund 2016), which has the philosophy of making data manipulation more intuitive.

We'll start with **`dplyr`**, which includes an array of data manipulation tools, including **`select`** for selecting variables, **`filter`** for subsetting observations, **`summarize`** for reducing variables to summary statistics, typically stratified by groups, and **`mutate`** for creating new variables from mathematical expressions from existing variables. Some `dplyr` tools such as data joins we'll look at later in the data transformation chapter.

3.1 The Tidyverse

The tidyverse refers to a suite of R packages developed at RStudio (see https://rstudio.com and <https://r4ds.had.co.nz>[1]) for facilitating data processing and analysis. While R itself is designed around exploratory data analysis, the tidyverse takes it further. Some of the packages in the tidyverse that are widely used are:

- **`dplyr`** : data manipulation like a database
- **`readr`** : better methods for reading and writing rectangular data
- **`tidyr`** : reorganization methods that extend dplyr's database capabilities
- **`purrr`** : expanded programming toolkit including enhanced "apply" methods
- **`tibble`** : improved data frame
- **`stringr`** : string manipulation library
- **`ggplot2`** : graphing system based on *the grammar of graphics*

In this chapter, we'll be mostly exploring **dplyr**, with a few other things thrown in like reading data frames with **readr**. For simplicity, we can just include `library(tidyverse)` to get everything.

3.2 Tibbles

Tibbles are an improved type of data frame

- part of the tidyverse
- serve the same purpose as a data frame, and all data frame operations work

Advantages

- display better

[1] https://r4ds.had.co.nz

- can be composed of more complex objects like lists, etc.
- can be grouped

There multiple ways to create a tibble:

- Reading from a CSV using `read_csv()`. *Note the underscore, a function naming convention in the tidyverse.*
- Using `tibble()` to either build from vectors or from scratch, or convert from a different type of data frame.
- Using `tribble()` to build in code from scratch.
- Using various tidyverse functions that return tibbles.

3.2.1 Building a tibble from vectors

We'll start by looking at a couple of built-in character vectors (there are lots of things like this in R):

- `letters` : lowercase letters
- `LETTERS` : uppercase letters

```
letters
```

```
##  [1] "a" "b" "c" "d" "e" "f" "g" "h" "i" "j" "k" "l" "m" "n" "o" "p" "q" "r" "s"
## [20] "t" "u" "v" "w" "x" "y" "z"
```

```
LETTERS
```

```
##  [1] "A" "B" "C" "D" "E" "F" "G" "H" "I" "J" "K" "L" "M" "N" "O" "P" "Q" "R" "S"
## [20] "T" "U" "V" "W" "X" "Y" "Z"
```

... then make a tibble of `letters`, `LETTERS`, and two random sets of 26 values, one normally distributed, the other uniform:

```
norm <- rnorm(26)
unif <- runif(26)
library(tidyverse)
tibble26 <- tibble(letters,LETTERS,norm,unif)
tibble26
```

```
## # A tibble: 26 x 4
##    letters LETTERS    norm   unif
##    <chr>   <chr>     <dbl>  <dbl>
##  1 a       A        0.368  0.0900
##  2 b       B        0.972  0.739
##  3 c       C       -0.0237 0.382
##  4 d       D        0.154  0.0366
##  5 e       E        1.21   0.402
##  6 f       F        0.479  0.151
```

TABLE 3.1 Peaks tibble

peak	elev	longitude	latitude
Mt. Whitney	4421	-118.2	36.5
Mt. Elbert	4401	-106.4	39.1
Mt. Hood	3428	-121.7	45.4
Mt. Rainier	4392	-121.8	46.9

```
##   7 g       G       -0.261  0.848
##   8 h       H        1.66   0.762
##   9 i       I        0.788  0.826
##  10 j       J       -0.0975 0.536
## # ... with 16 more rows
```

See section 10.2.3 for more on creating random (or rather *pseudo-random*) numbers in R.

3.2.2 tribble

As long as you don't let them multiply in your starship, tribbles are handy for creating tibbles. (Or rather the `tribble` function is a handy way to create tibbles in code.) You simply create the variable names with a series of entries starting with a tilde, then the data are entered one row at a time. If you line them all up in your code one row at a time, it's easy to enter the data accurately (Table 3.1).

```
peaks <- tribble(
  ~peak, ~elev, ~longitude, ~latitude,
  "Mt. Whitney", 4421, -118.2, 36.5,
  "Mt. Elbert", 4401, -106.4, 39.1,
  "Mt. Hood", 3428, -121.7, 45.4,
  "Mt. Rainier", 4392, -121.8, 46.9)
knitr::kable(peaks, caption = 'Peaks tibble')
```

3.2.3 read_csv

The `read_csv` function does somewhat the same thing as `read.csv` in base R, but creates a tibble instead of a data.frame, and has some other properties we'll look at below.

Note that the code below accesses data we'll be using a lot, from EPA Toxic Release Inventory (TRI) data. If you want to keep this data organized in a separate project, you might consider creating a new **air_quality** project. This is optional, and you can get by with staying in one project since all of our data will be accessed from the `igisci` package. But in your own work, you will find it useful to create separate projects to keep things organized with your code and data together.

```
library(tidyverse); library(igisci)
TRI87 <- read_csv(ex("TRI/TRI_1987_BaySites.csv"))
TRI87df <- read.csv(ex("TRI/TRI_1987_BaySites.csv"))
TRI87b <- tibble(TRI87df)
identical(TRI87, TRI87b)
```

```
## [1] FALSE
```

Note that they're not identical. *So what's the difference between `read_csv` and `read.csv`?* Why would we use one over the other? Since their names are so similar, you may accidentally choose one or the other. Some things to consider:

- To use read_csv, you need to load the `readr` or `tidyverse` library, or use `readr::read_csv`.
- The `read.csv` function "fixes" some things and sometimes that might be desired: problematic variable names like `MLY-TAVG-NORMAL` become `MLY.TAVG.NORMAL` – but this may create problems if those original names are a standard designation.
- With `read.csv`, numbers stored as characters are converted to numbers: "01" becomes 1, "02" becomes 2, etc.
- There are other known problems that read_csv avoids.

Recommendation: Use `read_csv` and `write_csv`.

You can still just call tibbles "data frames", since they are still data frames, and in this book we'll follow that practice.

FIGURE 3.2 Euc-Oak paired plot runoff and erosion study (Thompson, Davis, and Oliphant (2016))

3.3 Summarizing variable distributions

A simple statistical summary is very easy to do in R, and we'll use **eucoak** data in the `igisci` package from a study of comparative runoff and erosion under eucalyptus and oak canopies (Thompson, Davis, and Oliphant 2016). In this study (Figure 3.2), we looked at the amount of runoff and erosion captured in Gerlach troughs on paired eucalyptus and oak sites in the San Francisco Bay Area. *You might consider creating a `eucoak` project, since we'll be referencing this data set in several places.*

```
library(igisci)
summary(eucoakrainfallrunoffTDR)
```

```
##      site              site #          date              month
##  Length:90          Min.   :1.000   Length:90          Length:90
##  Class :character   1st Qu.:2.000   Class :character   Class :character
##  Mode  :character   Median :4.000   Mode  :character   Mode  :character
##                     Mean   :4.422
##                      3rd Qu.:6.000
##                     Max.   :8.000
##
##     rain_mm          rain_oak         rain_euc       runoffL_oak
##  Min.   : 1.00   Min.   : 1.00   Min.   : 1.00   Min.   : 0.000
##  1st Qu.:16.00   1st Qu.:16.00   1st Qu.:14.75   1st Qu.: 0.000
```

```
##  Median :28.50   Median :30.50   Median :30.00   Median : 0.450
##  Mean   :37.99   Mean   :35.08   Mean   :34.60   Mean   : 2.032
##  3rd Qu.:63.25   3rd Qu.:50.50   3rd Qu.:50.00   3rd Qu.: 2.800
##  Max.   :99.00   Max.   :98.00   Max.   :96.00   Max.   :14.000
##  NA's   :18      NA's   :2       NA's   :2       NA's   :5
##   runoffL_euc       slope_oak       slope_euc       aspect_oak
##  Min.   : 0.00   Min.   : 9.00   Min.   : 9.00   Min.   :100.0
##  1st Qu.: 0.07   1st Qu.:12.00   1st Qu.:12.00   1st Qu.:143.0
##  Median : 1.20   Median :24.50   Median :23.00   Median :189.0
##  Mean   : 2.45   Mean   :21.62   Mean   :19.34   Mean   :181.9
##  3rd Qu.: 3.30   3rd Qu.:30.50   3rd Qu.:25.00   3rd Qu.:220.0
##  Max.   :16.00   Max.   :32.00   Max.   :31.00   Max.   :264.0
##  NA's   :3
##    aspect_euc    surface_tension_oak surface_tension_euc
##  Min.   :106.0   Min.   :37.40       Min.   :28.51
##  1st Qu.:175.0   1st Qu.:72.75       1st Qu.:32.79
##  Median :196.5   Median :72.75       Median :37.40
##  Mean   :191.2   Mean   :68.35       Mean   :43.11
##  3rd Qu.:224.0   3rd Qu.:72.75       3rd Qu.:56.41
##  Max.   :296.0   Max.   :72.75       Max.   :72.75
##                  NA's   :22          NA's   :22
##  runoff_rainfall_ratio_oak runoff_rainfall_ratio_euc
##  Min.   :0.00000           Min.   :0.000000
##  1st Qu.:0.00000           1st Qu.:0.003027
##  Median :0.02046           Median :0.047619
##  Mean   :0.05357           Mean   :0.065902
##  3rd Qu.:0.08485           3rd Qu.:0.083603
##  Max.   :0.42000           Max.   :0.335652
##  NA's   :5                 NA's   :3
```

In the summary output, how are character variables handled differently from numeric ones?

Remembering what we discussed in the previous chapter, consider the `site` variable (Figure 3.3), and in particular its Length. Looking at the table, what does that length represent?

There are a couple of ways of seeing what unique values exist in a character variable like `site` which can be considered a categorical variable (factor). Consider what these return:

```
unique(eucoakrainfallrunoffTDR$site)
```

```
## [1] "AB1" "AB2" "KM1" "PR1" "TP1" "TP2" "TP3" "TP4"
```

```
factor(eucoakrainfallrunoffTDR$site)
```

```
##  [1] AB1 AB1 AB1 AB1 AB1 AB1 AB1 AB1 AB1 AB1 AB1 AB1 AB2 AB2 AB2 AB2 AB2 AB2 AB2
## [20] AB2 AB2 AB2 AB2 AB2 KM1 KM1 KM1 KM1 KM1 KM1 KM1 KM1 KM1 KM1 KM1 KM1 PR1 PR1
## [39] PR1 PR1 PR1 PR1 PR1 PR1 PR1 PR1 TP1 TP1 TP1 TP1 TP1 TP1 TP1 TP1 TP1 TP1 TP1
## [58] TP2 TP2 TP2 TP2 TP2 TP2 TP2 TP2 TP2 TP2 TP2 TP3 TP3 TP3 TP3 TP3 TP3 TP3 TP3
## [77] TP3 TP3 TP3 TP4 TP4 TP4 TP4 TP4 TP4 TP4 TP4 TP4 TP4 TP4
## Levels: AB1 AB2 KM1 PR1 TP1 TP2 TP3 TP4
```

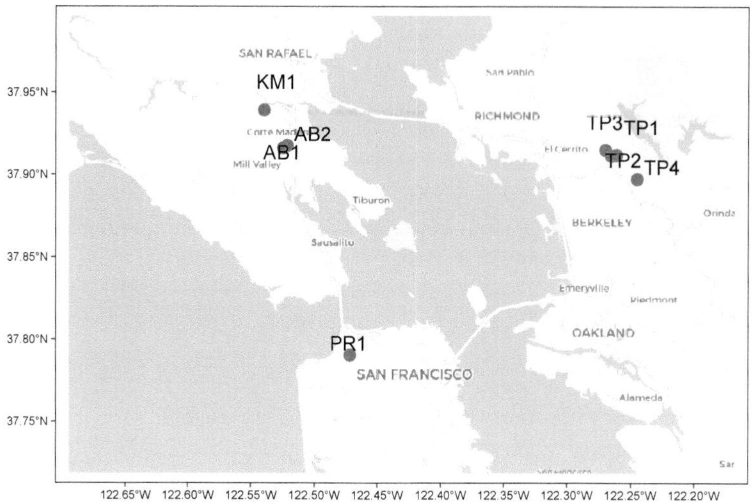

FIGURE 3.3 Eucalyptus/Oak paired site locations

3.3.1 Stratifying variables by site using a Tukey box plot

A good way to look at variable distributions stratified by a sample site factor is the Tukey box plot (Figure 3.4). We'll be looking more at this and other visualization methods in the next chapter.

```
ggplot(data = eucoakrainfallrunoffTDR) +
  geom_boxplot(mapping = aes(x=site, y=runoffL_euc))
```

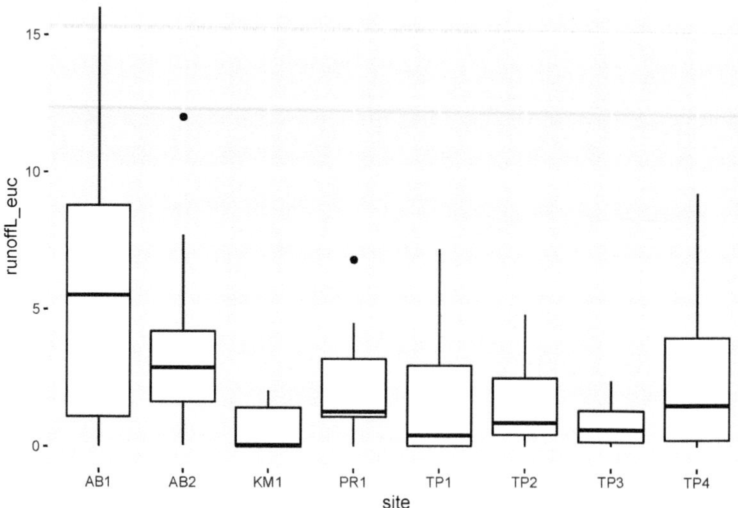

FIGURE 3.4 Tukey boxplot of runoff under eucalyptus canopy

TABLE 3.2 EucOak data reorganized a bit, first 6

site	Date	rain_mm	rain_subcanopy	runoffL_oak	runoffL_euc	slope_oak	slope_euc
AB1	2006-11-08	29	29.0	4.7900	6.70000	32	31
AB1	2006-11-12	22	18.5	3.2000	4.30000	32	31
AB1	2006-11-29	85	65.0	9.7000	16.00000	32	31
AB1	2006-12-12	82	87.5	14.0000	14.20000	32	31
AB1	2006-12-28	43	54.0	9.7472	4.32532	32	31
AB1	2007-01-29	7	54.0	1.4000	0.00000	32	31

3.4 Database operations with `dplyr`

As part of exploring our data, we'll typically simplify or reduce it for our purposes. The following methods are quickly discovered to be essential as part of exploring and analyzing data.

- **select rows** using logic, such as `population \> 10000`, with `filter`
- **select variable columns** you want to retain with `select`
- **add** new variables and assign their values with `mutate`
- **sort** rows based on a field with `arrange`
- **summarize** by group

3.4.1 Select, mutate, and the pipe

Read the pipe operator **`%>%`** as "and then..." This is bigger than it sounds and opens up many possibilities.

See the example below, and observe how the expression becomes several lines long. In the process, we'll see examples of new variables with mutate and selecting (and in the process *ordering*) variables (Table 3.2).

```
runoff <- eucoakrainfallrunoffTDR %>%
  mutate(Date = as.Date(date,"%m/%d/%Y"),
         rain_subcanopy = (rain_oak + rain_euc)/2) %>%
  dplyr::select(site, Date, rain_mm, rain_subcanopy,
         runoffL_oak, runoffL_euc, slope_oak, slope_euc)
```

Another way of thinking of the pipe that is very useful is that whatever goes before it becomes the first parameter for any functions that follow. So in the example above:

1. The parameter `eucoakrainfallrunoffTDR` becomes the first for `mutate()`, then
2. The result of the `mutate()` becomes the first parameter for `dplyr::select()`

To just rename a variable, use `rename` instead of `mutate`. It will stay in position.

3.4.1.1 Review: creating penguins from penguins_raw

To review some of these methods, it's useful to consider how the penguins data frame was created from the more complex penguins_raw data frame, both of which are part of the palmerpenguins package (Horst, Hill, and Gorman (2020)). First let's look at palmerpenguins::penguins_raw:

```
library(palmerpenguins)
library(tidyverse)
library(lubridate)
summary(penguins_raw)
```

```
##   studyName          Sample Number       Species             Region
##  Length:344         Min.   :  1.00   Length:344         Length:344
##  Class :character   1st Qu.: 29.00   Class :character   Class :character
##  Mode  :character   Median : 58.00   Mode  :character   Mode  :character
##                     Mean   : 63.15
##                     3rd Qu.: 95.25
##                     Max.   :152.00
##
##     Island             Stage           Individual ID      Clutch Completion
##  Length:344         Length:344         Length:344         Length:344
##  Class :character   Class :character   Class :character   Class :character
##  Mode  :character   Mode  :character   Mode  :character   Mode  :character
##
##
##
##
##     Date Egg           Culmen Length (mm) Culmen Depth (mm) Flipper Length (mm)
##  Min.   :2007-11-09   Min.   :32.10       Min.   :13.10      Min.   :172.0
##  1st Qu.:2007-11-28   1st Qu.:39.23       1st Qu.:15.60      1st Qu.:190.0
##  Median :2008-11-09   Median :44.45       Median :17.30      Median :197.0
##  Mean   :2008-11-27   Mean   :43.92       Mean   :17.15      Mean   :200.9
##  3rd Qu.:2009-11-16   3rd Qu.:48.50       3rd Qu.:18.70      3rd Qu.:213.0
##  Max.   :2009-12-01   Max.   :59.60       Max.   :21.50      Max.   :231.0
##                       NA's   :2           NA's   :2          NA's   :2
##  Body Mass (g)       Sex            Delta 15 N (o/oo) Delta 13 C (o/oo)
##  Min.   :2700   Length:344         Min.   : 7.632    Min.   :-27.02
##  1st Qu.:3550   Class :character   1st Qu.: 8.300    1st Qu.:-26.32
##  Median :4050   Mode  :character   Median : 8.652    Median :-25.83
##  Mean   :4202                      Mean   : 8.733    Mean   :-25.69
##  3rd Qu.:4750                      3rd Qu.: 9.172    3rd Qu.:-25.06
```

```
##  Max.   :6300                        Max.   :10.025    Max.   :-23.79
##  NA's   :2                           NA's   :14        NA's   :13
##    Comments
##  Length:344
##  Class :character
##  Mode  :character
##
##
##
##
```

Now let's create the simpler penguins data frame. We'll use rename for a couple, but most variables require mutation to manipulate strings (we'll get to that later), create factors, or convert to integers. And we'll rename some variables to avoid using backticks (the backward single quotation mark accessed just to the left of the `1` key and below the `Esc` key, and what you can use in markdown to create a monospaced font as I just used for `1` and `Esc`).

```
penguins <- penguins_raw %>%
  rename(bill_length_mm = `Culmen Length (mm)`,
         bill_depth_mm = `Culmen Depth (mm)`) %>%
  mutate(species = factor(word(Species)),
         island = factor(Island),
         flipper_length_mm = as.integer(`Flipper Length (mm)`),
         body_mass_g = as.integer(`Body Mass (g)`),
         sex = factor(str_to_lower(Sex)),
         year = as.integer(year(ymd(`Date Egg`)))) %>%
  dplyr::select(species, island, bill_length_mm, bill_depth_mm,
                flipper_length_mm, body_mass_g, sex, year)
summary(penguins)
```

```
##       species          island     bill_length_mm  bill_depth_mm
##  Adelie    :152   Biscoe   :168   Min.   :32.10   Min.   :13.10
##  Chinstrap: 68   Dream    :124   1st Qu.:39.23   1st Qu.:15.60
##  Gentoo    :124   Torgersen: 52   Median :44.45   Median :17.30
##                                   Mean   :43.92   Mean   :17.15
##                                   3rd Qu.:48.50   3rd Qu.:18.70
##                                   Max.   :59.60   Max.   :21.50
##                                   NA's   :2       NA's   :2
##  flipper_length_mm  body_mass_g       sex           year
##  Min.   :172.0     Min.   :2700   female:165   Min.   :2007
##  1st Qu.:190.0     1st Qu.:3550   male   :168   1st Qu.:2007
##  Median :197.0     Median :4050   NA's   : 11   Median :2008
##  Mean   :200.9     Mean   :4202                 Mean   :2008
##  3rd Qu.:213.0     3rd Qu.:4750                 3rd Qu.:2009
##  Max.   :231.0     Max.   :6300                 Max.   :2009
##  NA's   :2         NA's   :2
```

Unfortunately, they don't end up as *exactly* identical, though all of the variables are identical as vectors:

TABLE 3.3 Date-filtered EucOak data

site	Date	rain_mm	rain_subcanopy	runoffL_oak	runoffL_euc	slope_oak	slope_euc
AB1	2007-04-23	NA	33.5	6.94488	9.19892	32.0	31
AB1	2007-05-05	NA	31.0	6.33568	7.43224	32.0	31
AB2	2007-04-23	23	35.5	4.32000	2.88000	24.0	25
AB2	2007-05-05	11	25.5	4.98000	3.30000	24.0	25
KM1	2007-04-23	NA	37.0	1.56000	2.04000	30.5	25
KM1	2007-05-05	28	22.0	1.32000	1.32000	30.5	25

```
identical(penguins, palmerpenguins::penguins)
```

```
## [1] FALSE
```

3.4.1.2 Helper functions for `dplyr::select()`

In the `select()` example above, we listed all of the variables, but there are a variety of helper functions for using logic to specify which variables to select:

- `contains("_")` or any substring of interest in the variable name
- `starts_with("runoff")`
- `ends_with("euc")`
- `everything()`
- `matches()` a regular expression
- `num_range("x",1:5)` for the common situation where a series of variable names combine a string and a number
- `one_of(myList)` for when you have a group of variable names
- *range* of variables: e.g. `runoffL_oak:slope_euc` could have followed `rain_subcanopy` above
- *all but*: preface a variable or a set of variable names with `-` to select all others

3.4.2 filter

`filter` lets you select observations that meet criteria, similar to an SQL `WHERE` clause (Table 3.3).

```
runoff2007 <- runoff %>%
  filter(Date >= as.Date("04/01/2007", "%m/%d/%Y"))
```

3.4.2.1 Filtering out `NA` with `!is.na`

Here's a really important one. There are many times you need to avoid `NA`s. We thus commonly see summary statistics using `na.rm = TRUE` in order to *ignore* `NA`s when calculating a statistic like `mean`.

To simply filter out NAs from a vector or a variable use a filter:

```
feb_filt <- sierraFeb %>% filter(!is.na(TEMPERATURE))
```

3.4.3 Writing a data frame to a csv

Let's say you have created a data frame, maybe with read_csv

```
runoff20062007 <- read_csv(ex("eucoak/eucoakrainfallrunoffTDR.csv"))
```

Then you do some processing to change it, maybe adding variables, reorganizing, etc., and you want to write out your new `eucoak`, so you just need to use `write_csv`

```
write_csv(eucoak, "data/tidy_eucoak.csv")
```

Note the use of a data folder `data`: Remember that your default workspace (`wd` for *working directory*) is where your project file resides (check what it is with `getwd()`), so by default you're saving things in that wd. To keep things organized the above code is placing data in a data folder within the wd.

3.4.4 Summarize by group

You'll find that you need to use this all the time with real data. Let's say you have a bunch of data where some categorical variable is defining a grouping, like our site field in the `eucoak` data. This is a form of *stratifying* our data. We'd like to just create average slope, rainfall, and runoff for each site. Note that it involves two steps, first defining which field defines the group, then the various summary statistics we'd like to store. In this case, all of the slopes under `oak` remain the same for a given site – it's a *site* characteristic – and the same applies to the `euc` site, so we can just grab the first value (mean would have also worked of course) (Table 3.4).

```
eucoakSiteAvg <- runoff %>%
  group_by(site) %>%
  summarize(
    rain = mean(rain_mm, na.rm = TRUE),
    rain_subcanopy = mean(rain_subcanopy, na.rm = TRUE),
    runoffL_oak = mean(runoffL_oak, na.rm = TRUE),
    runoffL_euc = mean(runoffL_euc, na.rm = TRUE),
    slope_oak = first(slope_oak),
```

TABLE 3.4 EucOak data summarized by site

site	rain	rain_subcanopy	runoffL_oak	runoffL_euc	slope_oak	slope_euc
AB1	48.37500	43.08333	6.8018364	6.026523	32.0	31
AB2	34.08333	35.37500	4.9113636	3.654545	24.0	25
KM1	48.00000	36.12500	1.9362500	0.592500	30.5	25
PR1	56.50000	37.56250	0.4585714	2.310000	27.0	23
TP1	38.36364	30.04545	0.8772727	1.657273	9.0	9
TP2	34.33333	32.86364	0.0954545	1.525454	12.0	10

```
    slope_euc = first(slope_euc)
  )
```

Summarizing by group with TRI data

```
library(igisci)
TRI_BySite <- read_csv(ex("TRI/TRI_2017_CA.csv")) %>%
  mutate(all_air = `5.1_FUGITIVE_AIR` + `5.2_STACK_AIR`) %>%
  filter(all_air > 0) %>%
  group_by(FACILITY_NAME) %>%
  summarize(
    FACILITY_NAME = first(FACILITY_NAME),
    air_releases = sum(all_air, na.rm = TRUE),
    mean_fugitive = mean(`5.1_FUGITIVE_AIR`, na.rm = TRUE),
    LATITUDE = first(LATITUDE), LONGITUDE = first(LONGITUDE))
```

3.4.5 Count

The count function is a simple variant on summarizing by group, since the only statistic is the count of events.

```
tidy_eucoak %>% count(tree)
```

```
## # A tibble: 2 x 2
##   tree      n
##   <chr> <int>
## 1 euc      90
## 2 oak      90
```

Another way to create a count is to use n():

```
tidy_eucoak %>%
  group_by(tree) %>%
  summarize(n = n())
```

```
## # A tibble: 2 x 2
##   tree      n
##   <chr> <int>
## 1 euc      90
## 2 oak      90
```

3.4.6 Sorting after summarizing

Using the marine debris data from the *Marine Debris Monitoring and Assessment Project* (*Marine Debris Program*, n.d.), we can use `arrange` to sort by latitude, so we can see the beaches from south to north along the Pacific coast.

```
shorelineLatLong <- ConcentrationReport %>%
  group_by(`Shoreline Name`) %>%
  summarize(
    latitude = mean((`Latitude Start`+`Latitude End`)/2),
    longitude = mean((`Longitude Start`+`Longitude End`)/2)
  ) %>%
  arrange(latitude)
shorelineLatLong
```

```
## # A tibble: 38 x 3
##    `Shoreline Name`   latitude longitude
##    <chr>                 <dbl>     <dbl>
##  1 Aimee Arvidson         33.6     -118.
##  2 Balboa Pier #2         33.6     -118.
##  3 Bolsa Chica            33.7     -118.
##  4 Junipero Beach         33.8     -118.
##  5 Malaga Cove            33.8     -118.
##  6 Zuma Beach, Malibu     34.0     -119.
##  7 Zuma Beach             34.0     -119.
##  8 Will Rodgers           34.0     -119.
##  9 Carbon Beach           34.0     -119.
## 10 Nicholas Canyon        34.0     -119.
## # ... with 28 more rows
```

3.4.7 The dot operator

The dot (`.`) operator refers to *here*, similar to UNIX syntax for accessing files in the current folder where the path is "`./filename`". In a piped sequence, you might need to reference the object (like the data frame) the pipe is using, as you can see in the following code.

- The advantage of the pipe is you don't have to keep referencing the data frame.
- The dot is then used to connect to items inside the data frame

```
TRI87 <- read_csv(ex("TRI/TRI_1987_BaySites.csv"))
stackrate <- TRI87 %>%
  mutate(stackrate = stack_air/air_releases) %>%
  .$stackrate
head(stackrate)
```

```
## [1] 0.0000000 0.0000000 0.0000000 0.0000000 0.6666667 1.0000000
```

3.5 String abstraction

Character string manipulation is surprisingly critical to data analysis, and so the **stringr** package was developed to provide a wider array of string processing tools than what is in base R, including functions for detecting matches, subsetting strings, managing lengths, replacing substrings with other text, and joining, splitting, and sorting strings.

We'll look at some of the `stringr` functions, but a good way to learn about the wide array of functions is through the cheat sheet that can be downloaded from https://www.rstudio.com/resources/cheatsheets/

3.5.1 Detecting matches

These functions look for patterns within existing strings, which can then be used subset observations based on those patterns.

- **`str_detect`** detects patterns in a string, returns true or false if detected
- **`str_locate`** detects patterns in a string, returns **start** *and* **end** position if detected, or NA if not
- **`str_which`** returns the indices of strings that match a pattern
- **`str_count`** counts the number of matches in each string

```
str_detect(fruit,"qu")
str_which(fruit, "qu")
fruit[str_which(fruit,"qu")]
```

```
##  [1] FALSE FALSE FALSE FALSE FALSE FALSE FALSE FALSE FALSE FALSE FALSE FALSE
## [13] FALSE FALSE FALSE FALSE FALSE FALSE FALSE FALSE FALSE FALSE FALSE FALSE
## [25] FALSE FALSE FALSE FALSE FALSE FALSE FALSE FALSE FALSE FALSE FALSE FALSE
## [37] FALSE FALSE FALSE FALSE FALSE FALSE  TRUE FALSE FALSE  TRUE FALSE FALSE
## [49] FALSE FALSE FALSE FALSE FALSE FALSE FALSE FALSE FALSE FALSE FALSE FALSE
## [61] FALSE FALSE FALSE FALSE FALSE FALSE  TRUE FALSE FALSE FALSE FALSE FALSE
## [73] FALSE FALSE FALSE FALSE FALSE FALSE FALSE FALSE
## [1] 43 46 67
```

```
## [1] "kumquat" "loquat"  "quince"
```

Note that `str_locate` returns a matrix with rows for each item where the substring is located with both `start` and `end` locations.

```
fruit_a <- fruit[str_count(fruit,"a")>1]
fruit_a
str_locate(fruit_a,"a")
```

```
##  [1] "avocado"      "banana"       "blackcurrant" "canary melon" "cantaloupe"
##  [6] "guava"        "mandarine"    "papaya"       "pomegranate"  "rambutan"
## [11] "salal berry"  "satsuma"      "tamarillo"
##       start end
##  [1,]     1   1
##  [2,]     2   2
##  [3,]     3   3
##  [4,]     2   2
##  [5,]     2   2
##  [6,]     3   3
##  [7,]     2   2
##  [8,]     2   2
##  [9,]     7   7
## [10,]     2   2
## [11,]     2   2
## [12,]     2   2
## [13,]     2   2
```

The `str_locate_all` function carries this further and finds all locations of the substrings within the string.

```
fruit3a <- fruit[str_count(fruit,"a")==3]
fruit3a
str_locate_all(fruit3a,"a")
```

```
## [1] "banana" "papaya"
## [[1]]
##      start end
## [1,]     2   2
## [2,]     4   4
## [3,]     6   6
##
## [[2]]
##      start end
## [1,]     2   2
## [2,]     4   4
## [3,]     6   6
```

```
qufruit <- fruit[str_which(fruit,"qu")]
```

```
qufruit
str_locate(qufruit, "qu")
```

```
## [1] "kumquat" "loquat"  "quince"
##      start end
## [1,]     4   5
## [2,]     3   4
## [3,]     1   2
```

3.5.2 Subsetting strings

Subsetting in this case includes its normal use of abstracting the observations specified by a match (similar to a filter for data frames), or just a specified part of a string specified by start and end character positions, or the part of the string that matches an expression.

- **`str_sub`** extracts a part of a string from a start to an end character position
- **`str_subset`** returns the strings that contain a pattern match
- **`str_extract`** returns the first (or if `str_extract_all` then all matches) pattern matches
- **`str_match`** returns the first (or `_all`) pattern match as a matrix
- **`str_remove`** removes a specific substring (not on the cheat sheet, but handy)

```
qfruit <- str_subset(fruit, "q")
qfruit
str_sub(qfruit,1,2)
```

```
## [1] "kumquat" "loquat"  "quince"
## [1] "ku" "lo" "qu"
```

```
str_sub("94132",1,2)
str_extract(qfruit,"[aeiou]")
str_remove(str_subset(fruit,"fruit"), "fruit")
```

```
## [1] "94"
## [1] "u" "o" "u"
## [1] "bread"    "dragon"   "grape"    "jack"     "kiwi "    "passion" "star "
## [8] "ugli "
```

3.5.3 String length

The length of strings is often useful in an analysis process, either just knowing the length as an integer, or purposefully increasing or reducing it.

- **`str_length`** simply returns the length of the string as an integer

```
qfruit <- str_subset(fruit,"q")
qfruit
str_length(qfruit)
```

```
## [1] "kumquat" "loquat"  "quince"
## [1] 7 6 6
```

Using `str_locate` with `str_length`:

```
name <- "Inigo Montoya"
str_length(name)
firstname <- str_sub(name,1,str_locate(name," ")[1]-1)
firstname
lastname <- str_sub(name,str_locate(name," ")[1]+1,str_length(name))
lastname
```

```
## [1] 13
## [1] "Inigo"
## [1] "Montoya"
```

Padding and trimming lengths:

- **`str_pad`** adds a specified character (typically a space " ") to either end of a string
- **`str_trim`** removes white space from either end of a string

Here's an example `str_pad` and `str_trim` (in this case, reversing the former):

```
str_pad(qfruit,10,"both")
str_trim(str_pad(qfruit,10,"both"),"both")
```

```
## [1] " kumquat  " "  loquat  " "  quince  "
## [1] "kumquat" "loquat"  "quince"
```

3.5.4 Replacing substrings with other text ("mutating" strings)

These methods range from converting case to replacing substrings.

- **`str_to_lower`** converts strings to lowercase
- **`str_to_upper`** converts strings to uppercase
- **`str_to_title`** capitalizes strings (makes the first character of each word uppercase)

```
str_to_lower(name)
str_to_upper(name)
str_to_title("for whom the bell tolls")
```

```
## [1] "inigo montoya"
```

```
## [1] "INIGO MONTOYA"
## [1] "For Whom The Bell Tolls"
```

- **str_replace** replaces the first matched pattern (or all with `str_replace_all`) with a specified string
- **str_sub** a special use of this function to replace substrings with a specified string

```
str_replace(qfruit,"q","z")
str_sub(name,1,str_locate(name," ")[1]-1) <- "Diego"
name
```

```
## [1] "kumzuat" "lozuat"  "zuince"
## [1] "Diego Montoya"
```

The replacement mode of `str_sub` used above may seem pretty tortuous and confusing. Why did we have to access `[1]` in `str_locate(name," ")[1]`? Experiment with what you get with `str_locate(name," ")` itself, and maybe consult the help system with `?str_locate` (and run the examples), but you'll find this is one that returns a matrix, and even if there's only one item located will still have a `start` and `end` location. See what happens when you use it *without* accessing `[1]`. You're seeing one effect of vectorization – **reuse**. It can be very confusing, but reuse of inputs is one of the (very useful but still confusing) properties of vectorization. In this case it creates duplication, creating a vector of names.

3.5.5 Concatenating and splitting

One very common string function is to concatenate strings, and somewhat less common is splitting them using a key separator like space, comma, or line end. One use of using str_c in the example below is to create a comparable join field based on a numeric character string that might need a zero or something at the left or right.

- **str_c** The `paste()` function in base R will work, but you might want the default separator setting to be "" (as does `paste0()`) instead of `" "`, so `str_c` is just `paste` with a default "" separator, but you can also use `" "`.
- **str_split** splits a string into parts based upon the detection of a specified separator like space, comma, or line end. However, this ends up creating a list, which can be confusing to use. If you're wanting to access a particular member of a list of words, the `word` function is easier.
- **word** lets you pick the first, second, or nth word in a string of words.

```
library(stringr)
phrase <- str_c("for","whom","the","bell","tolls",sep=" ")
phrase
```

```
## [1] "for whom the bell tolls"
```

```
str_split(phrase, " ")
```

```
## [[1]]
## [1] "for"   "whom"  "the"   "bell"  "tolls"
```

```
word(phrase,4)
```

```
## [1] "bell"
```

... and of course, these methods can all be vectorized:

```
sentences[1:4]
```

```
## [1] "The birch canoe slid on the smooth planks."
## [2] "Glue the sheet to the dark blue background."
## [3] "It's easy to tell the depth of a well."
## [4] "These days a chicken leg is a rare dish."
```

```
word(sentences[1:4],2)
```

```
## [1] "birch" "the"   "easy"  "days"
```

Example of `str_c` use to modify a variable needed for a join:

```
csvPath <- system.file("extdata","CA/CA_MdInc.csv",package="igisci")
CA_MdInc <- read_csv(csvPath)
join_id <- str_c("0",CA_MdInc$NAME)
# could also use str_pad(CA_MdInc$NAME,1,side="left",pad="0")
head(CA_MdInc)
```

```
## # A tibble: 6 x 3
##         trID     NAME HHinc2016
##        <dbl>    <dbl>     <dbl>
## 1 6001400100 60014001    177417
## 2 6001400200 60014002    153125
## 3 6001400300 60014003     85313
## 4 6001400400 60014004     99539
## 5 6001400500 60014005     83650
## 6 6001400600 60014006     61597
```

```
head(join_id)
```

```
## [1] "060014001" "060014002" "060014003" "060014004" "060014005" "060014006"
```

3.6 Dates and times with `lubridate`

Simply stated, the `lubridate` package just makes it easier to work with dates and times. Dates and times are more difficult to work with than numbers and character strings, which is why `lubridate` was developed. The `lubridate` package is excellent at detecting and reading in dates and times – it can "parse" many forms. Spend some time to make sure you know how to use it. We'll be using dates and times a lot, since many data sets include time stamps. See the cheat sheet for more information but the following examples may demonstrate that it's pretty easy to use, and does a good job of making your job easier.

```
library(lubridate)
dmy("20 September 2020")
dmy_hm("20 September 2020 10:45")
mdy_hms("September 20, 2020 10:48")
mdy_hm("9/20/20 10:50")
mdy("9.20.20")
```

```
## [1] "2020-09-20"
## [1] "2020-09-20 10:45:00 UTC"
## [1] "2020-09-20 20:10:48 UTC"
## [1] "2020-09-20 10:50:00 UTC"
## [1] "2020-09-20"
```

```
start704 <- dmy_hm("24 August 2020 16:00")
end704 <- mdy_hm("12/18/2020 4:45 pm")
year(start704)
month(start704)
day(end704)
hour(end704)
```

```
## [1] 2020
## [1] 8
## [1] 18
## [1] 16
```

A bit of date math:

```
end704-start704
```

```
as_date(end704)
hms::as_hms(end704)
```

```
## Time difference of 116.0312 days
## [1] "2020-12-18"
## 16:45:00
```

You will find that dates are typically displayed as yyyy-mm-dd, e.g. "2020-09-20" above. You can display them other ways of course, but it's useful to write dates this way since they sort better, with the highest order coming first.

For obvious reasons, we'll use `lubridate` a lot when we get to time series.

3.7 Calling functions explicitly with ::

Sometimes you need to specify the package and function name this way, for instance, if more than one package has a function of the same name. You can also use this method to call a function without having loaded its library. Due to multiple packages having certain common names (like `select`), it's common to use this syntax, and you'll find that we'll use `dplyr::select(...)` throughout this book.

3.8 Exercises: Data Abstraction

Exercise 3.1. Create a tibble with 20 rows of two variables `norm` and `unif` with `norm` created with `rnorm()` and `unif` created with `runif()`.

Exercise 3.2. Read in "TRI/TRI_2017_CA.csv" in two ways, as a normal data frame assigned to df and as a tibble assigned to tbl. What field names result for what's listed in the CSV as `5.1_FUGITIVE_AIR`?

Exercise 3.3. Use the summary function to investigate the variables in either the data.frame or tibble you just created. What type of field and what values are assigned to `BIA_CODE`?

Exercise 3.4. Create a boxplot of `body_mass_g` by `species` from the `penguins` data frame in the palmerpenguins package (Horst, Hill, and Gorman 2020).

Exercise 3.5. Use select, mutate, and the pipe to create a `penguinMass` tibble where the only original variable retained is `species`, but with `body_mass_kg` created as $\frac{1}{1000}$ the `body_mass_g`. The statement should start with `penguinMass <- penguins` and use a pipe plus the other functions after that.

Exercise 3.6. Now, also with `penguins`, create `FemaleChinstaps` to include only the female Chinstrap penguins. Start with `FemaleChinstraps <- penguins %>%`

Exercise 3.7. Now, summarize by `species` groups to create mean and standard deviation variables from `bill_length_mm`, `bill_depth_mm`, `flipper_length_mm`, and `body_mass_g`. Preface the variable names with either `avg.` or `sd.` Include `na.rm=T` with all statistics function calls.

Exercise 3.8. Sort the penguins by `body_mass_g`.

Exercise 3.9. Using `stringr` methods, detect and print out a vector of sites in the `ConcentrationReport` that are parks (include the substring `"Park"` in the variable `Shoreline Name`), then detect and print out the longest shoreline name.

Exercise 3.10. The `sierraStations` data frame has weather stations only in California, but includes "`CA US`" at the end of the name. Use dplyr and stringr methods to create a new `STATION_NAME` that truncates the name, removing that part at the end.

Exercise 3.11. Use `dplyr` and `lubridate` methods to create a new date-type variable `sampleDate` from the existing `DATE` character variable in `soilCO2_97`, then use this to plot a graph of soil CO2 over time, with `sampleDate` as x, and `CO2pct` as y. Then use code to answer the question: What's the length of time (time difference) from beginning to end for this data set [Hint: two approaches might use sorting or `max()` and `min()`]?

4

Visualization

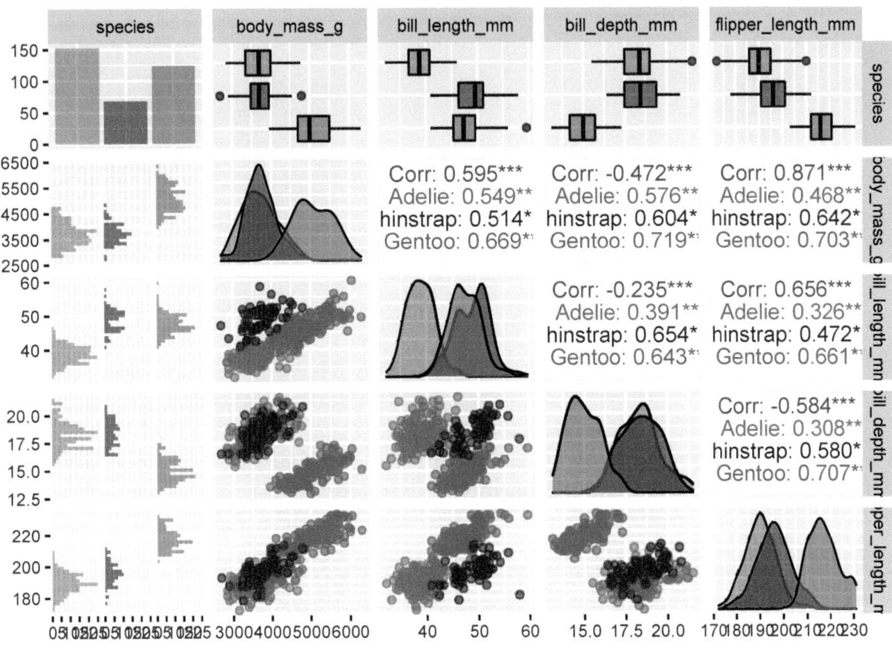

In this section, we'll explore visualization methods in R. Visualization has been a key element of R since its inception, since visualization is central to the exploratory philosophy of the language.

4.1 plot in base R

In this chapter, we'll spend a lot more time looking at `ggplot2` due to its clear syntax using the *grammar of graphics*. However, the base R `plot` system generally does a good job of coming up with the most likely graphical output based on the data you provide, so we'll use it from time to time. Through the book you'll see various references to its use, including setting graphical parameters with `par`, such as the `cex` for defining the relative size of text characters and point symbols, `lty` and `lwd` for line type and width, and *many* others. Users are encouraged to explore these with `?par` to learn about parameters (even things like creating multiple plots using `mfrow` as you can see in Figure 4.1), and `?plot` to learn more overall about the generic X-Y plotting system.

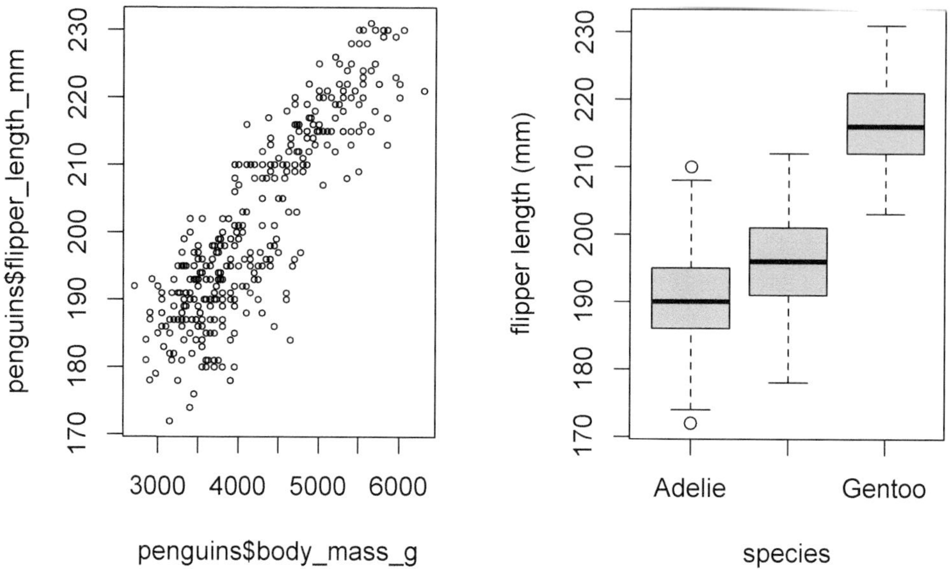

FIGURE 4.1 Flipper length by mass and by species, base plot system. The Antarctic peninsula penguin data set is from @palmer.

```
par(mfrow=c(1,2))
plot(penguins$body_mass_g, penguins$flipper_length_mm,
     cex=0.5) # half-size point symbol
plot(penguins$species, penguins$flipper_length_mm,
     ylab="flipper length (mm)",
     xlab="species")
```

4.2 ggplot2

We'll mostly focus however on gpplot2, based on the *Grammar of Graphics* because it provides considerable control over your graphics while remaining fairly easily readable, as long as you buy into its grammar.

The `ggplot2` app (and its primary function `ggplot`) looks at three aspects of a graph:

- data : where are the data coming from?
- geometry : what type of graph are we creating?

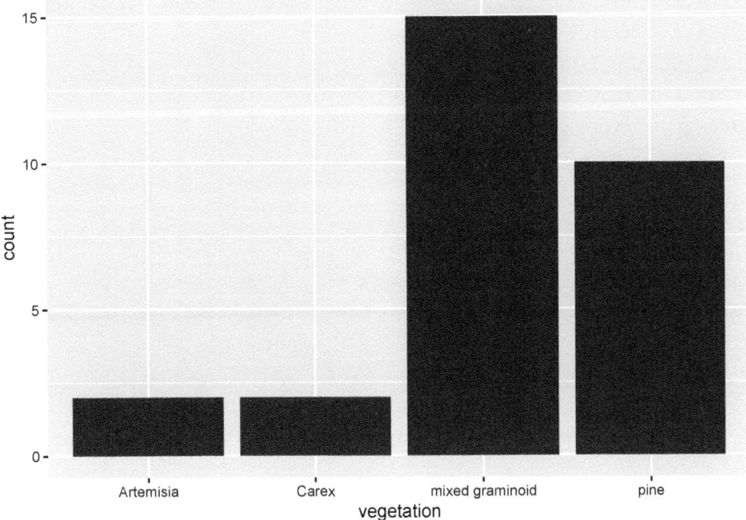

FIGURE 4.2 Simple bar graph of meadow vegetation samples

- aesthetics : what choices can we make about symbology and how do we connect symbology to data?

As with other tidyverse and RStudio packages, find the ggplot2 cheat sheet at https://www.rstudio.com/resources/cheatsheets/

4.3 Plotting one variable

The `ggplot` function provides plots of single and multiple variables, using various coordinate systems (including geographic). We'll start with just plotting one variable, which might be *continuous* – where we might want to see a histogram, density plot, or dot plot – or *discrete* – where we might want to see something like a a bar graph, like the first example below (Figure 4.2).

We'll look at a study of Normalized Difference Vegetation Index from a transect across a montane meadow in the northern Sierra Nevada, derived from multispectral drone imagery (JD Davis et al. 2020).

```
library(igisci)
library(tidyverse)
summary(XSptsNDVI)
```

```
##     DistNtoS        elevation     vegetation          geometry
##  Min.   :  0.0   Min.   :1510   Length:29          Length:29
##  1st Qu.: 37.0   1st Qu.:1510   Class :character   Class :character
```

```
## Median :175.0   Median :1511   Mode  :character   Mode  :character
## Mean   :164.7   Mean   :1511
## 3rd Qu.:275.5   3rd Qu.:1511
## Max.   :298.8   Max.   :1511
##  NDVIgrowing      NDVIsenescent
## Min.   :0.3255   Min.   :0.1402
## 1st Qu.:0.5052   1st Qu.:0.2418
## Median :0.6169   Median :0.2817
## Mean   :0.5901   Mean   :0.3662
## 3rd Qu.:0.6768   3rd Qu.:0.5407
## Max.   :0.7683   Max.   :0.7578
```

```
ggplot(XSptsNDVI, aes(vegetation)) +
  geom_bar()
```

4.3.1 Histogram

Histograms are very useful for looking at the distribution of continuous variables (Figure 4.3). We'll start by using a pivot table (these will be discussed in the next chapter, on data transformation.)

```
XSptsPheno <- XSptsNDVI %>%
  filter(vegetation != "pine") %>%
  pivot_longer(cols = starts_with("NDVI"),
               names_to = "phenology",
               values_to = "NDVI") %>%
  mutate(phenology = str_sub(phenology, 5, str_length(phenology)))
```

```
XSptsPheno %>%
  ggplot(aes(NDVI)) +
  geom_histogram(binwidth=0.05)
```

Histograms can be created in a couple of ways, one the conventional histogram that provides the most familiar view, for instance of the "bell curve" of a normal distribution (Figure 4.4).

```
sierraData %>%
    ggplot(aes(TEMPERATURE)) +
  geom_histogram(fill="dark green")
```

Alternatively, we can look at a *cumulative* histogram, which makes it easier to see percentiles and the median (50th percentile), by using the `cumsum()` function (Figure 4.5).

```
n <- length(sierraData$TEMPERATURE)
sierraData %>%
```

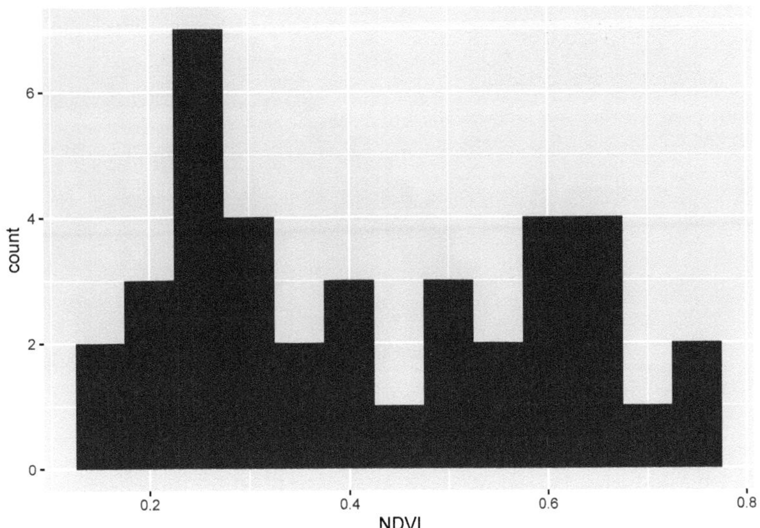

FIGURE 4.3 Distribution of NDVI, Knuthson Meadow

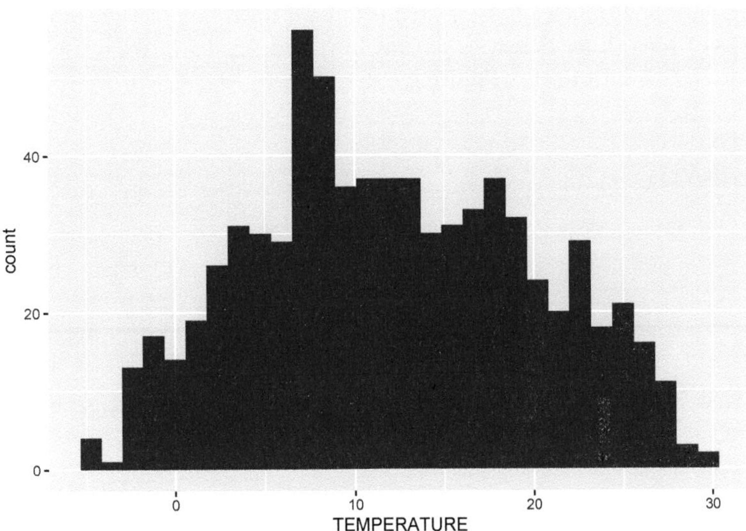

FIGURE 4.4 Distribution of Average Monthly Temperatures, Sierra Nevada

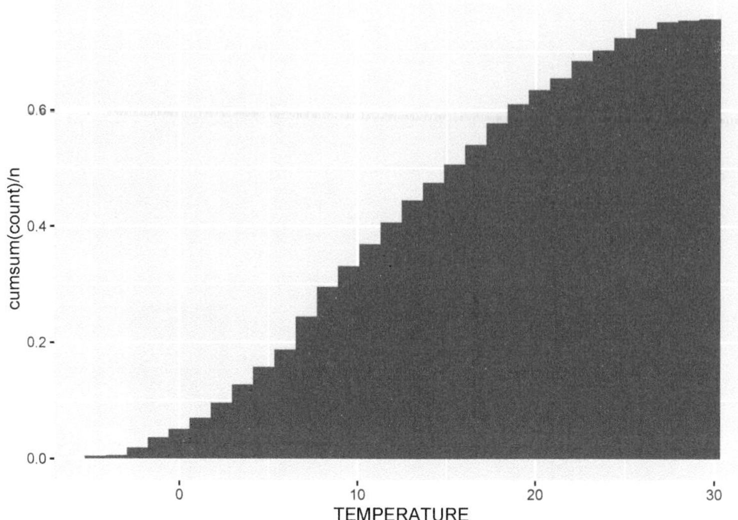

FIGURE 4.5 Cumulative Distribution of Average Monthly Temperatures, Sierra Nevada

```
  ggplot(aes(TEMPERATURE)) +
  geom_histogram(aes(y=cumsum(..count..)/n), fill="dark goldenrod")
```

4.3.2 Density plot

Density represents how much of the distribution we have out of the total. The total area (sum of widths of bins times densities of that bin) adds up to 1. We'll use a density plot to looking at our NDVI data again (Figure 4.6).

```
XSptsPheno %>%
  ggplot(aes(NDVI)) +
  geom_density()
```

Note that NDVI values are <1 so bins are very small numbers, so in this case densities can be >1.

To communicate more information, we might want to use color and transparency (alpha) settings. The following graph (Figure 4.7) will separate the data by phenology ("growing" vs. "senescence" seasons) using color, and use the alpha setting to allow these overlapping distributions to both be seen. To do this, we'll:

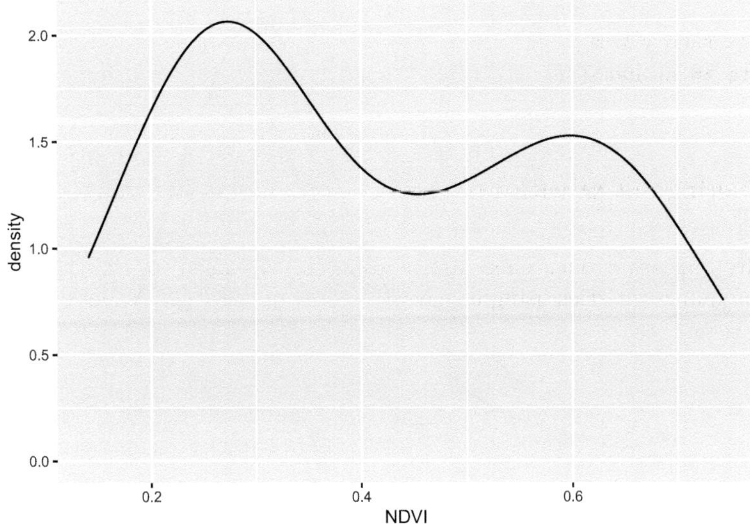

FIGURE 4.6 Density plot of NDVI, Knuthson Meadow

- "map" a variable (phenology) to an aesthetic property (fill color of the density polygon)
- set a a property (alpha = 0.2) to all polygons of the density plot. The alpha channel of colors defines its opacity, from invisible (0) to opaque (1) so is commonly used to set as its reverse, transparency.

```
XSptsPheno %>%
  ggplot(aes(NDVI, fill=phenology)) +
  geom_density(alpha=0.2)
```

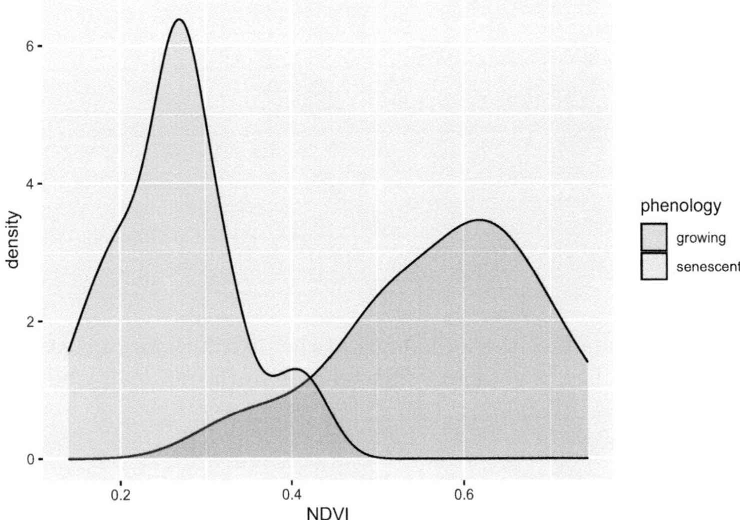

FIGURE 4.7 Comparative density plot using alpha setting

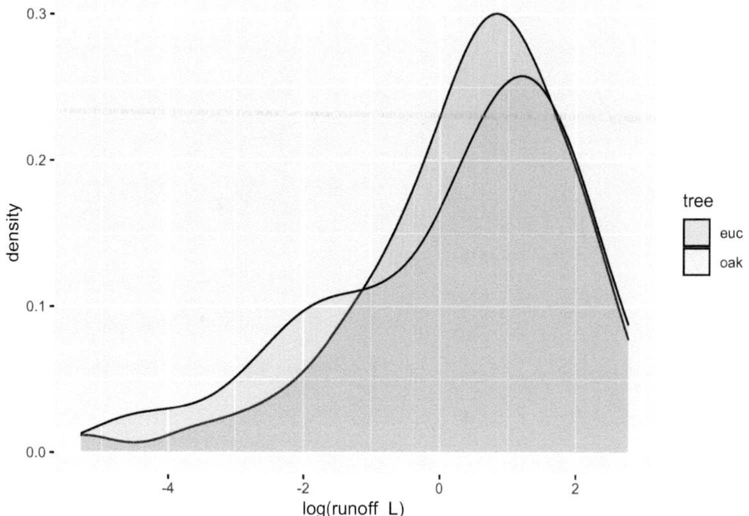

FIGURE 4.8 Runoff under eucalyptus and oak in Bay Area sites

Why is the color called **fill**? For polygons, "color" is used for the boundary.

Similarly for the eucalyptus and oak study, we can overlay the two distributions (Figure 4.8). Since the two distributions overlap considerably, the benefit of the alpha settings is clear.

```
tidy_eucoak %>%
  ggplot(aes(log(runoff_L),fill=tree)) +
  geom_density(alpha=0.2)
```

4.3.3 Boxplot

Tukey boxplots provide another way of looking at the distributions of continuous variables. Typically these are stratified by a factor, such as `site` in the euc/oak study (Figure 4.9):

```
ggplot(data = tidy_eucoak) +
  geom_boxplot(aes(x = site, y = runoff_L))
```

And then as we did above, we can communicate more by coloring by tree type. Note that this is called *within* the `aes()` function (Figure 4.10).

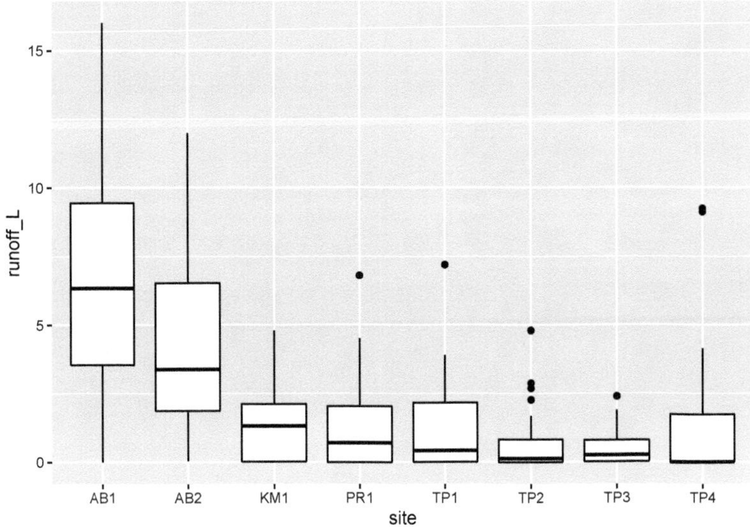

FIGURE 4.9 Boxplot of runoff by site

```
ggplot(data = tidy_eucoak) +
  geom_boxplot(aes(x=site, y=runoff_L, color=tree))
```

Visualizing soil CO_2 data with a box plot

In a study of soil CO_2 in the Marble Mountains of California (JD Davis, Amato, and Kiefer 2001), we sampled extracted soil air (Figure 4.11) in a 11-point transect across Marble Valley in 1997 (Figure 4.12). Again, a Tukey boxplot is useful for visualization (Figure 4.13).

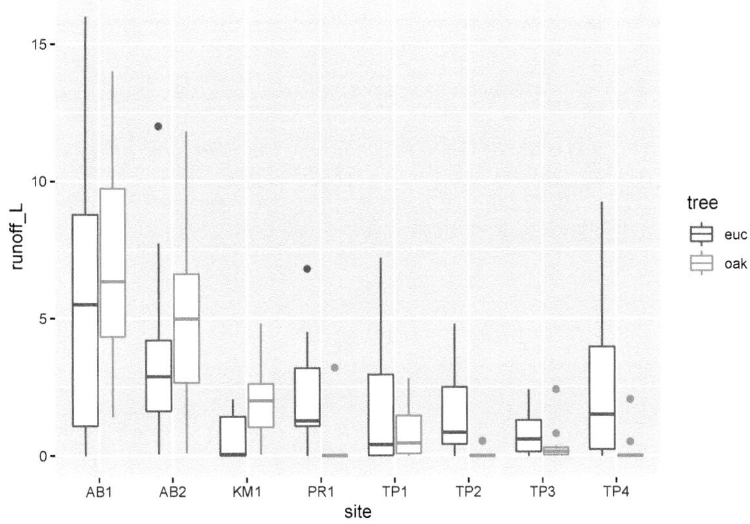

FIGURE 4.10 Runoff at Bay Area Sites, colored as eucalyptus and oak

FIGURE 4.11 Marble Valley, Marble Mountains Wilderness, California

---

Note that in this book you'll often see CO_2 written as CO2. These are both meant to refer to carbon dioxide, but I've learned that subscripts in figure headings don't always get passed through to the LaTeX compiler for the pdf/printed version, so I'm forced to write it without the subscript. Similarly CH_4 might be written as CH4, etc. The same applies often to variable names and axis labels in graphs, though there are some workarounds.

```
soilCO2 <- soilCO2_97
soilCO2$SITE <- factor(soilCO2$SITE)  # in order to make the numeric field a factor
ggplot(data = soilCO2, mapping = aes(x = SITE, y = CO2pct)) +
  geom_boxplot()
```

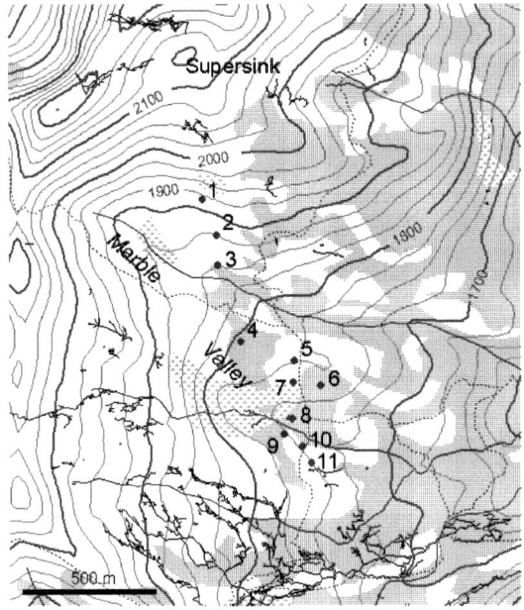

FIGURE 4.12 Marble Mountains soil gas sampling sites, with surface topographic features and cave passages

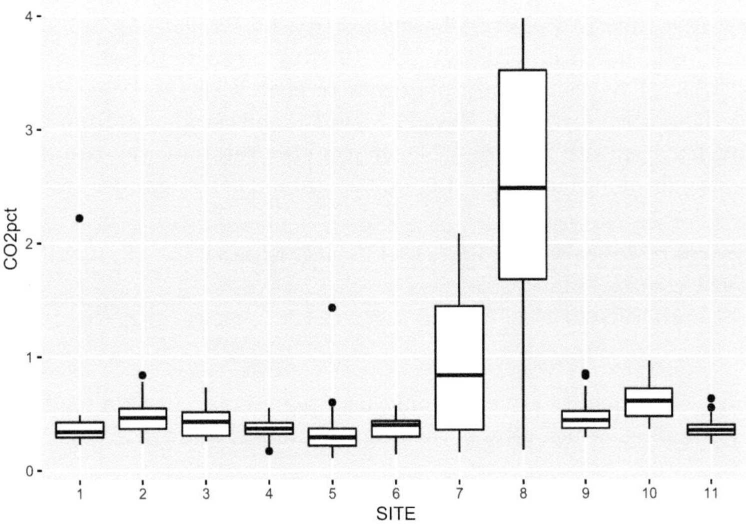

FIGURE 4.13 Visualizing soil CO2 data with a Tukey box plot

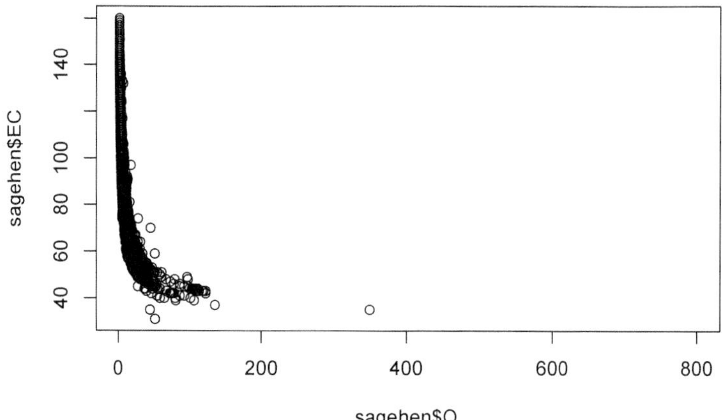

FIGURE 4.14 Scatter plot of discharge (Q) and specific electrical conductance (EC) for Sagehen Creek, California

4.4 Plotting Two Variables

4.4.1 Two continuous variables

We've looked at this before – the scatterplot – but let's try some new data: daily discharge (Q) and other data (like EC, electrical conductivity, a surrogate for solute concentration) from Sagehen Creek, north of Truckee, CA, 1970 to present. I downloaded the data for this location (which I've visited multiple times to chat with the USGS hydrologist about calibration methods with my Sierra Nevada Field Campus students) from the https://waterdata.usgs.gov/nwis/sw site (Figure 4.14). If you visit this site, you can download similar data from thousands of surface water (as well as groundwater) gauges around the country.

```
library(tidyverse); library(lubridate); library(igisci)
sagehen <- read_csv(ex("sierra/sagehen_dv.csv"))
plot(sagehen$Q,sagehen$EC)
```

Streamflow and water quality are commonly best represented using a log transform, or we can just use log scaling, which retains the original units (Figure 4.15).

```
ggplot(data=sagehen, aes(x=Q, y=EC)) + geom_point() +
  scale_x_log10() + scale_y_log10()
```

- For both graphs, the `aes` ("aesthetics") function specifies the variables to use as x and y coordinates
- geom_point creates a scatter plot of those coordinate points

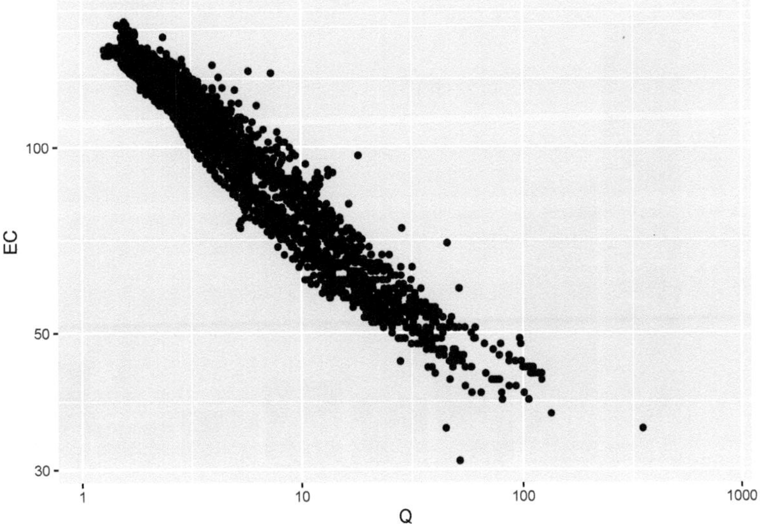

FIGURE 4.15 Q and EC for Sagehen Creek, using log10 scaling on both axes

Set color for all (*not* in `aes()`)

Sometimes all you want is a simple graph with one color for all points (Figure 4.16). Note that:

- color is defined outside of `aes`, so it applies to all points.
- mapping is first argument of `geom_point`, so `mapping =` is not needed.

```
ggplot(data=sierraFeb) +
  geom_point(aes(TEMPERATURE, ELEVATION), color="blue")
```

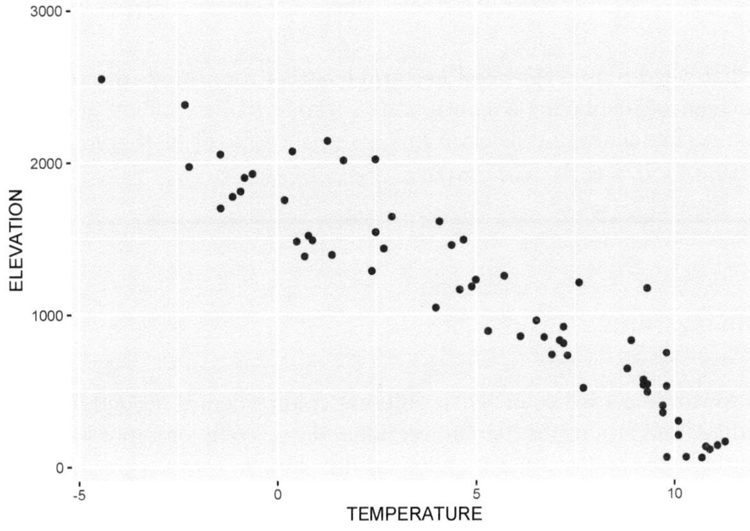

FIGURE 4.16 Setting one color for all points

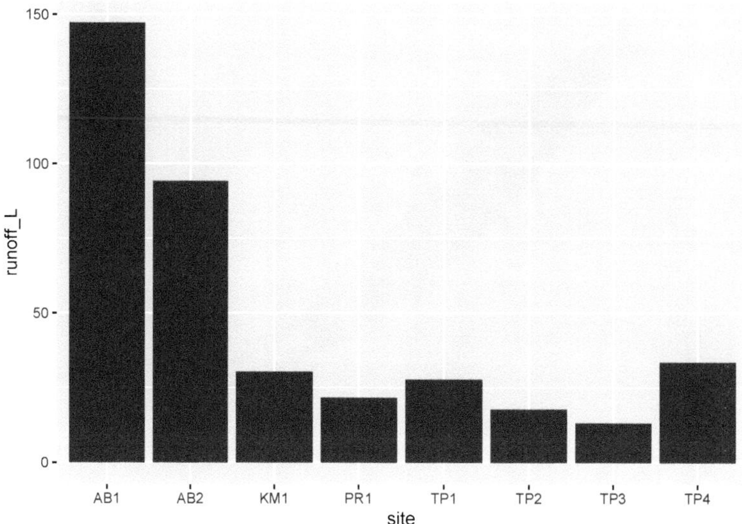

FIGURE 4.17 Two variables, one discrete

4.4.2 Two variables, one discrete

We've already looked at bringing in a factor to a histogram and a boxplot, but there's the even simpler bar graph that does this, if we're just interested in comparing values. This graph compares runoff by site (Figure 4.17).

```
ggplot(tidy_eucoak) +
  geom_bar(aes(site, runoff_L), stat="identity")
```

Note that we also used the `geom_bar` graph earlier for a single discrete variable, but instead of displaying a continuous variable like runoff, it just displayed the count (frequency) of observations, so was a type of histogram.

4.4.3 Color systems

There's a lot to working with color, with different color schemes needed for continuous data vs discrete values, and situations like bidirectional data. We'll look into some basics, but the reader is recommended to learn more at sources like https://cran.r-project.org/web/packages/RColorBrewer/index.html or https://colorbrewer2.org/ or just Googling "rcolorbrewer" or "colorbrewer" or even "R colors".

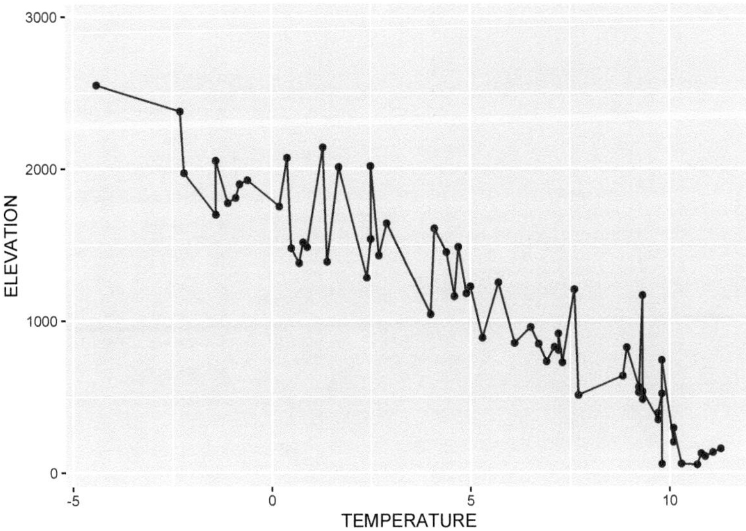

FIGURE 4.18 Using aesthetics settings for both points and lines

4.4.3.1 Specifying colors to use for a graphical element

When a color is requested for an entire graphical element, like geom_point or geom_line, and *not* in the aesthetics, all feature get that color. In the following graph the same x and y values are used to display as points in blue and as lines in red (Figure 4.18).

```
sierraFeb %>%
  ggplot(aes(TEMPERATURE,ELEVATION)) +
  geom_point(color="blue") +
  geom_line(color="red")
```

Note the use of pipe to start with the data then apply ggplot. This is one approach for creating graphs, and provides a fairly straightforward way to progress from data to visualization.

4.4.3.2 Color from variable, in aesthetics

If color is connected to a variable within the `aes()` call, a color scheme is chosen to assign either a range (for continuous) or a set of unique colors (for discrete). In this graph, color is defined inside `aes`, so is based on COUNTY (Figure 4.19).

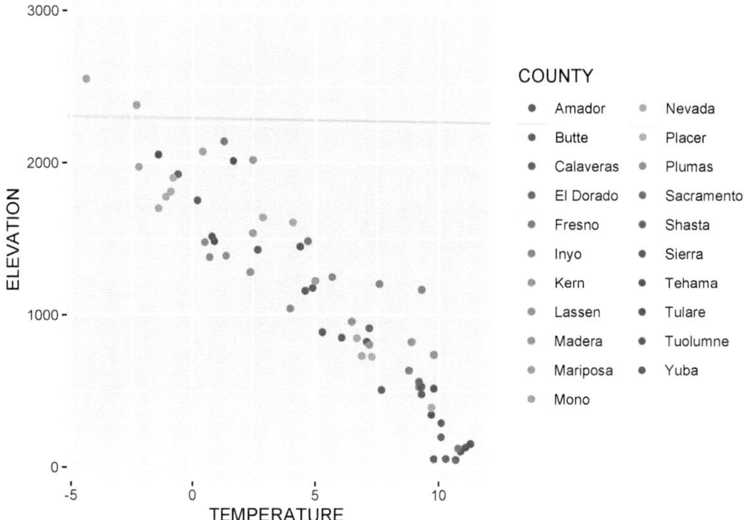

FIGURE 4.19 Color set within aes()

```
ggplot(data=sierraFeb) +
  geom_point(aes(TEMPERATURE, ELEVATION, color=COUNTY))
```

Note that counties represent discrete data, and this is detected by ggplot to assign an appropriate color scheme. Continuous data will require a different palette (Figure 4.20).

```
ggplot(data=sagehen, aes(x=Q, y=EC, col=waterTmax)) + geom_point() +
  scale_x_log10() + scale_y_log10()
```

River map and profile

We'll build a `riverData` dataframe with x and y location values and elevation. We'll need to start by creating empty vectors we'll populate with values in a loop: `d`, `longd`, and `s` are assigned an empty value `double()`, then slope `s` (since it only occurs between two points) needs one `NA` value assigned for the last point `s[length(x]` to have the same length as other vectors.

```
library(tidyverse)
x <- c(1000, 1100, 1300, 1500, 1600, 1800, 1900)
y <- c(500, 780, 820, 950, 1250, 1320, 1500)
elev <- c(0, 1, 2, 5, 25, 75, 150)
d <- double()      # creates an empty numeric vector
longd <- double()  # ("double" means double-precision floating point)
s <- double()
for(i in 1:length(x)){
  if(i==1){longd[i] <- 0; d[i] <-0}
  else{
```

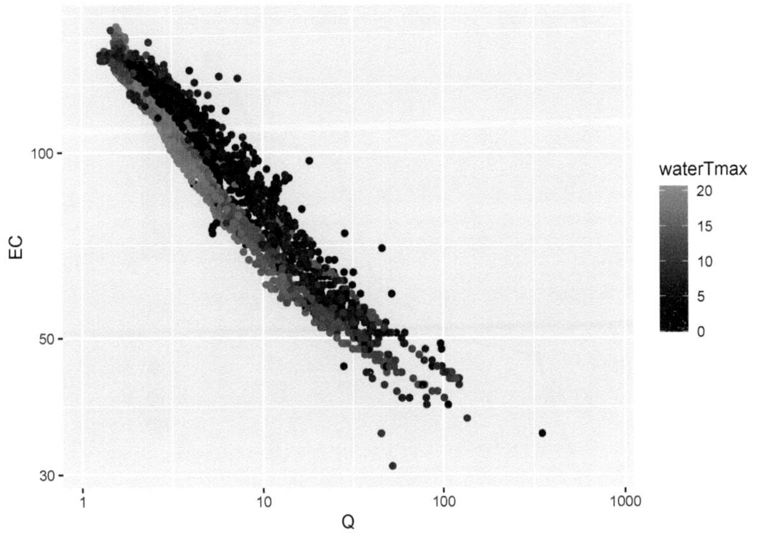

FIGURE 4.20 Streamflow (Q) and specific electrical conductance (EC) for Sagehen Creek, colored by temperature

```
    d[i] <- sqrt((x[i]-x[i-1])^2 + (y[i]-y[i-1])^2)
    longd[i] <- longd[i-1] + d[i]
    s[i-1] <- (elev[i]-elev[i-1])/d[i]}}
s[length(x)] <- NA  # make the last slope value NA since we have no data past it,
                    # and so the vector lengths are all the same
riverData <- bind_cols(x=x,y=y,elev=elev,d=d,longd=longd,s=s)
riverData
```

```
## # A tibble: 7 x 6
##       x     y  elev     d longd        s
##   <dbl> <dbl> <dbl> <dbl> <dbl>    <dbl>
## 1  1000   500     0     0     0  0.00336
## 2  1100   780     1  297.  297.  0.00490
## 3  1300   820     2  204.  501.  0.0126
## 4  1500   950     5  239.  740.  0.0632
## 5  1600  1250    25  316. 1056.  0.236
## 6  1800  1320    75  212. 1268.  0.364
## 7  1900  1500   150  206. 1474. NA
```

For this continuous data, a range of values is detected and a continous color scheme is assigned (Figure 4.21). The ggplot `scale_color_gradient` function is used to establish end points of a color range that the data are stretched between (Figure 4.22). We can use this for many continuous variables, such as slope (Figure 4.23). The `scale_color_gradient2` lets you use a `mid` color. Note that there's a comparable `scale_fill_gradient` and `scale_fill_gradient2` for use when specifying a fill (e.g. for a polygon) instead of a color (for polygons linked to the border).

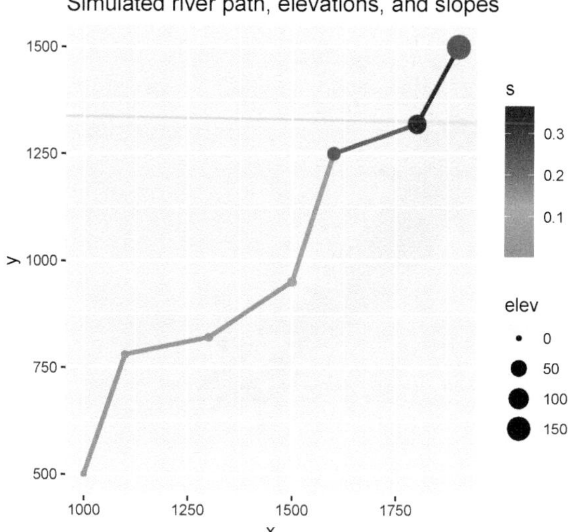

FIGURE 4.21 Channel slope as range from green to red, vertices sized by elevation

```
ggplot(riverData, aes(x,y)) +
  geom_line(mapping=aes(col=s), size=1.2) +
  geom_point(mapping=aes(col=s, size=elev)) +
  coord_fixed(ratio=1) + scale_color_gradient(low="green", high="red") +
  ggtitle("Simulated river path, elevations, and slopes")
```

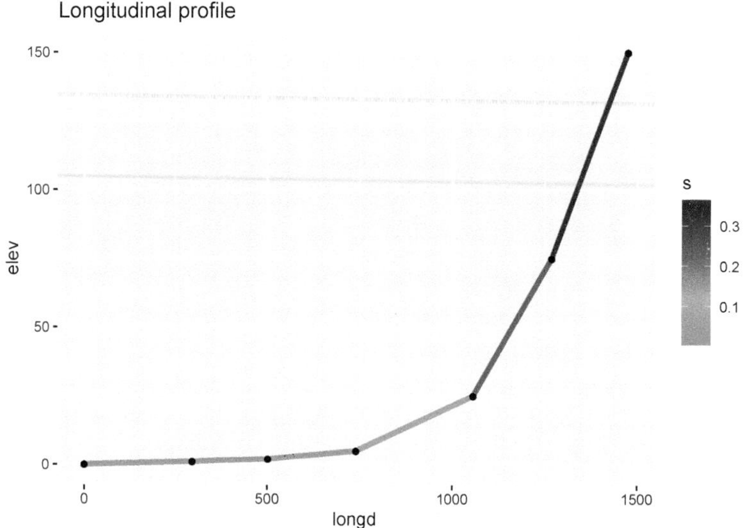

FIGURE 4.22 Channel slope as range of line colors on a longitudinal profile

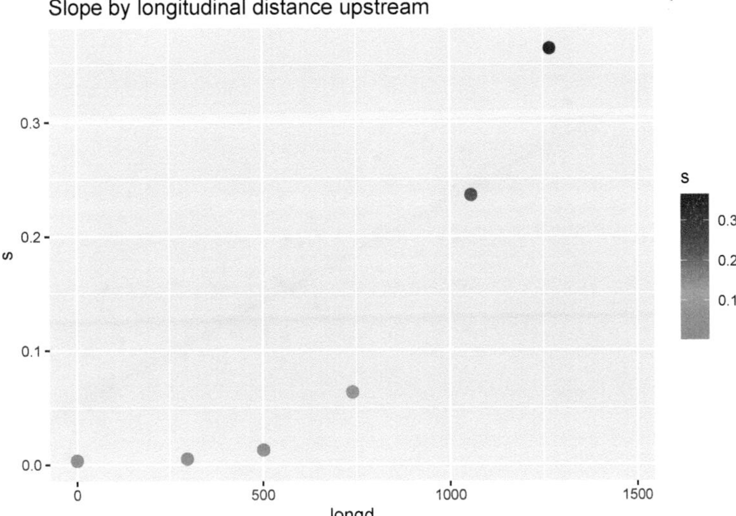

FIGURE 4.23 Channel slope by longitudinal distance as scatter points colored by slope

```
ggplot(riverData, aes(longd,elev)) + geom_line(aes(col=s), size=1.5) + geom_point()  +
  scale_color_gradient(low="green", high="red") +
  ggtitle("Longitudinal profile")
```

```
ggplot(riverData, aes(longd,s)) + geom_point(aes(col=s), size=3) +
  scale_color_gradient(low="green", high="red") +
  ggtitle("Slope by longitudinal distance upstream")
```

4.4.4 Trend line

When we get to statistical models, the first one we'll look at is a simple linear model. It's often useful to display this as a *trend line*, and this can be done with ggplot2's `geom_smooth()` function, specifying the linear model "lm" method. By default, the graph displays the standard error as a gray pattern (Figure 4.24).

```
sierraFeb %>%
  ggplot(aes(TEMPERATURE,ELEVATION)) +
  geom_point(color="blue") +
  geom_smooth(color="red", method="lm")
```

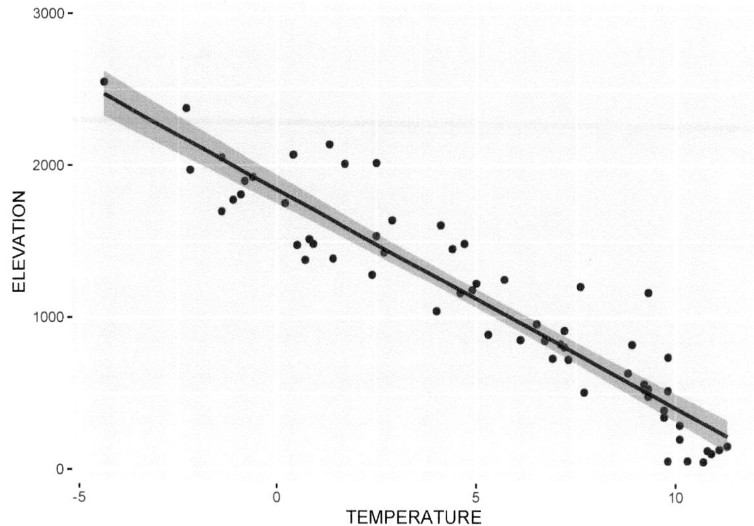

FIGURE 4.24 Trend line with a linear model

4.5 General Symbology

There's a lot to learn about symbology in graphs. We've included the basics, but readers are encouraged to also explore further. A useful vignette accessed by `vignette("ggplot2-specs")` lets you see aesthetic specifications for symbols, including:

- Color and fill
- Lines
  - line type, size, ends
- Polygon
  - border color, linetype, size
  - fill
- Points
  - shape
  - size
  - color and fill
  - stroke
- Text
  - font face and size
  - justification

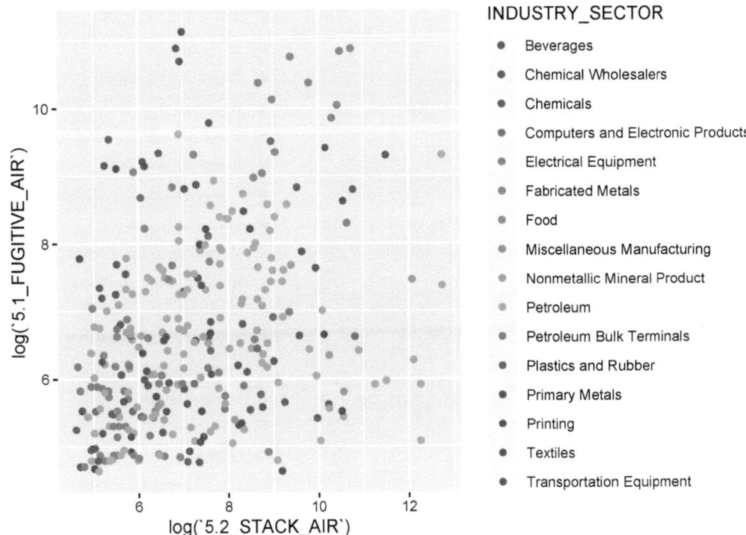

FIGURE 4.25 EPA TRI, categorical symbology for industry sector

4.5.1 Categorical symbology

One example of a "big data" resource is EPA's Toxic Release Inventory that tracks releases from a wide array of sources, from oil refineries on down. One way of dealing with big data in terms of exploring meaning is to use symbology to try to make sense of it (Figure 4.25).

```
library(igisci)
TRI <- read_csv(ex("TRI/TRI_2017_CA.csv")) %>%
  filter(`5.1_FUGITIVE_AIR` > 100 &
         `5.2_STACK_AIR` > 100 &
         `INDUSTRY_SECTOR` != "Other")
ggplot(data = TRI, aes(log(`5.2_STACK_AIR`), log(`5.1_FUGITIVE_AIR`),
                       color = INDUSTRY_SECTOR)) +
       geom_point()
```

4.5.2 Log scales instead of transform

In the above graph, we used the log() function in aes to use natural logarithms instead of the actual value. That's a simple way to do this. But what if we want to display the original data, just using a logarithmic grid? This might communicate better since readers would see the actual values (Figure 4.26).

```
ggplot(data=TRI, aes(`5.2_STACK_AIR`,`5.1_FUGITIVE_AIR`,color=INDUSTRY_SECTOR)) +
       geom_point() + scale_x_log10() + scale_y_log10()
```

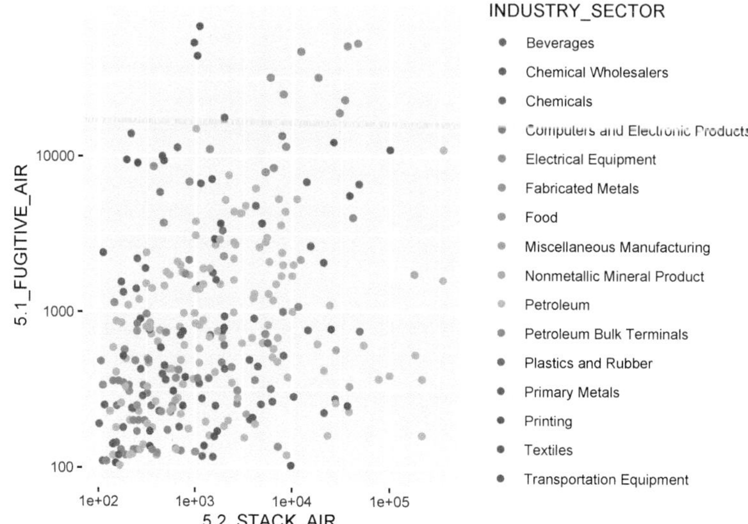

FIGURE 4.26 Using log scales instead of transforming

4.6 Graphs from Grouped Data

Earlier in this chapter, we used a pivot table to create a data frame `XSptsPheno`, which has NDVI values for phenology factors. [You may need to run that again if you haven't run it yet this session.] We can create graphs showing the relationship of NDVI and elevation grouped by phenology (Figure 4.27).

```
XSptsPheno %>%
  ggplot() +
  geom_point(aes(elevation, NDVI, shape=vegetation,
                 color = phenology), size = 3) +
  geom_smooth(aes(elevation, NDVI,
                 color = phenology), method="lm")
```

And similarly, we can create graphs of rainfall vs. runoff for eucs and oaks from the `tidy_eucoak` dataframe from the Thompson, Davis, and Oliphant (2016) study [and you may need to run that again to prep the data] (Figure 4.28).

```
ggplot(data = tidy_eucoak) +
  geom_point(mapping = aes(x = rain_mm, y = runoff_L, color = tree)) +
  geom_smooth(mapping = aes(x = rain_mm, y= runoff_L, color = tree),
            method = "lm") +
  scale_color_manual(values = c("seagreen4", "orange3"))
```

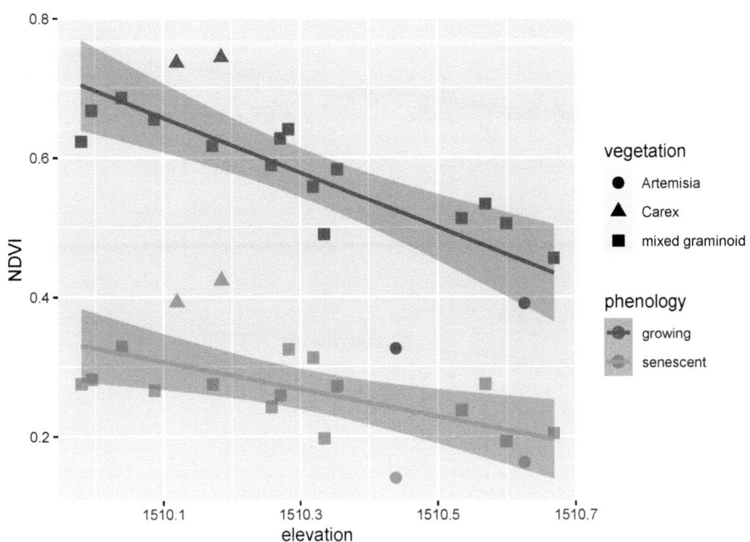

FIGURE 4.27 NDVI symbolized by vegetation in two seasons

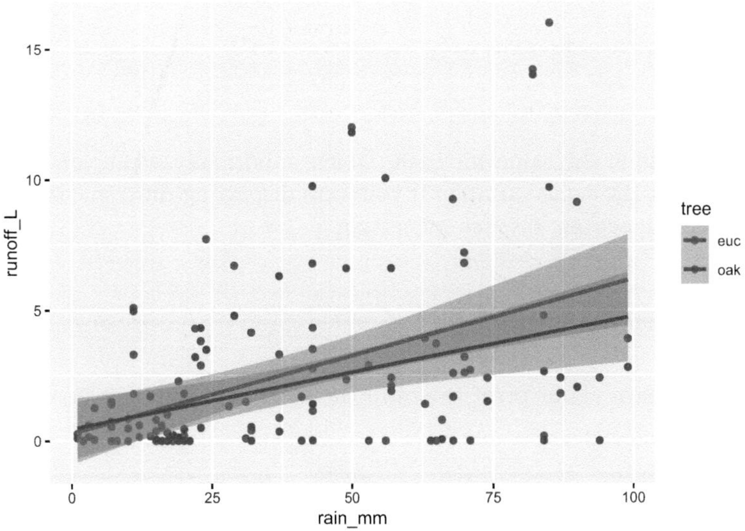

FIGURE 4.28 Eucalyptus and oak: rainfall and runoff

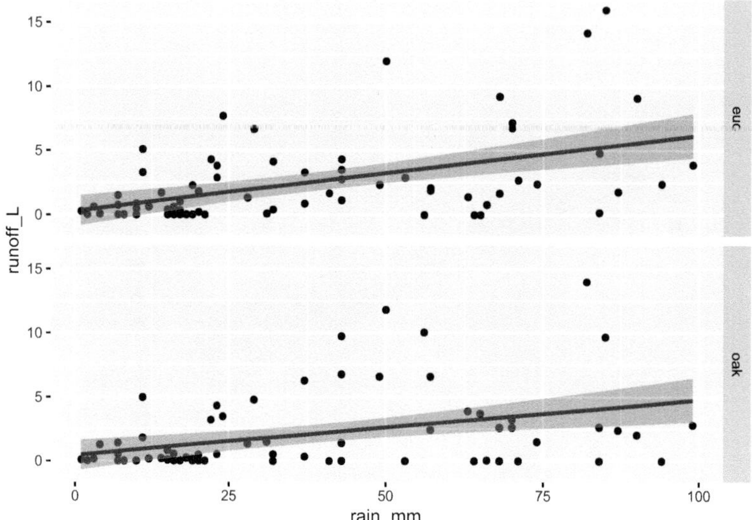

FIGURE 4.29 Faceted graph alternative to color grouping (note that the y scale is the same for each)

4.6.1 Faceted graphs

A theme we've already seen in this chapter is communicating more by comparing data on the same graph. We've been using symbology for that, but another approach is to create parallel groups of graphs called "faceted graphs" (Figure 4.29).

```
ggplot(data = tidy_eucoak) +
  geom_point(aes(x=rain_mm,y=runoff_L)) +
  geom_smooth(aes(x=rain_mm,y=runoff_L), method="lm") +
  facet_grid(tree ~ .)
```

Note that the y scale is the same for each, which is normally what you want since each graph is representing the same variable. If you were displaying different variables, however, you'd want to use the `scales = "free_y"` setting.

Again, we'll learn about pivot tables in the next chapter to set up our data.

4.7 Titles and Subtitles

All graphs need titles, and ggplot2 uses its `labs()` function for this (Figure 4.30).

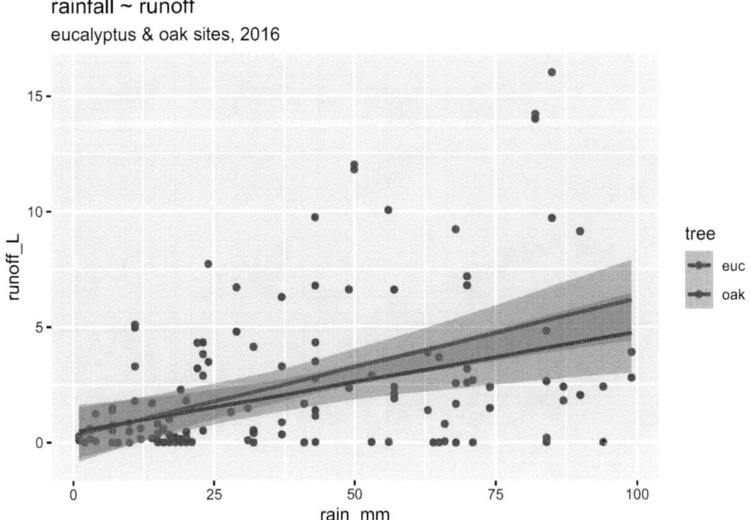

FIGURE 4.30 Titles added

```
ggplot(data = tidy_eucoak) +
  geom_point(aes(x=rain_mm,y=runoff_L, color=tree)) +
  geom_smooth(aes(x=rain_mm,y=runoff_L, color=tree), method="lm") +
  scale_color_manual(values=c("seagreen4","orange3")) +
  labs(title="rainfall ~ runoff",
       subtitle="eucalyptus & oak sites, 2016")
```

4.8 Pairs Plot

Pairs plots are an excellent exploratory tool to see which variables are correlated. Since only continuous data are useful for this, and since pairs plots can quickly get overly complex, it's good to use `dplyr::select` to select the continuous variables, or maybe use a helper function like `is.numeric` with `dplyr::select_if` (Figure 4.31).

```
sierraFeb %>%
     dplyr::select(ELEVATION:TEMPERATURE) %>%
     pairs()
```

The GGally package has a very nice-looking pairs plot that takes it even further. In addition to scatterplots, it has boxplots, histograms, density plots, and correlations, all stratified and colored by species. The code for this figure is mostly borrowed from Horst, Hill, and Gorman (2020) (Figure 4.32).

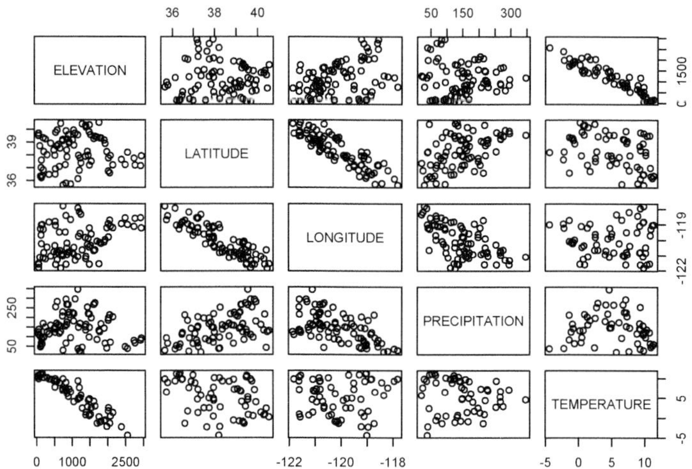

FIGURE 4.31 Pairs plot for Sierra Nevada stations variables

```
library(palmerpenguins)
penguins %>%
  dplyr::select(species, body_mass_g, ends_with("_mm")) %>%
  GGally::ggpairs(aes(color = species, alpha = 0.8)) +
  scale_colour_manual(values = c("darkorange","purple","cyan4")) +
  scale_fill_manual(values = c("darkorange","purple","cyan4"))
```

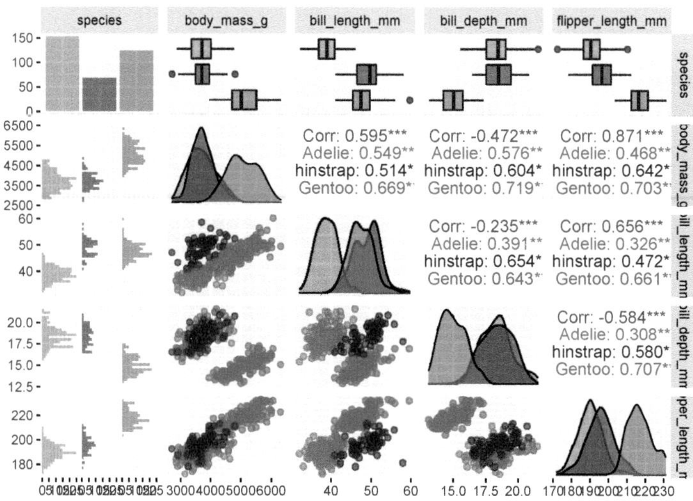

FIGURE 4.32 Enhanced GGally pairs plot for palmerpenguin data

4.9 Exercises: Visualization

Exercise 4.1. Create a bar graph of the counts of the species in the **`penguins`** data frame (Horst, Hill, and Gorman 2020). What can you say about what it shows?

Exercise 4.2. Use `bind_cols` in `dplyr` to create a tibble from built-in vectors `state.abb` and `state.region`, then use `ggplot` with `geom_bar` to create a bar graph of the four regions.

Exercise 4.3. Convert the built-in time series `treering` into a tibble `tr` using the `tibble()` function with the single variable assigned as `treering = treering` (or just specifying `treering` will also work for this simple example), then create a histogram, using that tibble and variable for the `data` and `x` settings needed. Attach a screen capture of the histogram. (Also, learn about the treering data by entering `?treering` in the console and read the Help displayed.)

Exercise 4.4. Create a new tibble `st` using `bind_cols` with `Name=state.name`, `Abb=state.abb`, `Region=state.region`, and `as_tibble(state.x77)`. *Note that this works since all of the parts are sorted by state.* From `st`, create a density plot from the variable `Frost` (number of days with frost for that state). What is the approximate modal value?

Exercise 4.5. From `st` create a a boxplot of `Area` by `Region`. Which region has the highest and which has the lowest median Area? Do the same for `Frost`.

Exercise 4.6. From `st`, compare murder rate (y is `Murder`) to `Frost` (as x) in a scatter plot, colored by `Region`.

Exercise 4.7. Add a trend line (smooth) with method="lm" to your scatterplot, not colored by `Region` (but keep the points colored by region). What can you say about what this graph is showing you?

Exercise 4.8. Add a title to your graph.

Exercise 4.9. Change your scatterplot to place labels using the `Abb` variable (still colored by Region) using `geom_label(aes(label=Abb, col=Region))`. Any observations about outliers?

Exercise 4.10. Change the boxplot of CO2 soil samples by site to use a log10 scale grid but display the original numbers (i.e. not in aes()).

5

Data Transformation

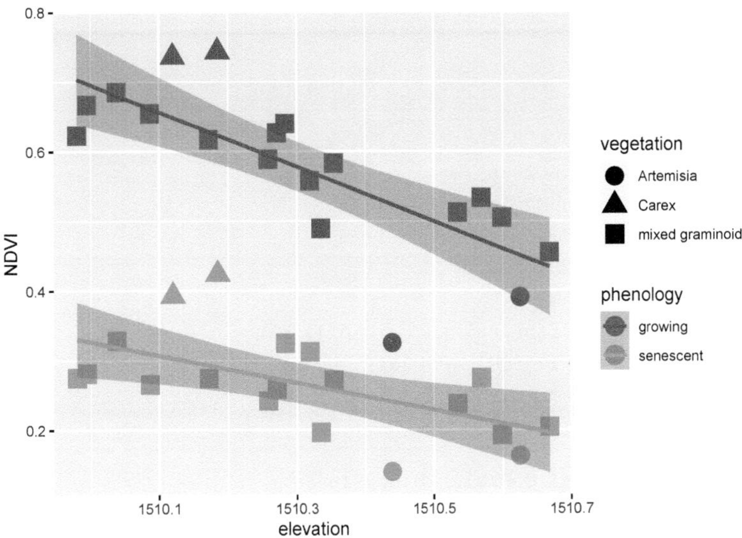

The goal of this section is to continue where we started in the earlier chapter on data abstraction with **`dplyr`** to look at the more transformational functions applied to data in a database, and **`tidyr`** adds other tools like pivot tables.

- **`dplyr`** tools:
  - joins: `left_join`, `right_join`, `inner_join`, `full_join`, `semi_join`, `anti_join`
  - set operations: `intersect`, `union`, `setdiff`
  - binding rows and columns: `bind_cols`, `bind_rows`
- **`tidyr`** tools:
  - pivot tables: `pivot_longer`, `pivot_wider`

The term "data wrangling" has been used for what we're doing with these tools, and the relevant cheat sheet used to be called that, but now they have names like "Data transformation with dplyr" and "Data tidying with tidyr", but those could change too. See what's current at https://www.rstudio.com/resources/cheatsheets/

5.1 Data joins

To bring in variables from another data frame based on a common join field. There are multiple types of joins. Probably the most common is **left_join** since it starts from the data frame (or sf) you want to continue working with and bring in data from an additional source. You'll retain all records of the first data set. For any non-matches, `NA` is assigned.

```
library(tidyverse)
library(igisci)
library(sf)
income <- read_csv(ex("CA/CA_MdInc.csv")) %>%
  dplyr::select(trID, HHinc2016) %>%
  mutate(HHinc2016 = as.numeric(HHinc2016),
         joinid = str_c("0", trID)) %>%
  dplyr::select(joinid, HHinc2016)
census <- BayAreaTracts %>%
  left_join(income, by = c("FIPS" = "joinid")) %>%
  dplyr::select(FIPS, POP12_SQMI, POP2012, HHinc2016)
head(census %>% st_set_geometry(NULL))
```

```
##          FIPS POP12_SQMI POP2012 HHinc2016
## 1 06001400100   1118.797    2976    177417
## 2 06001400200   9130.435    2100    153125
## 3 06001400300  11440.476    4805     85313
## 4 06001400400  14573.077    3789     99539
## 5 06001400500  15582.609    3584     83650
## 6 06001400600  13516.667    1622     61597
```

Other joins are:

- **right_join** where you end up retaining all the rows of the second data set and NA is assigned to non-matches
- **inner_join** where you only retain records for matches
- **full_join** where records are retained for both sides, and NAs are assigned to non-matches

Right join example We need to join NCDC monthly climate data for all California weather stations to a selection of 82 stations that are in the Sierra.

- The monthly data has 12 rows (1/month) for each station
- The right_join gets all months for all stations, so we weed out the non-Sierra stations by removing NAs from a field only with Sierra station data

```
sierra <- right_join(sierraStations, CA_ClimateNormals, by="STATION") %>%
  filter(!is.na(STATION_NA)) %>% dplyr::select(-STATION_NA)
head(sierra %>% filter(DATE == "01") %>%
  dplyr::select(NAME, ELEVATION, `MLY-TAVG-NORMAL`), n=10)
```

```
## # A tibble: 10 x 3
##    NAME                              ELEVATION `MLY-TAVG-NORMAL`
##    <chr>                                 <dbl>             <dbl>
##  1 GROVELAND 2, CA US                    853.                5.6
##  2 CANYON DAM, CA US                    1390.                0.2
##  3 KERN RIVER PH 3, CA US                824.                7.6
##  4 DONNER MEMORIAL ST PARK, CA US       1810.               -1.9
##  5 BOWMAN DAM, CA US                    1641.                3
##  6 BRUSH CREEK RANGER STATION, CA US    1085.               NA
##  7 GRANT GROVE, CA US                   2012.                1.9
##  8 LEE VINING, CA US                    2072.               -1.2
##  9 OROVILLE MUNICIPAL AIRPORT, CA US      57.9               7.7
## 10 LEMON COVE, CA US                     156.                8.6
```

The exact same thing, however, could be accomplished with an inner_join, and it doesn't required removing the NAs:

```
sierraAlso <- inner_join(sierraStations, CA_ClimateNormals, by="STATION") %>%
   dplyr::select(-STATION_NA)
```

5.2 Set operations

Set operations compare two data frames (or vectors) to handle observations or rows that are the same for each, or not the same. The three set methods are:

- `dplyr::intersect(x,y)` retains rows that appear in *both* x and y
- `dplyr::union(x,y)` retains rows that appear in either or both of x and y
- `dplyr::setdiff(x,y)` retains rows that appear in x but not in y

```
squares <- (1:10)^2
evens <- seq(0,100,2)
squares
```

```
##  [1]   1   4   9  16  25  36  49  64  81 100
```

```
evens
```

```
##  [1]   0   2   4   6   8  10  12  14  16  18  20  22  24  26  28  30  32  34  36
## [20]  38  40  42  44  46  48  50  52  54  56  58  60  62  64  66  68  70  72  74
## [39]  76  78  80  82  84  86  88  90  92  94  96  98 100
```

```
intersect(squares,evens)
```

```
## [1]   4  16  36  64 100
```

```
sort(union(squares,evens))
```

```
## [1]   0   1   2   4   6   8   9  10  12  14  16  18  20  22  24  25  26  28  30
## [20]  32  34  36  38  40  42  44  46  48  49  50  52  54  56  58  60  62  64  66
## [39]  68  70  72  74  76  78  80  81  82  84  86  88  90  92  94  96  98 100
```

```
sort(setdiff(squares,evens))
```

```
## [1]  1  9 25 49 81
```

5.3 Binding rows and columns

These `dplyr` functions are similar to `cbind` and `rbind` in base R, but always create data frames. For instance, `cbind` usually creates matrices and makes all vectors the same class. Note that in `bind_cols`, the order of data in rows must be the same.

```
states <- bind_cols(abb=state.abb,
                    name=state.name,
                    region=state.region,
                    state.x77)
head(states)
```

```
## # A tibble: 6 x 11
##   abb   name   region Popul~1 Income Illit~2 Life ~3 Murder HS Gr~4 Frost   Area
##   <chr> <chr>  <fct>    <dbl>  <dbl>   <dbl>   <dbl>  <dbl>   <dbl> <dbl>  <dbl>
## 1 AL    Alaba~ South     3615   3624     2.1    69.0   15.1    41.3    20  50708
## 2 AK    Alaska West       365   6315     1.5    69.3   11.3    66.7   152 566432
## 3 AZ    Arizo~ West      2212   4530     1.8    70.6    7.8    58.1    15 113417
## 4 AR    Arkan~ South     2110   3378     1.9    70.7   10.1    39.9    65  51945
## 5 CA    Calif~ West     21198   5114     1.1    71.7   10.3    62.6    20 156361
## 6 CO    Color~ West      2541   4884     0.7    72.1    6.8    63.9   166 103766
## # ... with abbreviated variable names 1: Population, 2: Illiteracy,
## #   3: `Life Exp`, 4: `HS Grad`
```

To compare, note that `cbind` converts numeric fields to character type when any other field is character, and character fields are converted to character integers where there are any repeats, which would require manipulating them into factors:

```
states <- as_tibble(cbind(abb=state.abb,
                          name=state.name,
                          region=state.region,
                          division=state.division,
```

```
                                   state.x77))
head(states)
```

```
## # A tibble: 6 x 12
##   abb   name  region divis~1 Popul~2 Income Illit~3 Life ~4 Murder HS Gr~5 Frost
##   <chr> <chr> <chr>  <chr>   <chr>   <chr>  <chr>   <chr>   <chr>  <chr>   <chr>
## 1 AL    Alab~ 2      4       3615    3624   2.1     69.05   15.1   41.3    20
## 2 AK    Alas~ 4      9       365     6315   1.5     69.31   11.3   66.7    152
## 3 AZ    Ariz~ 4      8       2212    4530   1.8     70.55   7.8    58.1    15
## 4 AR    Arka~ 2      5       2110    3378   1.9     70.66   10.1   39.9    65
## 5 CA    Cali~ 4      9       21198   5114   1.1     71.71   10.3   62.6    20
## 6 CO    Colo~ 4      8       2541    4884   0.7     72.06   6.8    63.9    166
## # ... with 1 more variable: Area <chr>, and abbreviated variable names
## #   1: division, 2: Population, 3: Illiteracy, 4: `Life Exp`, 5: `HS Grad`
```

5.4 Pivoting data frames

Pivot tables are a popular tool in Excel, allowing you to transform your data to be more useful in a particular analysis. A common need to pivot is 2+ variables with the same data where the variable name should be a factor. `Tidyr` has **`pivot_wider`** and **`pivot_longer`**.

- **`pivot_wider`** pivots rows into variables.
- **`pivot_longer`** pivots variables into rows, creating factors.

5.4.1 `pivot_longer`

In our meadows study cross-section (JD Davis et al. 2020) created by intersecting normalized difference vegetation index (NDVI) values from multispectral drone imagery with surveyed elevation and vegetation types (xeric, mesic, and hydric), we have fields `NDVIgrowing` from a July 2019 growing season and `NDVIsenescent` from a September 2020 dry season, but would like "growing" and "senescent" to be factors with a single `NDVI` variable. This is how we used `pivot_longer` to accomplish this, using data from the `igisci` data package:

```
XSptsPheno <- XSptsNDVI %>%
      filter(vegetation != "pine") %>%     # trees removed
      pivot_longer(cols = starts_with("NDVI"),
                   names_to = "phenology", values_to = "NDVI") %>%
      mutate(phenology = str_sub(phenology, 5, str_length(phenology)))
```

To see what the reverse would be, we'd use `pivot_wider` to return to the original, but note that we're not writing over our `XSptsPheno` data frame.

```
XSptsPheno %>%
  pivot_wider(names_from = phenology, names_prefix = "NDVI",
              values_from = NDVI)
```

```
## # A tibble: 19 x 6
##    DistNtoS elevation vegetation      geometry                       NDVIg~1 NDVIs~2
##       <dbl>     <dbl> <chr>           <chr>                            <dbl>   <dbl>
##  1      0       1510. Artemisia       c(718649.456, 4397466.714)       0.326   0.140
##  2     16.7     1510. mixed graminoid c(718649.4309, 4397450.07~       0.627   0.259
##  3     28.6     1510. mixed graminoid c(718649.413, 4397438.222)       0.686   0.329
##  4     30.5     1510. mixed graminoid c(718649.4101, 4397436.33)       0.668   0.282
##  5     31.1     1510. mixed graminoid c(718649.4092, 4397435.73~       0.655   0.266
##  6     33.4     1510. mixed graminoid c(718649.4058, 4397433.44~       0.617   0.274
##  7     35.6     1510. mixed graminoid c(718649.4025, 4397431.24~       0.623   0.275
##  8     37       1510. mixed graminoid c(718649.4004, 4397429.85~       0.589   0.242
##  9     74       1510. mixed graminoid c(718649.3448, 4397392.99~       0.641   0.325
## 10    101       1510. mixed graminoid c(718649.3042, 4397366.09~       0.558   0.312
## 11    126.      1511. Artemisia       c(718649.2672, 4397341.59)       0.391   0.163
## 12    137       1510. mixed graminoid c(718649.25, 4397330.233)        0.583   0.272
## 13    149.      1510. Carex           c(718649.2317, 4397318.07~       0.736   0.392
## 14    154.      1510. Carex           c(718649.224, 4397312.999)       0.744   0.424
## 15    175       1511. mixed graminoid c(718649.1929, 4397292.37~       0.455   0.204
## 16    195.      1511. mixed graminoid c(718649.1633, 4397272.75~       0.512   0.237
## 17    197.      1510. mixed graminoid c(718649.1597, 4397270.36~       0.489   0.197
## 18    201.      1511. mixed graminoid c(718649.1544, 4397266.88~       0.533   0.275
## 19    259.      1511. mixed graminoid c(718649.0663, 4397208.49~       0.505   0.193
## # ... with abbreviated variable names 1: NDVIgrowing, 2: NDVIsenescent
```

```
XSptsPheno
```

```
## # A tibble: 38 x 6
##    DistNtoS elevation vegetation      geometry                    phenol~1  NDVI
##       <dbl>     <dbl> <chr>           <chr>                       <chr>    <dbl>
##  1      0       1510. Artemisia       c(718649.456, 4397466.714)  growing  0.326
##  2      0       1510. Artemisia       c(718649.456, 4397466.714)  senesce~ 0.140
##  3     16.7     1510. mixed graminoid c(718649.4309, 4397450.077) growing  0.627
##  4     16.7     1510. mixed graminoid c(718649.4309, 4397450.077) senesce~ 0.259
##  5     28.6     1510. mixed graminoid c(718649.413, 4397438.222)  growing  0.686
##  6     28.6     1510. mixed graminoid c(718649.413, 4397438.222)  senesce~ 0.329
##  7     30.5     1510. mixed graminoid c(718649.4101, 4397436.33)  growing  0.668
##  8     30.5     1510. mixed graminoid c(718649.4101, 4397436.33)  senesce~ 0.282
##  9     31.1     1510. mixed graminoid c(718649.4092, 4397435.732) growing  0.655
## 10     31.1     1510. mixed graminoid c(718649.4092, 4397435.732) senesce~ 0.266
## # ... with 28 more rows, and abbreviated variable name 1: phenology
```

We'll use the `pivot_longer` result (the one we actually assigned to `XSptsPheno`) to allow us to create the graph we're after (Figure 5.1).

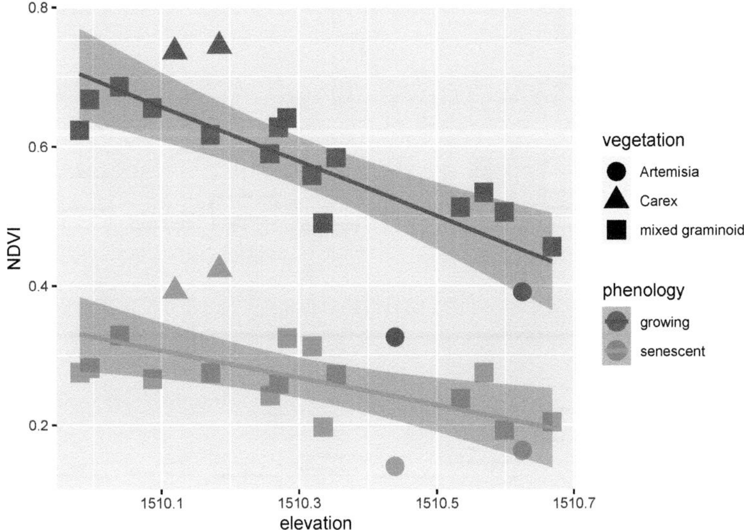

FIGURE 5.1 Color classified by phenology, data created by a pivot

```
XSptsPheno %>%
  ggplot() +
  geom_point(aes(elevation, NDVI, shape=vegetation,
                 color = phenology), size = 5) +
  geom_smooth(aes(elevation, NDVI,
                 color = phenology), method="lm")
```

Pivots turn out to be commonly useful. We've already seen their use in the Visualization chapter, such as when we graphed runoff from the Eucalyptus/Oak study (Thompson, Davis, and Oliphant 2016), where we used a `pivot_longer` (Figure 5.2).

```
eucoakrainfallrunoffTDR %>%
  pivot_longer(cols = starts_with("runoffL"),
               names_to = "tree", values_to = "runoffL") %>%
  mutate(tree = str_sub(tree, str_length(tree)-2, str_length(tree))) %>%
  ggplot() + geom_boxplot(aes(site, runoffL)) +
    facet_grid(tree ~ .)
```

Combining a pivot with bind_rows to create a runoff/rainfall scatterplot colored by tree

With a bit more code, we can combine pivoting with binding rows to set up a useful scatter plot (Figure 5.3).

```
runoffPivot <- eucoakrainfallrunoffTDR %>%
  pivot_longer(cols = starts_with("runoffL"),
               names_to = "tree", values_to = "runoffL") %>%
```

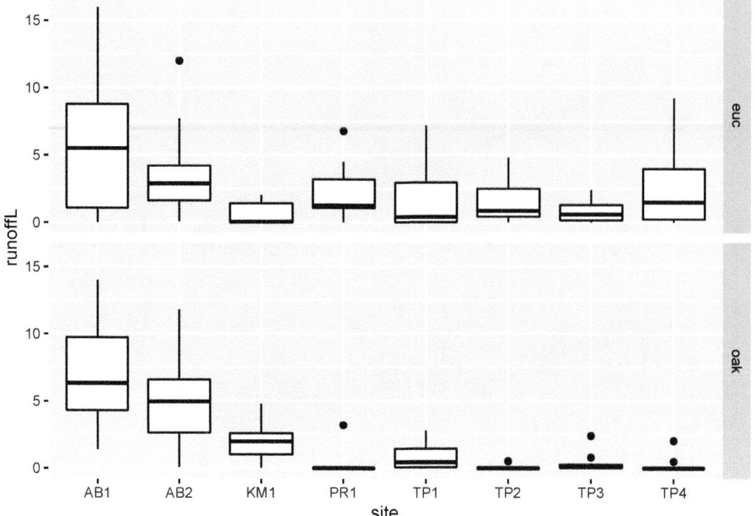

FIGURE 5.2 Euc vs oak graphs created using a pivot

```
  mutate(tree = str_sub(tree, str_length(tree)-2, str_length(tree)),
         Date = as.Date(date, "%m/%d/%Y"))
euc <- runoffPivot %>%
  filter(tree == "euc") %>%
  mutate(rain_subcanopy = rain_euc,
         slope = slope_euc,     aspect = aspect_euc,
         surface_tension = surface_tension_euc,
```

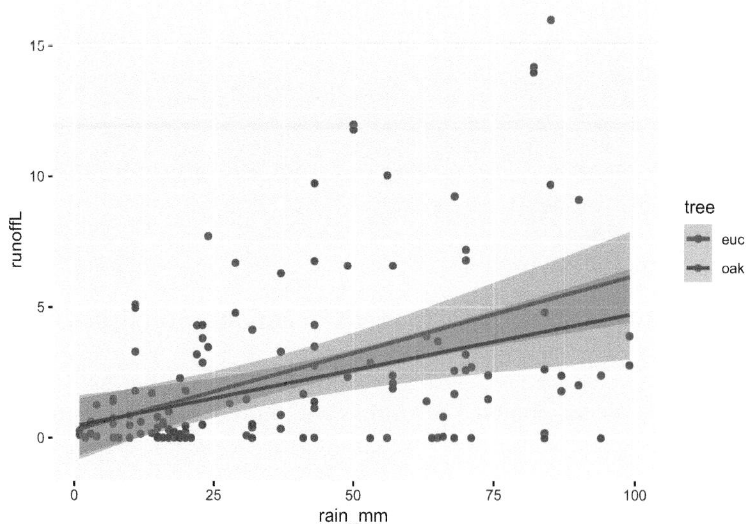

FIGURE 5.3 Runoff/rainfall scatterplot colored by tree, created by pivot and binding rows

```
         runoff_rainfall_ratio = runoff_rainfall_ratio_euc) %>%
  dplyr::select(site, `site #`, tree, Date, month, rain_mm,
         rain_subcanopy, slope, aspect, runoffL,
         surface_tension, runoff_rainfall_ratio)
oak <- runoffPivot %>%
  filter(tree == "oak") %>%
  mutate(rain_subcanopy = rain_oak,
         slope = slope_oak, aspect = aspect_oak,
         surface_tension = surface_tension_oak,
         runoff_rainfall_ratio = runoff_rainfall_ratio_oak) %>%
  dplyr::select(site, `site #`, tree, Date, month, rain_mm,
         rain_subcanopy, slope, aspect, runoffL,
         surface_tension, runoff_rainfall_ratio)
bind_rows(euc, oak) %>%
  ggplot() +
  geom_point(mapping = aes(x = rain_mm, y = runoffL, color = tree)) +
  geom_smooth(mapping = aes(x = rain_mm, y= runoffL, color = tree),
              method = "lm") +
  scale_color_manual(values = c("seagreen4", "orange3"))
```

5.4.2 **pivot_wider**

The opposite of `pivot_longer`, `pivot_wider` is less commonly used for tidying data, but can be useful for creating tables of that desired format. An environmental application of `pivot_wider` can be found in `vignette("pivot")` (modified below) for studying fish detected by automatic monitors, with `fish_encounters` data contributed by Johnston and Rudis (n.d.) . This pivot makes it easier to see fish encounters by station. See Johnston and Rudis (n.d.) and `vignette("pivot")` for more information on the dataset and how the pivot provides this view.

```
library(tidyverse)
fish_encounters <- read_csv(ex("fishdata.csv"))
fish_encounters
```

```
## # A tibble: 209 x 3
##     TagID Station value
##     <dbl> <chr>   <dbl>
##   1  4842 Release     1
##   2  4843 Release     1
##   3  4844 Release     1
##   4  4845 Release     1
##   5  4847 Release     1
##   6  4848 Release     1
##   7  4849 Release     1
##   8  4850 Release     1
##   9  4851 Release     1
##  10  4854 Release     1
## # ... with 199 more rows
```

```
fishEncountersWide <- fish_encounters %>%
  pivot_wider(names_from = Station, values_from = value, values_fill = 0)
fishEncountersWide
```

```
## # A tibble: 19 x 12
##     TagID Release I80_1 Lisbon  Rstr Base_TD   BCE   BCW  BCE2  BCW2   MAE   MAW
##     <dbl>   <dbl> <dbl>  <dbl> <dbl>   <dbl> <dbl> <dbl> <dbl> <dbl> <dbl> <dbl>
##  1  4842       1     1      1     1       1     1     1     1     1     1     1
##  2  4843       1     1      1     1       1     1     1     1     1     1     1
##  3  4844       1     1      1     1       1     1     1     1     1     1     1
##  4  4845       1     1      1     1       1     0     0     0     0     0     0
##  5  4847       1     1      1     0       0     0     0     0     0     0     0
##  6  4848       1     1      1     1       0     0     0     0     0     0     0
##  7  4849       1     1      0     0       0     0     0     0     0     0     0
##  8  4850       1     1      0     1       1     1     1     0     0     0     0
##  9  4851       1     1      0     0       0     0     0     0     0     0     0
## 10  4854       1     1      0     0       0     0     0     0     0     0     0
## 11  4855       1     1      1     1       1     0     0     0     0     0     0
## 12  4857       1     1      1     1       1     1     1     1     1     0     0
## 13  4858       1     1      1     1       1     1     1     1     1     1     1
## 14  4859       1     1      1     1       1     0     0     0     0     0     0
## 15  4861       1     1      1     1       1     1     1     1     1     1     1
## 16  4862       1     1      1     1       1     1     1     1     1     0     0
## 17  4863       1     1      0     0       0     0     0     0     0     0     0
## 18  4864       1     1      0     0       0     0     0     0     0     0     0
## 19  4865       1     1      1     0       0     0     0     0     0     0     0
```

5.4.3 A `free_y` faceted graph using a pivot

Creating parallel multi-parameter graphs over a time series can be challenging. We need to link the graphs with a common x axis, but we may need to vary the scaling on the y axis, unlike the faceted graph of grouped data we looked at in the visualization chapter. We'll look at time series in a later chapter, but this is a good time to explore the use of a pivot_longer to set up the data for a graph like this, and at the same time to expand upon our visualization toolkit.

Flux tower data

We'll look at this data set in more depth in the time series chapter, but here's a quick introduction. To look at micrometeorological changes related to phenological changes in vegetation from seasonal hydrologic changes from snowmelt through summer drying, we captured an array of variables at a flux tower in Loney Meadow in the South Yuba River watershed of the northern Sierra during the 2016 season (Figure 5.4).

A spreadsheet of 30-minute summaries from 17 May to 6 September can be found in the igisci extdata folder as "meadows/LoneyMeadow_30minCO2fluxes_Geog604.xls", and among other parameters includes data on CO_2 flux, net radiation (Qnet), air temperature (Tair), and relative humidity (RH). There's clearly a lot more we can do with these data

FIGURE 5.4 Flux tower installed at Loney Meadow, 2016. Photo credit: Darren Blackburn

(see Blackburn, Oliphant, and Davis (2021)), but we'll look at this selection of parameters to see how we can use a pivot to create a multi-parameter graph.

First, we'll read in the data and simplify some of the variables. Since the second row of data contains measurement units, we needed to wrangle the data a bit to capture the variable names then add those back after removing the first two rows:

```
library(readxl); library(tidyverse); library(lubridate)
vnames <- read_xls(ex("meadows/LoneyMeadow_30minCO2fluxes_Geog604.xls"),
                  n_max=0) %>% names()
vunits <- read_xls(ex("meadows/LoneyMeadow_30minCO2fluxes_Geog604.xls"),
                  n_max=0, skip=1) %>% names()
Loney <- read_xls(ex("meadows/LoneyMeadow_30minCO2fluxes_Geog604.xls"),
                 skip=2, col_names=vnames) %>%
         rename(YDay = `Day of Year`, CO2flux = `CO2 Flux`)
```

The time unit we'll want to use for time series is going to be days, and we can also then look at the data over time, and a group_by summarization by days will give us a generalized picture of changes over the collection period reflecting phenological changes from first exposure after snowmelt through the maximum growth period and through the major senescence period of late summer. We'll create the data to graph by using a group_by summarize to create a daily picture of a selection of four micrometeorological parameters:

```
LoneyDaily <- Loney %>%
  group_by(YDay) %>%
  summarize(CO2flux = mean(CO2flux),
            Qnet = mean(Qnet),
            Tair = mean(Tair),
            RH = mean(RH))
```

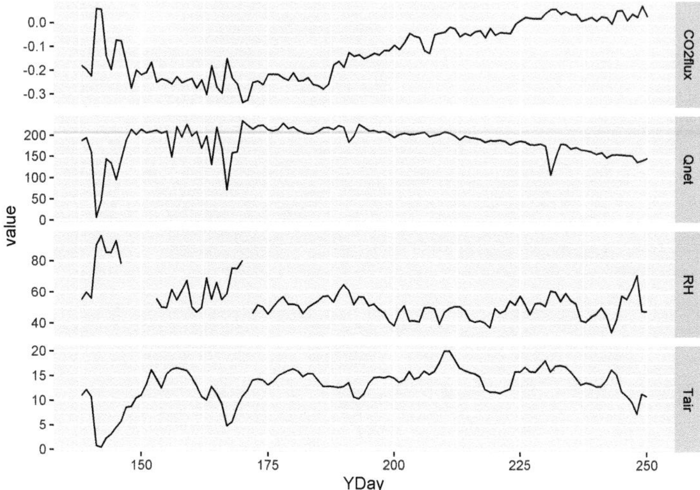

FIGURE 5.5 free-y facet graph supported by pivot (note the y axis scaling varies among variables)

Then from this daily data frame, we pivot_longer to store all of the parameter names in a `parameter` variable and all parameter values in a `value` variable:

```
LoneyDailyLong <- LoneyDaily %>%
  pivot_longer(cols = CO2flux:RH,
               names_to="parameter",
               values_to="value") %>%
  filter(parameter %in% c("CO2flux", "Qnet", "Tair", "RH"))
```

Now we have what we need to create a multi-parameter graph using facet_grid – note the scales = "free_y" setting to allow each variable's y axis to correspond to its value range (Figure 5.5).

```
p <- ggplot(data = LoneyDailyLong, aes(x=YDay, y=value)) +
  geom_line()
p + facet_grid(parameter ~ ., scales = "free_y")
```

The need to create a similar graph from soil CO_2 data inspired me years ago to learn how to program Excel. It was otherwise impossible to get Excel to make all the x scales the same, so I learned how to force it with code...

5.5 Exercise: Transformation

by Josh von Nonn (2021)

The impetus behind this exercise came from the movie *Dark Waters* (https://www.youtube.com/watch?v=RvAOuhyunhY), inspired by a true story of negligent chemical waste disposal by Dupont.

First create a new RStudio project, named `GAMA` and save this .Rmd file there, and create a folder `GAMA_water_data` in the project folder; the path to the data can then be specified as `"GAMA_water_data/gama_pfas_statewide_v2.txt"` assuming that the latter name matches what is unzipped from your download. Change to match the actual name if it differs.

Then download from the California Water Boards, GAMA groundwater website: https://gamagroundwater.waterboards.ca.gov/gama/datadownload

Then select "Statewide Downloads" then "Statewide PFOS Data" and extract the zip file into the `GAMA_water_data` folder. This is a large txt file so if you open it with notepad it may take some time to load. Notice that this is a space delimited file.

Note that this data structure is similar to what was discussed in the introductory chapter in how you should use RStudio projects to organize your data, allowing *relative paths* to your data, such as "GAMA_water_data/gama_pfas_statewide_v2.txt", which will work wherever you copy your project folder. An absolute path to somewhere on your computer in contrast won't work for anyone else trying to run your code; absolute paths should only be used for servers that other users have access to and URLs on the web.

Required packages:

1. Read in the downloaded file `"gama_pfas_statewide_v2.txt"` and call it `cal_pfas` and have a look at the data set. You can select if from the Environment pane or use `view(cal_pfas)`.

2. Before we clean up this data, let's preserve the locations of all the wells. Create a new data frame, `cal_pfas_loc`, Select `GM_WELL_ID`, `GM_LATITUDE`, and `GM_LONGITUDE` and remove all the duplicate wells (hint: dplyr cheat sheet provides a function to do this).

3. Now to trim down the data. Create a new data frame, `cal_pfas_trim`; add a new column, `DATE`, using the associated `lubridate` function on `GM_SAMP_COLLECTION_DATE` (this will allow ggplot to recognize it as a date and not a character string), select `GM_WELL_ID,GM_CHEMICAL_VVL,GM_RESULT`, and the newly created `DATE`. Sort (`arrange`) the data by `GM_WELL_ID`.

4. Use `pivot_wider` to create new columns from the chemical names and values from the result column and store the new data frame as `cal_pfas_wide`. Notice the warnings. Some of the wells have multiple samples on the same day so they will be put into a vector (ex. c(6.8,9,4.3,etc..)). Rerun the pivot but include the argument `values_fn = mean`. This will return only the average of all the samples. Once the pivot is working correctly, keep the columns `GM_WELL_ID,DATE,PFOS,PFOA`

and pipe a mutate to create a new column, `SUMPFS`, that is the sum of `PFOS` and `PFOA`.

The US EPA has established a lifetime Health Advisory Level (HAL) for PFOA and PFOS of 70 ng/L. When both PFOA and PFOS are found in drinking water, the combined concentrations of PFOA and PFOS should be compared with the 70 ng/L HAL. From the GROUNDWATER INFORMATION SHEET for PFOA (website:https://www.waterboards.ca.gov/water_issues/programs/gama/factsheets.html)

---

5. For the sake of creating an interesting time series plot, let's filter data for wells that have a `SUMPFS` greater than 70 and that have more than 10 sampling dates. Start by creating a new data frame- `cal_pfas_index` from the pivoted data frame. Hint: one way to do this is use `group_by`, `filter`, `count`, and `filter` again. The resulting data frame downloaded in 2021 had 11 observations and 2 variables.

6. Create a new data frame, `cal_pfas_max` to locate the well with the most sampling dates (n). The data wrangling function from base R, `subset`, can do this using the `max` function as an argument.

7. Now let's pull all the data on that well by joining the max indexed well with `cal_pfas_wide` and call the new data frame `cal_pfas_join`. Remove the "n" column using the select function.

8. Create a new data frame, `cal_pfs_long` and `pivot_longer` the `cal_pfs_join` data, creating new columns: `"chemical"` and `"ngl"`.

9. Plot the well using the wide data from `cal_pfs_join` with ggplot, using `DATE` on the x axis and plot the three variables (`PFOS,PFOA,SUMPFS`) with different colored lines of your choice. Add a horizontal reference line (`geom_hline(yintercept = 70)`) for the HAL limit at 70.

10. Plot the well using the long data from `cal_pfs_long` using DATE on the x axis and `ngl` on the y axis. Distinguish the chemicals by setting the line color to `Chemical` in the aesthetics. Add the horizontal reference at 70 (Figure 5.6).

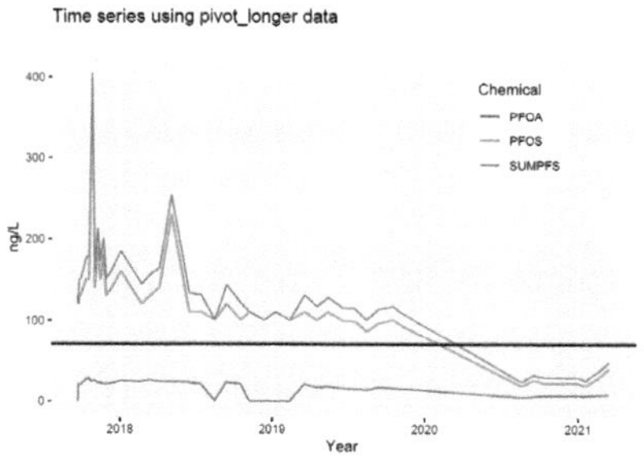

FIGURE 5.6 Goal

Part II

Spatial

6

Spatial Data and Maps

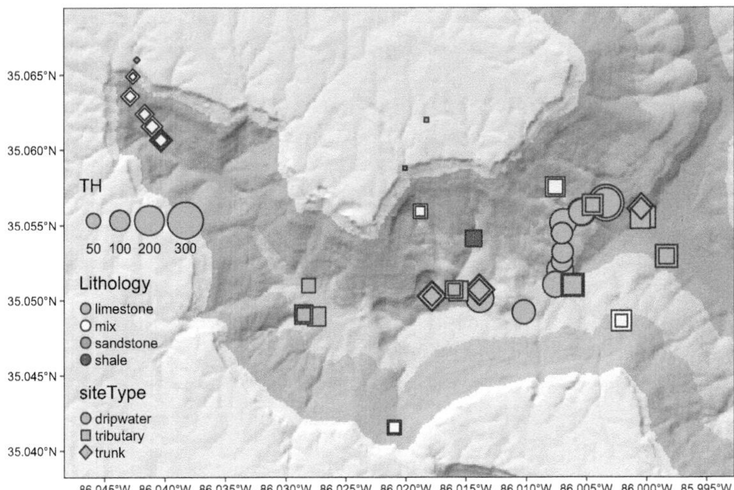

The Spatial section of this book adds the spatial dimension and *geographic* data science. Most environmental systems are significantly affected by location, so the geographic perspective is highly informative. In this chapter, we'll explore the basics of

- building and using feature-based (vector) data using the `sf` (simple features) package
- raster data using the `terra` package
- some common mapping methods in `plot` and `ggplot2`

Especially for the feature data, which is built on a database model (with geometry as one variable), the tools we've been learning in `dplyr` and `tidyr` will also be useful to working with attributes.

Caveat: Spatial data and analysis is very useful for environmental data analysis, and we'll only scratch the surface. Readers are encouraged to explore other resources on this important topic, such as:

- *Geocomputation with R* at https://geocompr.robinlovelace.net/
- *Simple Features for R* at https://r-spatial.github.io/sf/
- *Spatial Data Science with R* at https://rspatial.org

6.1 Spatial Data

Spatial data are data using a Cartesian coordinate system with x, y, z, and maybe more dimensions. Geospatial data are spatial data that we can map on our planet and relate to other geospatial data based on geographic coordinate systems (GCS) of longitude and latitude or known projections to planar coordinates like Mercator, Albers, or many others. You can also use local coordinate systems with the methods we'll learn, to describe geometric objects or a local coordinate system to "map out" an archaeological dig or describe the movement of ants on an anthill, but we'll primarily use geospatial data.

To work with spatial data requires extending R to deal with it using packages. Many have been developed, but the field is starting to mature using international open GIS standards.

`sp` (until recently, the dominant library of spatial tools)

- Includes functions for working with spatial data
- Includes `spplot` to create maps
- Also needs `rgdal` package for `readOGR` – reads spatial data frames
- Earliest archived version on CRAN is 0.7-3, 2005-04-28, with 1.0 released 2012-09-30. Authors: Edzer Pebesma and Roger Bivand

`sf` (Simple Features)

- Feature-based (vector) methods
- ISO 19125 standard for GIS geometries
- Also has functions for working with spatial data, but clearer to use
- Doesn't need many additional packages, though you may still need `rgdal` installed for some tools you want to use
- Replacing `sp` and `spplot` though you'll still find them in code
- Works with ggplot2 and tmap for nice looking maps
- Cheat sheet and other information at https://r-spatial.github.io/sf/
- Earliest CRAN archive is 0.2 on 2016-10-26, 1.0 not until 2021-06-29
- Author: Edzer Pebesma

`raster`

- The established raster package for R, with version 1.0 from 2010-03-20, Author: Robert J. Hijmans (Professor of Environmental Science and Policy at UC Davis)
- Depends on `sp`
- CRAN: "Reading, writing, manipulating, analyzing and modeling of spatial data. The package implements basic and high-level functions for raster data and for vector data operations such as intersections. See the manual and tutorials on https://rspatial.org/ to get started."

`terra`

- Intended to eventually replace `raster`, by the same author (Hijmans), current version 1.6-7 2022-08-07; does not depend on `sp`, but includes Bivand and Pebesma as contributors
- CRAN: "Methods for spatial data analysis with raster and vector data. Raster methods

TABLE 6.1 Some common sf functions for building geometries from coordinates

sf function	input
st_point()	numeric vector of length 2, 3, or 4, represent a single point
st_multipoint()	numeric matrix with points in rows
st_linestring()	numeric matrix with points in rows
st_multilinestring()	list of numeric matrices with points in rows
st_polygon()	list of numeric matrices with points in rows
st_multipolygon()	list of lists with numeric matrices

allow for low-level data manipulation as well as high-level global, local, zonal, and focal computation. The predict and interpolate methods facilitate the use of regression type (interpolation, machine learning) models for spatial prediction, including with satellite remote sensing data. Processing of very large files is supported. See the manual and tutorials on https://rspatial.org/terra/ to get started."

6.1.1 Simple geometry building in sf

Simple Features (`sf`) includes a set of simple geometry building tools that use matrices or lists of matrices of XYZ (or just XY) coordinates (or lists of lists for multipolygon) as input, and each basic geometry type having single and multiple versions (Table 6.1). Note that sf functions have the pattern `st_*` (st means "space and time").

```
library(tidyverse)
library(sf)
library(igisci)
```

6.1.1.1 Building simple sf geometries

We'll build some simple geometries and plot them, just to illustrate the types of geometries. Try the following code.

```
library(sf); library(tidyverse)
eyes <- st_multipoint(rbind(c(1,5), c(3,5)))
nose <- st_point(c(2,4))
mouth <- st_linestring(rbind(c(1,3),c(3, 3)))
border <- st_polygon(list(rbind(c(0,5), c(1,2), c(2,1), c(3,2),
                               c(4,5), c(3,7), c(1,7), c(0,5))))
plot(st_sfc(eyes, nose, mouth, border))
```

6.1.1.2 Building geospatial points as sf columns (`sfc`)

The face-building example above is really intended for helping to understand the concept and isn't really how we'd typically build data. For one thing, it doesn't have a real coordinate system that can be reprojected into locations anywhere on the planet. And most commonly we'll be accessing data that have been created by government agencies such as those from the USGS in The National Map, or NGO, university, or corporate sources who have an interest in collecting and providing data.

See Geocomputation with R at https://geocompr.robinlovelace.net/ or https://r-spatial.github.io/sf/ for more details, but when building spatial data from scratch, we would need to follow this by building a simple feature geometry list column from the collection with `st_sfc()`. We'll look at this using real geographic coordinates, building a generalized map of California and Nevada boundaries (and then later in the exercises, you'll expand this to a map of the west.) We'll use a shortcut to the coordinate referencing system (crs) by using a short numeric EPSG code (*EPSG Geodetic Parameter Dataset* (n.d.)), 4326, for the common geographic coordinate system (GCS) of latitude and longitude.

```
library(sf)
CA_matrix <- rbind(c(-124,42),c(-120,42),c(-120,39),c(-114.5,35),
  c(-114.1,34.3),c(-114.6,32.7),c(-117,32.5),c(-118.5,34),c(-120.5,34.5),
  c(-122,36.5),c(-121.8,36.8),c(-122,37),c(-122.4,37.3),c(-122.5,37.8),
  c(-123,38),c(-123.7,39),c(-124,40),c(-124.4,40.5),c(-124,41),c(-124,42))
NV_matrix <- rbind(c(-120,42),c(-114,42),c(-114,36),c(-114.5,36),
  c(-114.5,35),c(-120,39),c(-120,42))
CA_list <- list(CA_matrix);       NV_list <- list(NV_matrix)
CA_poly <- st_polygon(CA_list);   NV_poly <- st_polygon(NV_list)
sfc_2states <- st_sfc(CA_poly,NV_poly,crs=4326)  # crs=4326 specifies GCS
st_geometry_type(sfc_2states)
```

```
## [1] POLYGON POLYGON
## 18 Levels: GEOMETRY POINT LINESTRING POLYGON MULTIPOINT ... TRIANGLE
```

Then we'll use the `geom_sf()` function in ggplot2 to create a map (Figure 6.1).

```
library(tidyverse)
ggplot() + geom_sf(data = sfc_2states)
```

6.1.1.3 Building an sf class from sf columns.

So far, we've built an `sfg` (sf geometry) and then a collection of matrices in an `sfc` (sf column), and we were even able to make a map of it. But to be truly useful, we'll want other attributes other than just where the feature is, so we'll build an `sf` that has attributes, and that can also be used like a data frame – similar to a shapefile, since a data frame is like a database table `.dbf`. We'll build one with the `st_sf()` function from attributes and geometry (Figure 6.2).

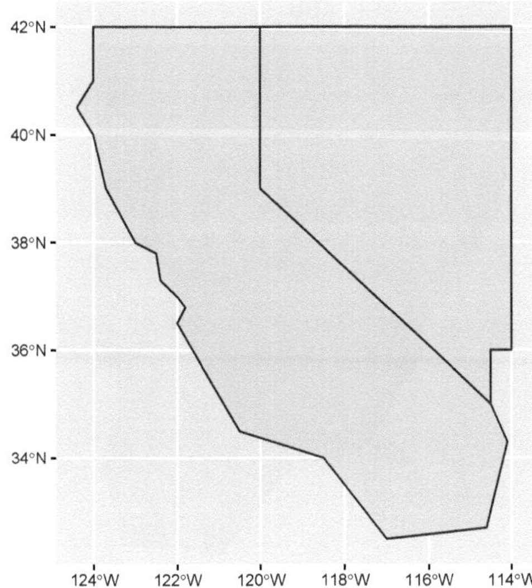

FIGURE 6.1 A simple ggplot2 map built from scratch with hard-coded data as simple feature columns

```
attributes <- bind_rows(c(abb="CA", area=423970, pop=39.56e6),
                        c(abb="NV", area=286382, pop=3.03e6)) %>%
  mutate(area=as.numeric(area),pop=as.numeric(pop))
twostates <- st_sf(attributes, geometry = sfc_2states)
ggplot(twostates) + geom_sf() + geom_sf_text(aes(label = abb))
```

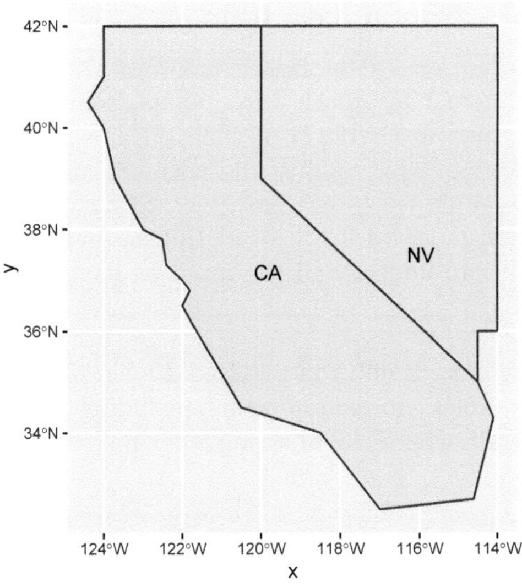

FIGURE 6.2 Using an sf class to build a map in ggplot2, displaying an attribute

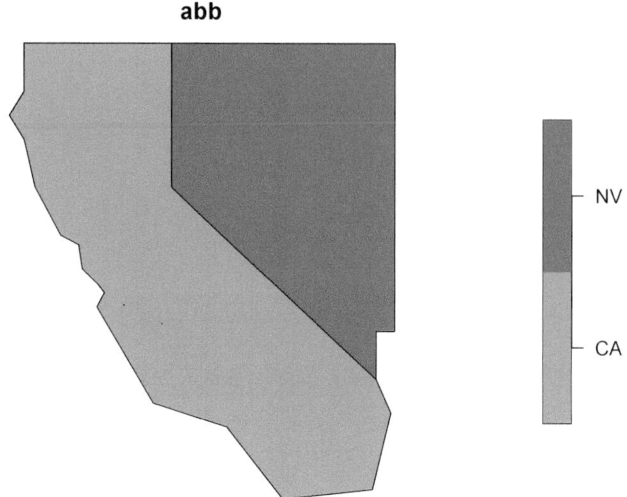

FIGURE 6.3 Base R plot of one attribute from two states

Or we could let the plot system try to figure out what to do with it. To help it out a bit, we can select just one variable, for which we need to use the `[]` accessor (Figure 6.3). Try the `$` accessor instead to see what you get, and consider why this might be.

```
plot(twostates["abb"])
```

6.1.2 Building points from a data frame

We're going to take a different approach with points, since if we're building them from scratch they commonly come in the form of a data frame with x and y columns, so the `st_as_sf` is much simpler than what we just did with `sfg` and `sfc` data.

The `sf::st_as_sf` function is useful for a lot of things, and follows the convention of R's many "as" functions: if it can understand the input as having "space and time" (st) data, it can convert it to an sf.

We'll start by creating a data frame representing 12 Sierra Nevada (and some adjacent) weather stations, with variables entered as vectors, including attributes. The order of vectors needs to be the same for all, where the first point has index 1, etc.

```
sta  <- c("OROVILLE","AUBURN","PLACERVILLE","COLFAX","NEVADA CITY","QUINCY",
          "YOSEMITE","PORTOLA","TRUCKEE","BRIDGEPORT","LEE VINING","BODIE")
long <- c(-121.55,-121.08,-120.82,-120.95,-121.00,-120.95,
          -119.59,-120.47,-120.17,-119.23,-119.12,-119.01)
lat <-  c(39.52,38.91,38.70,39.09,39.25,39.94,
          37.75,39.81,39.33,38.26,37.96,38.21)
elev <- c(52,394,564,725,848,1042,
```

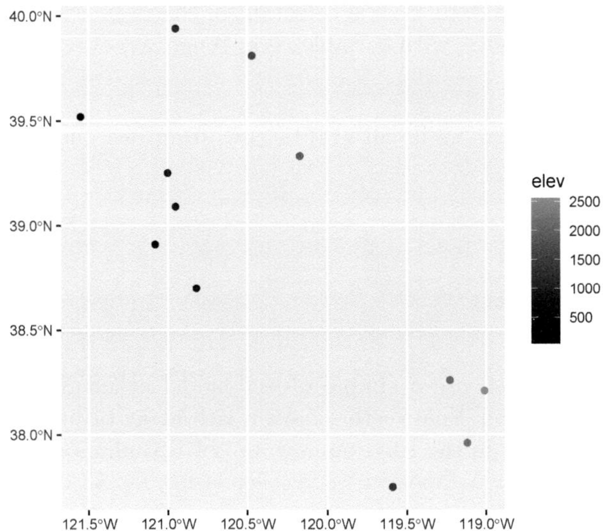

FIGURE 6.4 Points created from a dataframe with Simple Features

```
          1225,1478,1775,1972,2072,2551)
temp <- c(10.7,9.7,9.2,7.3,6.7,4.0,
          5.0,0.5,-1.1,-2.2,0.4,-4.4)
prec <- c(124,160,171,207,268,182,
          169,98,126,41,72,40)
```

To build the sf data, we start by building the data frame...

```
df <- data.frame(sta,elev,temp,prec,long,lat)
```

... then use the very handy `sf::st_as_sf` function to make an sf out of it, using the long and lat as a source of coordinates (Figure 6.4).

```
wst <- st_as_sf(df, coords=c("long","lat"), crs=4326)
ggplot(data=wst) + geom_sf(mapping = aes(col=elev))
```

6.1.3 SpatVectors in `terra`

We'll briefly look at another feature-based object format, the `SpatVector` from `terra`, which uses a formal S4 object definition that provides some advantages over S3 objects such as what's in most of R, at the expense of being incompatible with some existing packages, at least at the time of writing; that will probably change. See Hijmans (n.d.) for more about SpatVectors. We'll use terra's raster capabilities a lot more, but we'll take a brief look at building and using a `SpatVector`.

We'll start with the climate data from a selection of Sierra Nevada (and some adjacent) weather stations, created above, with variables `sta`, `long`, `lat`, `elev`, `temp`, and `prec` entered as vectors.

Then we'll build a `SpatVector` we'll call `station_pts` from just the longitude and latitude vectors we just created:

```
library(terra)
longlat <- cbind(long,lat)
station_pts <- vect(longlat)
```

Unlike other data sets which use a simple object model called S3, `SpatVector` data use the more formal S4 definition. SpatVectors have a rich array of properties and associated methods, which you can see in the Environment tab of RStudio, or access with `@ptr`. This is a good thing, but the `str()` function we've been using for S3 objects presents us with arguably too much information for easy understanding. But just by entering `station_pts` on the command line we get something similar to what `str()` does with S3 data, so this might be better for seeing a few key properties:

```
station_pts
```

```
##  class       : SpatVector
##  geometry    : points
##  dimensions  : 12, 0  (geometries, attributes)
##  extent      : -121.55, -119.01, 37.75, 39.94  (xmin, xmax, ymin, ymax)
##  coord. ref. :
```

Note that there's nothing listed for `coord. ref.` meaning the coordinate referencing system (CRS). We'll look at this more below, but for now let's change the code a bit to add this information using the PROJ-string method:

```
station_pts <- vect(longlat, crs = "+proj=longlat +datum=WGS84")
```

Of the many methods associated with these objects, one example is getting with geom():

```
geom(station_pts)
```

```
##        geom part       x     y hole
##  [1,]     1    1 -121.55 39.52    0
##  [2,]     2    1 -121.08 38.91    0
##  [3,]     3    1 -120.82 38.70    0
##  [4,]     4    1 -120.95 39.09    0
##  [5,]     5    1 -121.00 39.25    0
##  [6,]     6    1 -120.95 39.94    0
##  [7,]     7    1 -119.59 37.75    0
##  [8,]     8    1 -120.47 39.81    0
##  [9,]     9    1 -120.17 39.33    0
## [10,]    10    1 -119.23 38.26    0
```

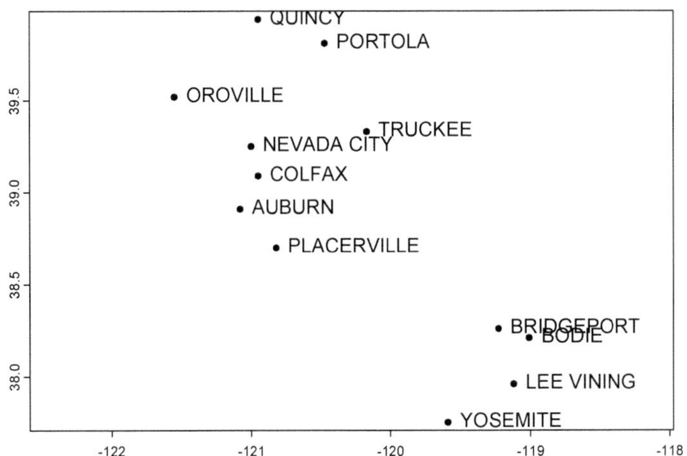

FIGURE 6.5 Simple plot of SpatVector point data with labels (note that overlapping labels may result, as seen here)

```
## [11,]   11    1 -119.12 37.96    0
## [12,]   12    1 -119.01 38.21    0
```

Let's add the attributes to the original `longlat` matrix to build a SpatVect with attributes we can see in a basic plot (Figure 6.5).

```
df <- data.frame(sta,elev,temp,prec)
clim <- vect(longlat, atts=df, crs="+proj=longlat +datum=WGS84")
clim
```

```
## class       : SpatVector
## geometry    : points
## dimensions  : 12, 4  (geometries, attributes)
## extent      : -121.55, -119.01, 37.75, 39.94  (xmin, xmax, ymin, ymax)
## coord. ref. : +proj=longlat +datum=WGS84 +no_defs
## names       :         sta  elev  temp  prec
## type        :       <chr> <num> <num> <num>
## values      :    OROVILLE    52  10.7   124
##                   AUBURN   394   9.7   160
##               PLACERVILLE   564   9.2   171
```

```
plot(clim)
text(clim, labels="sta", pos=4)
```

The SpatVector format looks promising, but sometimes you may want an sf instead. Fortunately, you can use the `sf::sf_as_sf` function to convert it, just as you can create a `SpatVector` from an sf with the `terra::vect` function. The following code demonstrates

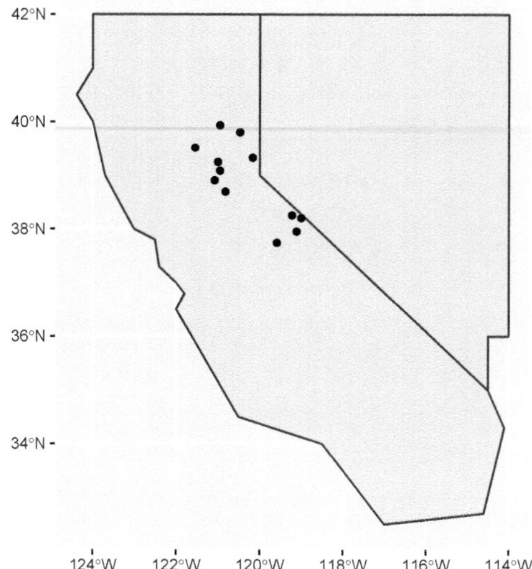

FIGURE 6.6 ggplot of twostates and stations

both, with a ggplot2 map of sf data (Figure 6.6) and then a base R plot map with SpatVector data (Figure 6.7).

```
library(sf); library(ggplot2)
climsf <- st_as_sf(clim)
str(climsf)
```

```
## Classes 'sf' and 'data.frame':   12 obs. of  5 variables:
##  $ sta     : chr  "OROVILLE" "AUBURN" "PLACERVILLE" "COLFAX" ...
##  $ elev    : num  52 394 564 725 848 ...
##  $ temp    : num  10.7 9.7 9.2 7.3 6.7 4 5 0.5 -1.1 -2.2 ...
##  $ prec    : num  124 160 171 207 268 182 169 98 126 41 ...
##  $ geometry:sfc_POINT of length 12; first list element:  'XY' num  -121.5 39.5
##  - attr(*, "sf_column")= chr "geometry"
##  - attr(*, "agr")= Factor w/ 3 levels "constant","aggregate",..: NA NA NA NA
##   ..- attr(*, "names")= chr [1:4] "sta" "elev" "temp" "prec"
```

```
ggplot() + geom_sf(data=twostates) + geom_sf(data=climsf)
```

```
twostatesV <- vect(twostates)
plot(twostatesV)
plot(clim, add = T)
```

```
twostatesV
```

```
##  class       : SpatVector
##  geometry    : polygons
```

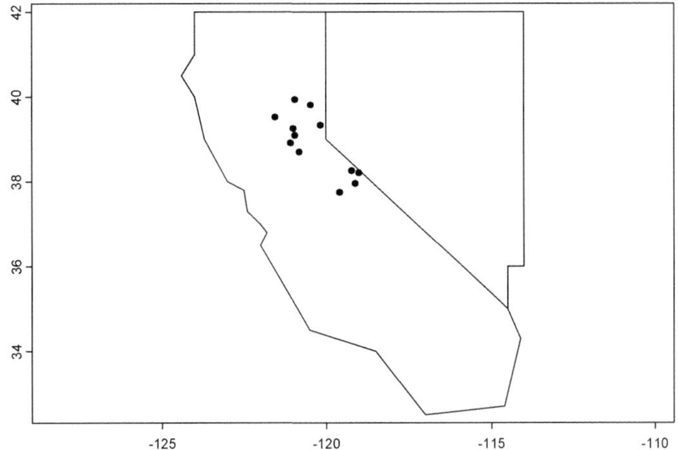

FIGURE 6.7 Base R plot of twostates and stations SpatVectors

```
## dimensions  : 2, 3  (geometries, attributes)
## extent      : -124.4, -114, 32.5, 42  (xmin, xmax, ymin, ymax)
## coord. ref. : lon/lat WGS 84 (EPSG:4326)
## names       :   abb       area        pop
## type        : <chr>     <num>      <num>
## values      :    CA  4.24e+05 3.956e+07
##                 NV 2.864e+05  3.03e+06
```

6.1.4 Creating features from shapefiles

Both sf's `st_read` and terra's `vect` read the open-GIS *shapefile* format developed by Esri for points, polylines, and polygons. You would normally have shapefiles (and all the files that go with them – .shx, etc.) stored on your computer, but we'll access one from the `igisci` external data folder, and use that `ex()` function we used earlier with CSVs. *Remember that we could just include that and the library calls just once at the top of our code like this...*

```
library(igisci)
library(sf)
```

If we just send a spatial dataset like an `sf` spatial data frame to the plot system, it will plot all of the variables by default (Figure 6.8).

```
BayAreaCounties <- st_read(ex("BayArea/BayAreaCounties.shp"))
```

```
plot(BayAreaCounties)
```

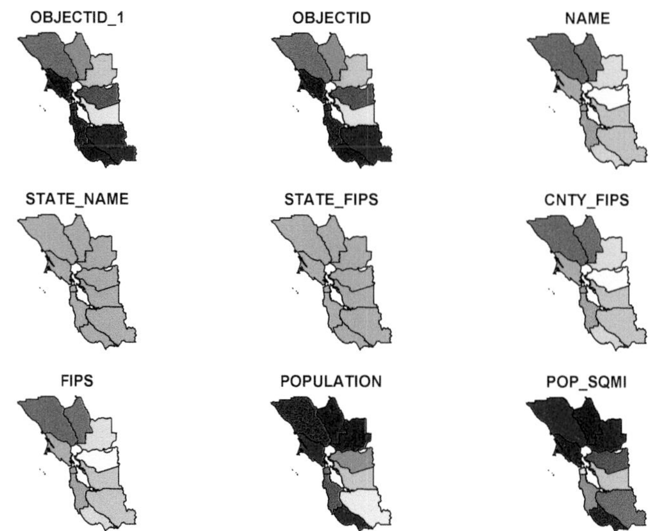

FIGURE 6.8 A simple plot of polygon data by default shows all variables

But with just one variable, of course, it just produces a single map (Figure 6.9).

```
plot(BayAreaCounties["POPULATION"])
```

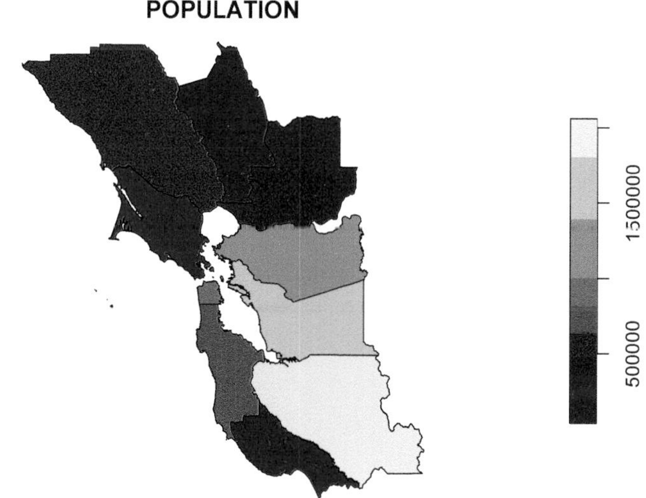

FIGURE 6.9 A single map with a legend is produced when a variable is specified

Notice that in the above map, we used the `[]` accessor. Why didn't we use the simple `$` accessor? Remember that plot() figures out what to do based on what you provide it. And there's an important difference in what you get with the two accessors, which we can check with class():

```
class(BayAreaCounties["POPULATION"])
```

```
## [1] "sf"         "data.frame"
```

```
class(BayAreaCounties$POPULATION)
```

```
## [1] "numeric"
```

You might see that what you get with `plot(BayAreaCounties$POPULATION)` is not very informative, since the object is just a numeric vector, while using `[]` accessor returns a spatial dataframe.

There's a lot more we could do with the base R plot system, so we'll learn some of these before exploring what we can do with `ggplot2`, `tmap`, and `leaflet`. But first we need to learn more about building geospatial data.

We'll use `st_as_sf()` for that, but we'll need to specify the coordinate referencing system (CRS), in this case GCS. We'll only briefly explore how to specify the CRS here. For a thorough coverage, please see Lovelace, Nowosad, and Muenchow (2019).

6.2 Coordinate Referencing Systems

Before we try the next method for bringing in spatial data – converting data frames – we need to look at coordinate referencing systems (CRS). First, there are quite a few, with some spherical like the geographic coordinate system (GCS) of longitude and latitude, and others planar projections of GCS using mathematically defined projections such as Mercator, Albers Conformal Conic, Azimuthal, etc., and including widely used government-developed systems such as UTM (universal transverse mercator) or state plane. Even for GCS, there are many varieties since geodetic datums can be chosen, and for very fine resolution work where centimetres or even millimetres matter, this decision can be important (and tectonic plate movement can play havoc with tying it down.)

There are also multiple ways of expressing the CRS, either to read it or to set it. The full specification of a CRS can be displayed for data already in a CRS, with either `sf::st_crs` or `terra::crs`.

```
st_crs(CA_counties)
```

```
## Coordinate Reference System:
##   User input: WGS 84
##   wkt:
## GEOGCRS["WGS 84",
##     DATUM["World Geodetic System 1984",
##         ELLIPSOID["WGS 84",6378137,298.257223563,
```

```
##             LENGTHUNIT["metre",1]]],
##     PRIMEM["Greenwich",0,
##         ANGLEUNIT["degree",0.0174532925199433]],
##     CS[ellipsoidal,2],
##         AXIS["latitude",north,
##             ORDER[1],
##             ANGLEUNIT["degree",0.0174532925199433]],
##         AXIS["longitude",east,
##             ORDER[2],
##             ANGLEUNIT["degree",0.0174532925199433]],
##     ID["EPSG",4326]]
```

... though there are other ways to see the crs in shorter forms, or its individual properties :

```
st_crs(CA_counties)$proj4string
```

```
## [1] "+proj=longlat +datum=WGS84 +no_defs"
```

```
st_crs(CA_counties)$units_gdal
```

```
## [1] "degree"
```

```
st_crs(CA_counties)$epsg
```

```
## [1] 4326
```

> There's also `st_crs()$Wkt`, which you should try, but it creates a long, continuous string that goes off the page, so doesn't work for the book formatting.

```
st_crs(CA_counties)$Wkt # [Creates too long of a string for this book]
```

So, to convert the sierra data into geospatial data with `st_as_sf`, we might either do it with the reasonably short PROJ format ...

```
GCS <- "+proj=longlat +datum=WGS84 +no_defs +ellps=WGS84 +towgs84=0,0,0"
wsta = st_as_sf(sierraFeb, coords = c("LONGITUDE","LATITUDE"), crs=GCS)
```

... or with the even shorter EPSG code (which we can find by Googling), which can either be provided as text `"epsg:4326"` or often just the number, which we'll use next.

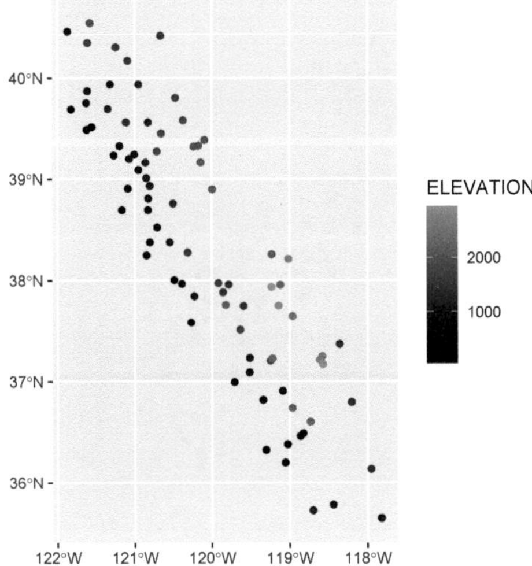

FIGURE 6.10 Points created from data frame with coordinate variables

6.3 Creating sf Data from Data Frames

As we saw earlier, if your data frame has geospatial coordinates like `LONGITUDE` and `LATITUDE` ...

```
names(sierraFeb)
```

```
## [1] "STATION_NAME"  "COUNTY"        "ELEVATION"     "LATITUDE"
## [5] "LONGITUDE"     "PRECIPITATION" "TEMPERATURE"
```

... we have what we need to create geospatial data from it. Earlier we read in a series of vectors built in code with `c()` functions from 12 selected weather stations; this time we'll use a data frame that has all of the Sierra Nevada weather stations (Figure 6.10).

```
wsta <- st_as_sf(sierraFeb, coords = c("LONGITUDE","LATITUDE"), crs=4326)
ggplot(data=wsta) + geom_sf(aes(col=ELEVATION))
```

6.3.1 *Removing* geometry

There are many instances where you want to remove geometry from a sf data frame.

- Some R functions run into problems with geometry and produce confusing error messages, like "non-numeric argument"
- To work with an sf data frame in a non-spatial way

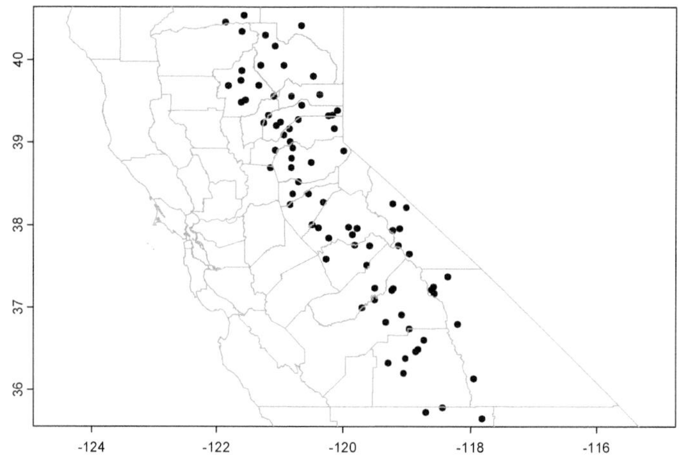

FIGURE 6.11 Plotting SpatVector data with base R plot system

One way to remove geometry :

```
myNonSFdf <- mySFdf %>% st_set_geometry(NULL)
```

6.4 Base R's `plot()` with `terra`

As we've seen in the examples above, the venerable plot system in base R can often do a reasonable job of displaying a map. The terra package extends this a bit by providing tools for overlaying features on other features (or rasters using plotRGB), so it's often easy to use (Figure 6.11).

```
library(terra)
plot(vect(wsta))
lines(vect(CA_counties), col="gray")
```

Note that we converted the sf data to `terra` `SpatVector` data with `vect()`, but the base R plot system can work with either `sf` data or `SpatVector` data from `terra`.

The `lines`, `points`, and `polys` functions (to name a few – see `?terra` and look at the Plotting section) will add to an existing plot. Alternatively, we could use plot's `add=TRUE` parameter to add onto a plot that we've previously scaled with a plot. In this case, we'll cheat and just use the first plot to set the scaling but then cover over that plot with other plots (Figure 6.12).

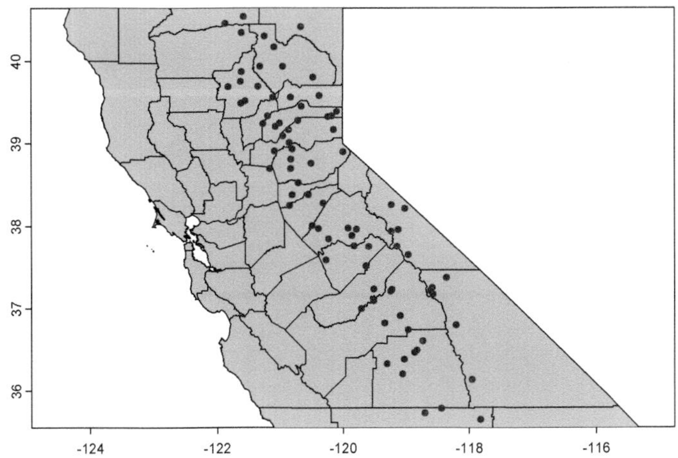

FIGURE 6.12 Features added to the map using the base R plot system

```
library(terra)
plot(vect(wsta[1])) # simple way of establishing the coordinates
# Then will be covered by the following:
plot(vect(CA_counties), col="lightgray", add=TRUE)
plot(vect(wsta), col="red", add=TRUE)
```

6.4.1 Using maptiles to create a basemap

For vector data, it's often nice to display over a basemap by accessing raster tiles that are served on the internet by various providers. We'll use the `maptiles` package to try displaying the `CA` and `NV` boundaries we created earlier. The `maptiles` package supports a variety of basemap providers, and I've gotten the following to work: `"OpenStreetMap"`, `"Stamen.Terrain"`, `"Stamen.TerrainBackground"`, `"Esri.WorldShadedRelief"`, `"Esri.NatGeoWorldMap"`, `"Esri.WorldGrayCanvas"`, `"CartoDB.Positron"`, `"CartoDB.Voyager"`, `"CartoDB.DarkMatter"`, `"OpenTopoMap"`, `"Wikimedia"`, *however, they don't work at all scales and locations – you'll often see an Error in grDevices, if so then try another provider – the default `"OpenStreetMap"` seems to work the most reliably* (Figure 6.13).

```
library(terra); library(maptiles)
# Get the raster that covers the extent of CANV:
calnevaBase <- get_tiles(twostates, provider="OpenTopoMap")
st_crs(twostates)$epsg
```

```
## [1] 4326
```

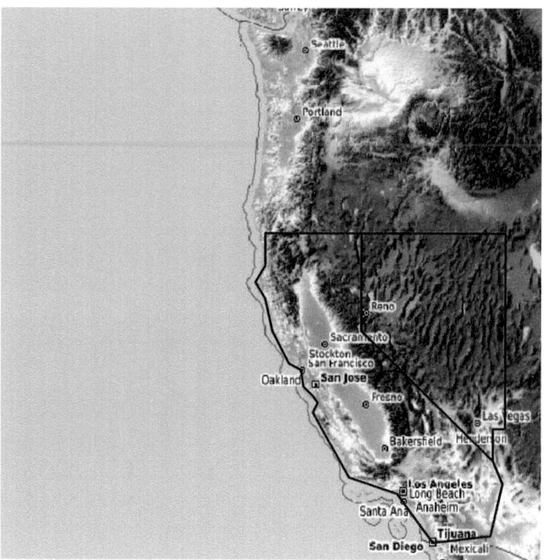

FIGURE 6.13 Using maptiles for a base map

```
plotRGB(calnevaBase)  # starts plot with a raster basemap
lines(vect(twostates), col="black", lwd=2)
```

In the code above, we can also see the use of `terra` functions and parameters. Learn more about these by reviewing the code and considering:

terra functions: The `terra::vect()` function creates a `SpatVector` that works with `terra::lines` to display the boundary lines in `plot()`. (More on `terra` later, in the Raster section.)

parameters: Note the use of the parameter `lwd` (line width) from the plot system. This is one of *many* parameter settings described in **`?par`**. It defaults to 1, so 2 makes it twice as thick. You could also use lty (line type) to change it to a dashed line with `lty="dashed"` or `lty=2`.

And for the `sierraFeb` data, we'll start with `st_as_sf` and the coordinate system (4326 for GCS), then use a `maptiles` basemap again, and the `terra::points` method to add the points (Figure 6.14).

```
library(terra); library(maptiles)
sierraFebpts <- st_as_sf(sierraFeb, coords = c("LONGITUDE", "LATITUDE"), crs=4326)
sierraBase <- get_tiles(sierraFebpts)
st_crs(sierraFebpts)$epsg
```

```
## [1] 4326
```

```
plotRGB(sierraBase)
points(vect(sierraFebpts))
```

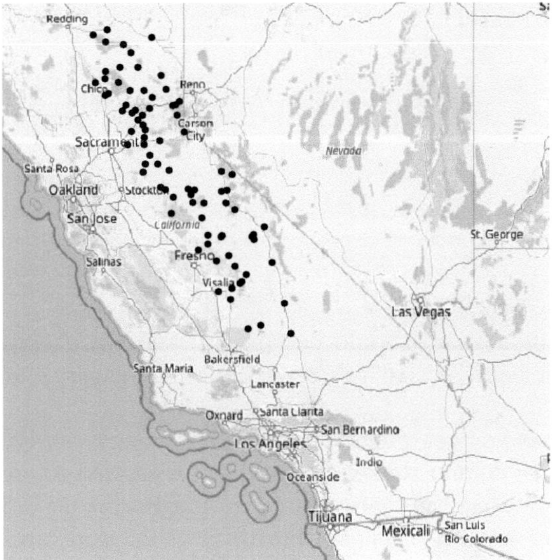

FIGURE 6.14 Converted sf data for map with tiles

6.5 Raster data

Simple *Features* are feature-based, of course, so it's not surprising that `sf` doesn't have support for rasters. So we'll want to either use the `raster` package or its imminent replacement, `terra` (which also has vector methods – see Hijmans (n.d.) and Lovelace, Nowosad, and Muenchow (2019)), and I'm increasingly using `terra` since it has some improvements that I find useful.

6.5.1 Building rasters

We can start by building one from scratch. The `terra` package has a `rast()` function for this, and creates an S4 (as opposed to the simpler S3) object type called a `SpatRaster`.

```
library(terra)
new_ras <- rast(ncol = 12, nrow = 6,
                xmin = -180, xmax = 180, ymin = -90, ymax = 90,
                vals = 1:72)
```

Formal S4 objects like `SpatRasters` have many associated properties and methods. To get a sense, type into the console `new_ras@ptr$` and see what suggestions you get from the autocomplete system. To learn more about them, see Hijmans (n.d.), but we'll look at a few of the key properties by simply entering the `SpatRaster` name in the console.

```
new_ras
```

```
## class       : SpatRaster
## dimensions  : 6, 12, 1  (nrow, ncol, nlyr)
## resolution  : 30, 30  (x, y)
## extent      : -180, 180, -90, 90  (xmin, xmax, ymin, ymax)
## coord. ref. : lon/lat WGS 84
## source      : memory
## name        : lyr.1
## min value   :     1
## max value   :    72
```

- The `nrow` and `ncol` dimensions should be familiar, and the `nlyr` tells us that this raster has a single layer (or band as they're referred to with imagery), and it gives it the name `lyr.1`.
- The resolution is what you get when you divide 360 (degrees of longitude) by 12 and 180 (degrees of latitude) by 6, and the extent is what we entered to create it.
- But how does it know that these are degrees of longitude and latitude, as it appears to from the `coord. ref.` property? The author of this tool seems to let it assume GCS if it doesn't exceed the limits of the planet: -180 to +180 longitude and -90 to +90 latitude.
- The source is in memory, since we didn't read it from a file, but entered it directly. New rasters we might create from raster operations will also be in memory.
- The minimum and maximum values are what we used to create it.

The name `lyr.1` isn't very useful, so let's change the name with the `names` function, and then access the name with a `@ptr` property (you could also use `names(new_ras)` to access it, but I thought it might be useful to see the `@ptr` in action.)

```
names(new_ras) <- "world30deg"
new_ras
```

```
## class       : SpatRaster
## dimensions  : 6, 12, 1  (nrow, ncol, nlyr)
## resolution  : 30, 30  (x, y)
## extent      : -180, 180, -90, 90  (xmin, xmax, ymin, ymax)
## coord. ref. : lon/lat WGS 84
## source      : memory
## name        : world30deg
## min value   :          1
## max value   :         72
```

Simple feature data or `SpatVectors` can be plotted along with rasters using terra's lines, polys, or points functions. Here we'll use the `twostates sf` we created earlier (make sure to run that code again if you need it). Look closely at the map for our two states (Figure 6.15).

```
plot(new_ras, main=paste("CANV on",new_ras@ptr$names))
CANV <- vect(twostates)
lines(CANV)
```

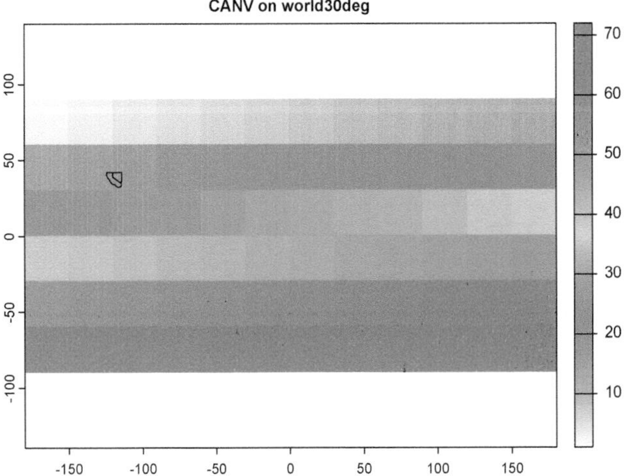

FIGURE 6.15 Simple plot of a worldwide SpatRaster of 30-degree cells, with SpatVector of CA and NV added

6.5.2 Vector to raster conversion

To convert rasters to vectors requires having a template raster with the desired cell size and extent, or an existing raster we can use as a template – we ignore what the values are – such as `elev.tif` in the following example (Figure 6.16):

```
streams <- vect(ex("marbles/streams.shp"))
elev <- rast(ex("marbles/elev.tif"))
streamras <- rasterize(streams,elev)
plot(streamras, col="blue")
```

6.5.2.1 What if we have no raster to use as a template?

In the above process, we referenced `elev` essentially as a template to get a raster structure and crs to use. But what if we wanted to do the conversion and we didn't have a raster to use as a template? We'd need to create a raster, specifying the cell size and the extent (`xmin`, `xmax`, `ymin`, `ymax`).

If you know those extent values, you could start from scratch. But we'll look at a common situation where we have existing features, from the `streams.shp` we are wishing to rasterize...

```
streams <- vect(ex("marbles/streams.shp"))
streams
```

```
## class       : SpatVector
## geometry    : lines
```

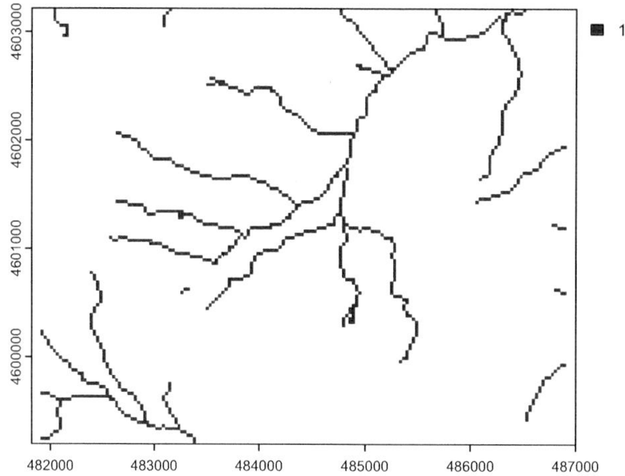

FIGURE 6.16 Stream raster converted from stream features, with 30 m cells from an elevation raster template

```
##  dimensions  : 48, 8  (geometries, attributes)
##  extent      : 481903.6, 486901.9, 4599199, 4603201  (xmin, xmax, ymin, ymax)
##  source      : streams.shp
##  coord. ref. : NAD83 / UTM zone 10N (EPSG:26910)
##  names       : FNODE_ TNODE_ LPOLY_ RPOLY_ LENGTH STREAMS_ STREAMS_ID
##  type        :  <num>  <num>  <num>  <num>  <num>    <num>      <num>
##  values      :      5      6      4      4  269.3        1          1
##                     2      7      2      2  153.7        2          5
##                     8      1      2      2  337.9        3         36
##  Shape_Leng
##       <num>
##       269.3
##       153.7
##       337.9
```

... and in code we could pull out the four extent properties with the `ext()` function, and use that also to work out the aspect ratio and (assuming that we know we want a raster cell size of 30 m based on the crs being in UTM metres) we can come up with the parameters (with integer numbers of columns and rows) for creating the raster template.

```
XMIN <- ext(streams)$xmin
XMAX <- ext(streams)$xmax
YMIN <- ext(streams)$ymin
YMAX <- ext(streams)$ymax
aspectRatio <- (YMAX-YMIN)/(XMAX-XMIN)
cellSize <- 30
NCOLS <- as.integer((XMAX-XMIN)/cellSize)
NROWS <- as.integer(NCOLS * aspectRatio)
```

Then we use those parameters to create the template, also borrowing the `crs` from the streams layer.

```
templateRas <- rast(ncol=NCOLS, nrow=NROWS,
                    xmin=XMIN, xmax=XMAX, ymin=YMIN, ymax=YMAX,
                    vals=1, crs=crs(streams))
```

Finally we can do something similar to original example, but with the template instead of `elev` as reference for the vector-raster conversion. The result will be identical to what we saw above.

```
strms <- rasterize(streams,templateRas)
```

Note that if you knew of a reference raster, you can also look at its properties with

```
elev
```

```
## class       : SpatRaster
## dimensions  : 134, 167, 1  (nrow, ncol, nlyr)
## resolution  : 30, 30  (x, y)
## extent      : 481905, 486915, 4599195, 4603215  (xmin, xmax, ymin, ymax)
## coord. ref. : NAD_1983_UTM_Zone_10N (EPSG:26910)
## source      : elev.tif
## name        : elev
## min value   : 1336
## max value   : 2260
```

and use or modify those properties to manually create a template raster, with code something like:

```
templateRas <- rast(ncol=167, nrow=134,
                    xmin=481905, xmax=486915, ymin=4599195, ymax=4603215,
                    vals=1, crs="EPSG:26910")
```

The purpose here is just to show how we can create and use a raster template. How you define your raster template will depend on the extent and scale of your data, and if you have other rasters (like `elev` above) you want to match with, you may want to use it as a template as we did first.

6.5.2.2 Plotting some existing downloaded raster data

Let's plot some Shuttle Radar Topography Mission (SRTM) elevation data for the Virgin River Canyon at Zion National Park (Figure 6.17), from Jakub Nowosad's `spDataLarge` repository, which we'll need to first install with:

```
install.packages("spDataLarge",repos="https://nowosad.github.io/drat/",type="source")
```

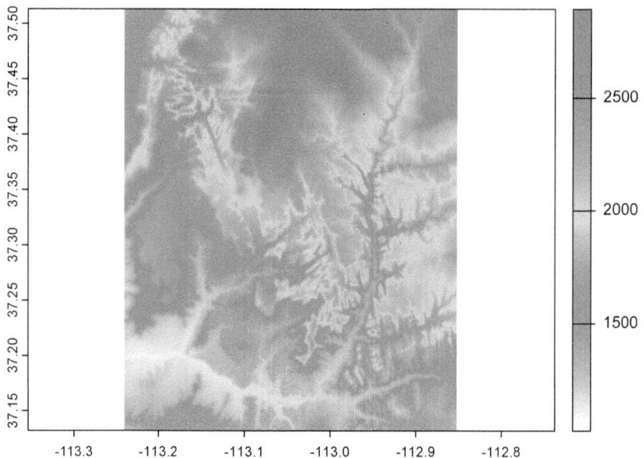

FIGURE 6.17 Shuttle Radar Topography Mission (SRTM) image of Virgin River Canyon area, southern Utah

```
library(terra)
plot(rast(system.file("raster/srtm.tif", package="spDataLarge")))
```

6.6 ggplot2 for Maps

The Grammar of Graphics is the gg of ggplot.

- Key concept is separating aesthetics from data
- Aesthetics can come from variables (using aes()setting) or be constant for the graph

Mapping tools that follow this lead

- ggplot, as we have seen, and it continues to be enhanced (Figure 6.18)
- tmap (Thematic Maps) https://github.com/mtennekes/tmap Tennekes, M., 2018, tmap: Thematic Maps in R, *Journal of Statistical Software* 84(6), 1-39

```
ggplot(CA_counties) + geom_sf()
```

Try `?geom_sf` and you'll find that its first parameter is mapping with `aes()` by default. The data property is inherited from the ggplot call, but commonly you'll want to specify data=something in your geom_sf call.

Another simple ggplot, with labels

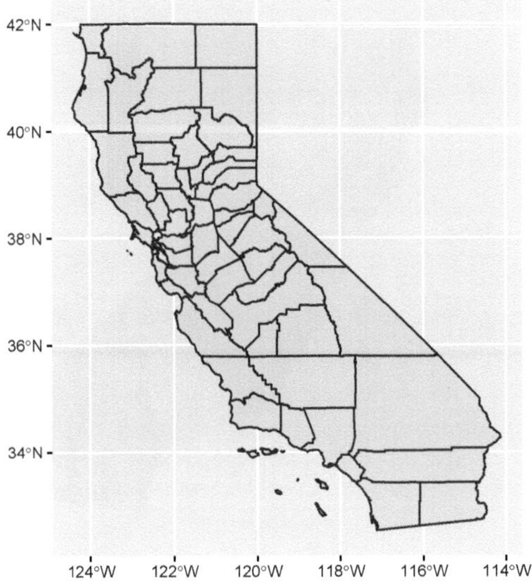

FIGURE 6.18 simple ggplot map

Adding labels is also pretty easy using `aes()` (Figure 6.19).

```
ggplot(CA_counties) + geom_sf() +
  geom_sf_text(aes(label = NAME), size = 1.5)
```

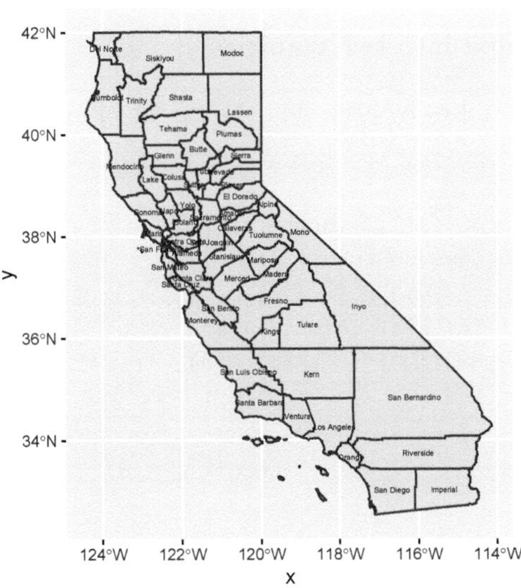

FIGURE 6.19 labels added

And now with fill color, repositioned legend, and no "x" or "y" labels

The x and y labels are unnecessary since the graticule is provided, and for many maps there's a better place to put the legend than what happens by default – for California's shape, the legend goes best in Nevada (Figure 6.20).

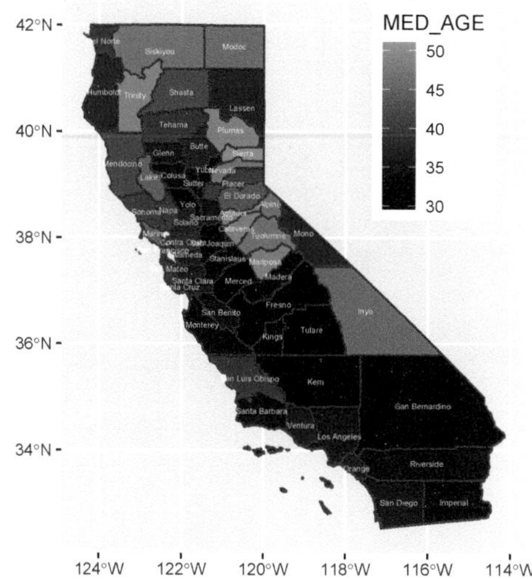

FIGURE 6.20 repositioned legend

```
ggplot(CA_counties) + geom_sf(aes(fill=MED_AGE)) +
  geom_sf_text(aes(label = NAME), col="white", size=1.5) +
  theme(legend.position = c(0.8, 0.8)) +
  labs(x="",y="")
```

Map in ggplot2, zoomed into two counties

We can zoom into two counties by accessing the extent of an existing spatial dataset using `st_bbox()` (Figure 6.21).

```
library(tidyverse); library(sf); library(igisci)
census <- st_make_valid(BayAreaTracts) %>%
   filter(CNTY_FIPS %in% c("013", "095"))
TRI <- read_csv(ex("TRI/TRI_2017_CA.csv")) %>%
  st_as_sf(coords = c("LONGITUDE", "LATITUDE"), crs=4326) %>%
  st_join(census) %>%
  filter(CNTY_FIPS %in% c("013", "095"),
         (`5.1_FUGITIVE_AIR` + `5.2_STACK_AIR`) > 0)
bnd = st_bbox(census)
ggplot() +
  geom_sf(data = BayAreaCounties, aes(fill = NAME)) +
  geom_sf(data = census, color="grey40", fill = NA) +
  geom_sf(data = TRI) +
  coord_sf(xlim = c(bnd[1], bnd[3]), ylim = c(bnd[2], bnd[4])) +
  labs(title="Census Tracts and TRI air-release sites") +
  theme(legend.position = "none")
```

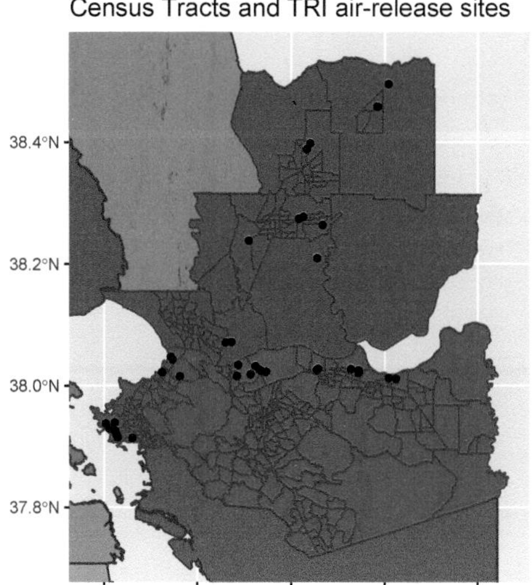

FIGURE 6.21 Using bbox to zoom into two counties

6.6.1 Rasters in ggplot2

Raster display in `ggplot2` is currently a little awkward, as are rasters in general in the world of database-centric GIS where rasters don't comfortably sit.

We can use a trick: converting rasters to a grid of points, to then be able to create Figure 6.22.

```
library(igisci)
library(tidyverse)
library(sf)
library(terra)
elevation <- rast(ex("marbles/elev.tif"))
trails <- st_read(ex("marbles/trails.shp"))
elevpts <- st_as_sf(as.points(elevation))
```

```
ggplot() +
  geom_sf(data = elevpts, aes(col=elev)) +
  geom_sf(data = trails)
```

Note that the raster name stored in the `elevation@ptr$names` property is derived from the original source raster `elev.tif`, not from the object name we gave it, so we needed to specify `elev` for the variable to color from the resulting point features.

```
names(elevation)
```

```
## [1] "elev"
```

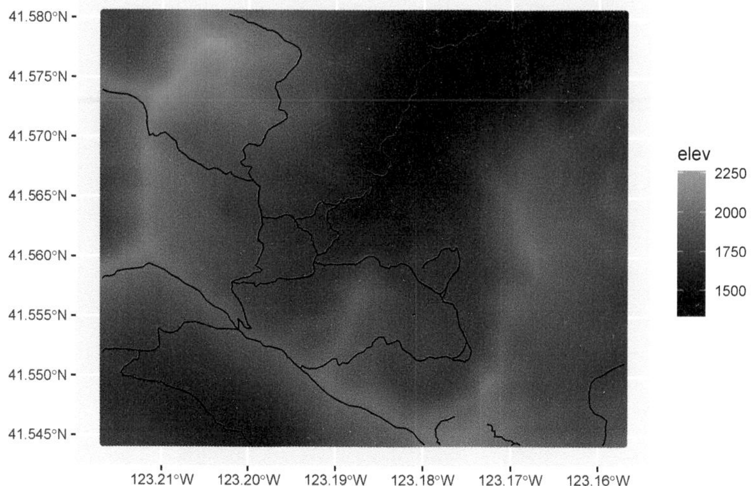

FIGURE 6.22 Rasters displayed in ggplot by converting to points

6.7 tmap

The tmap package provides some nice cartographic capabilities. We'll use some of its capabilities, but for more thorough coverage, see the "Making maps with R" section of *Geocomputation with R* (Lovelace, Nowosad, and Muenchow (2019)).

The basic building block is `tm_shape(data)` followed by various layer elements, such as tm_fill(). The tm_shape function can work with either features or rasters.

```
library(spData); library(tmap)
m <- tm_shape(world) + tm_fill() + tm_borders()
```

But we'll make a better map (Figure 6.23) by inserting a graticule before filling the polygons, then use `tm_layout` for some useful cartographic settings: make the background light blue and avoid excessive inner margins.

```
library(spData); library(tmap)
bounds <- st_bbox(world)
m <- tm_shape(world, bbox=bounds) + tm_graticules(col="seashell2") +
  tm_fill() + tm_borders() + tm_layout(bg.color="lightblue", inner.margins=0)
m
```

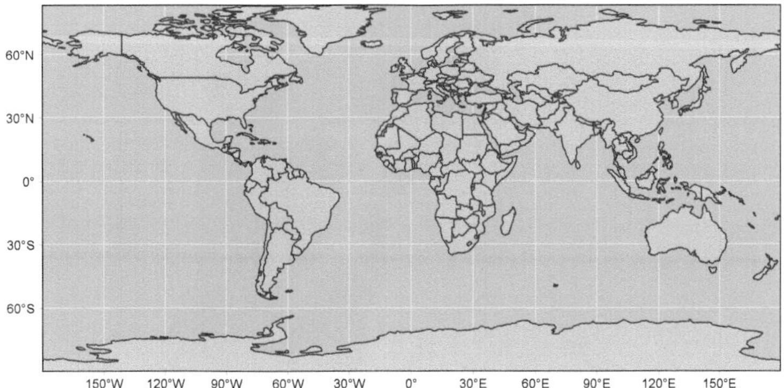

FIGURE 6.23 tmap of the world

Experiment by leaving settings out to see the effect, and explore other options with `?tm_layout`, etc. The tmap package has a wealth of options, just a few of which we'll explore. For instance, we might want to use a different map projection than the default.

Color by variable

As with plot and ggplot2, we can reference a variable to provide a range of colors for features we're plotting, such as coloring polygon fills to create a choropleth map (Figure 6.24).

```
library(sf); library(igisci)
library(tmap)
tm_shape(st_make_valid(BayAreaTracts)) + tm_graticules(col="#e5e7e9") + tm_fill(col = "MED_AGE")
```

tmap of sierraFeb with hillshade and point symbols

We'll use a raster hillshade as a basemap using `tm_raster`, zoom into an area with a bounding box (from `st_bbox`), include county boundaries with `tm_borders`, and color station points with temperatures with `tm_symbols` (Figure 6.25).

```
library(terra) # alt: library(raster)
tmap_mode("plot")
tmap_options(max.categories = 8)
sierra <- st_as_sf(sierraFeb, coords = c("LONGITUDE", "LATITUDE"), crs = 4326)
```

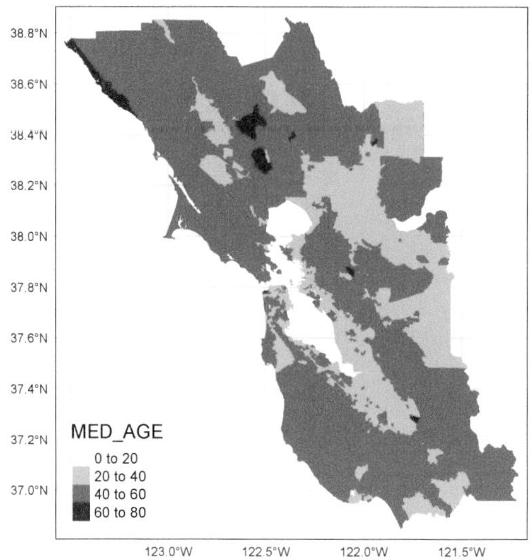

FIGURE 6.24 tmap fill colored by variable

```
hillsh <- rast(ex("CA/ca_hillsh_WGS84.tif"))   # alt: hillsh <- raster(...)
bounds <- st_bbox(sierra)
tm_shape(hillsh,bbox=bounds)+
  tm_graticules(col="azure2") +
  tm_raster(palette="-Greys",legend.show=FALSE,n=10) +
  tm_shape(sierra) +
  tm_symbols(col="TEMPERATURE", palette=c("blue","red"), style="cont",n=8) +
  tm_shape(st_make_valid(CA_counties)) + tm_borders() +
```

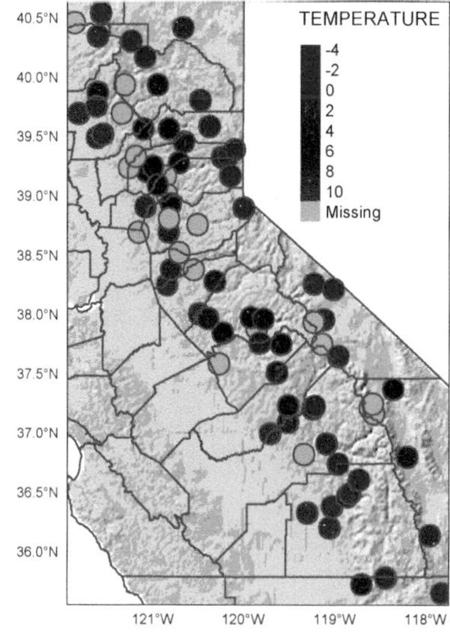

FIGURE 6.25 hillshade, borders and point symbols in tmap

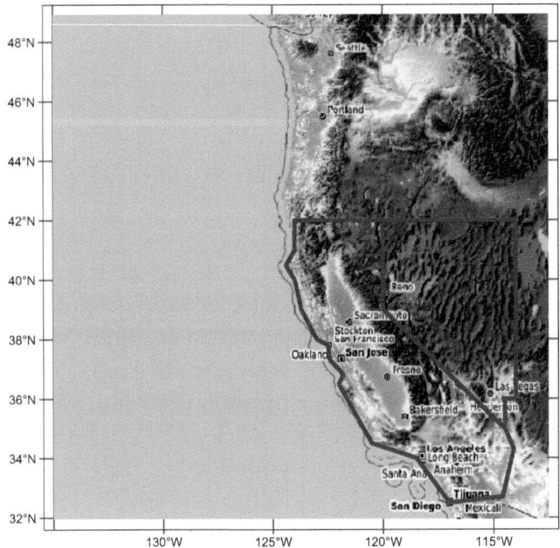

FIGURE 6.26 Two western states with a basemap in tmap

```
  tm_legend() +
  tm_layout(legend.position=c("RIGHT","TOP"))
```

Color (rgb) basemap

Earlier we looked at creating a basemap using `maptiles`. These are also supported in tmap using the `tm_rgb` (Figure 6.26). Note that here the graticule is covered by the raster tiles, as it was by polygons earlier. We could have put it after the basemap tiles, but it doesn't look good, but works ok just showing on the edges.

```
library(tmap); library(maptiles)
# The following gets the raster that covers the extent of CA & NV [twostates]
calnevaBase <- get_tiles(twostates, provider="OpenTopoMap")# g
tm_shape(calnevaBase) + tm_graticules() + tm_rgb()  +
  tm_shape(twostates) + tm_borders(lwd=3)
```

6.8 Interactive Maps

The word "static" in "static maps" isn't something you would have heard in a cartography class thirty years ago, since essentially *all* maps then were static. Very important in designing maps was, and still is, considering your audience and their perception of symbology.

- Figure-to-ground relationships assume "ground" is a white piece of paper (or possibly a standard white background in a pdf), so good cartographic color schemes tend to range from light for low values to dark for high values.
- Scale is fixed, and there are no "tools" for changing scale, so a lot of attention must be paid to providing scale information.
- Similarly, without the ability to see the map at different scales, inset maps are often needed to provide context.

Interactive maps change the game in having tools for changing scale and *always* being "printed" on a computer or device where the color of the background isn't necessarily white. We are increasingly used to using interactive maps on our phones or other devices, and often get frustrated not being able to zoom into a static map. However, as we'll see, there are trade-offs in that interactive maps don't provide the same level of control on symbology for what we want to put on the map, but instead depend a lot on basemaps for much of the cartographic design, generally limiting the symbology of data being mapped on top of it.

6.8.1 Leaflet

We'll come back to tmap to look at its interactive option, but we should start with a very brief look at the package that it uses when you choose interactive mode: **leaflet**. The R leaflet library itself translates to *Javascript* and its **Leaflet** library, which was designed to support "mobile-friendly interactive maps" (https://leafletjs.com). We'll also look at another R package that translates to leaflet: **mapview**.

Interactive maps tend to focus on using basemaps for most of the cartographic design work, and quite a few are available. The only code required to display the basemap is `addTiles()`, which will display the default `OpenStreetMap` in the area where you provide any features, typically with `addMarkers()`. This default basemap is pretty good to use for general purposes, especially in urban areas where OpenStreetMap contributors (including a lot of former students I think) have provided a lot of data.

You can specify additional choices using `addProviderTiles()`, and use a layers control with the choices provided as `baseGroups`. You have access to all of the basemap provider names with the vector `providers`, and to see what they look like in your area, explore http://leaflet-extras.github.io/leaflet-providers/preview/index.html, where you can zoom into an area of interest and select the base map to see how it would look in a given zoom (Figure 6.27).

```
library(leaflet)
leaflet() %>%
  addTiles(group="OpenStreetMap") %>%
  addProviderTiles("OpenTopoMap",group="OpenTopoMap") %>%
  addProviderTiles("Esri.NatGeoWorldMap",group="Esri.NatGeoWorldMap") %>%
  addProviderTiles("Esri.WorldImagery",group="Esri.WorldImagery") %>%
  addMarkers(lng=-122.4756, lat=37.72222,
           popup="Institute for Geographic Information Science, SFSU") %>%
  addLayersControl(
    baseGroups = c("OpenStreetMap","OpenTopoMap",
                 "Esri.WorldImagery","Esri.NatGeoWorldMap"))
```

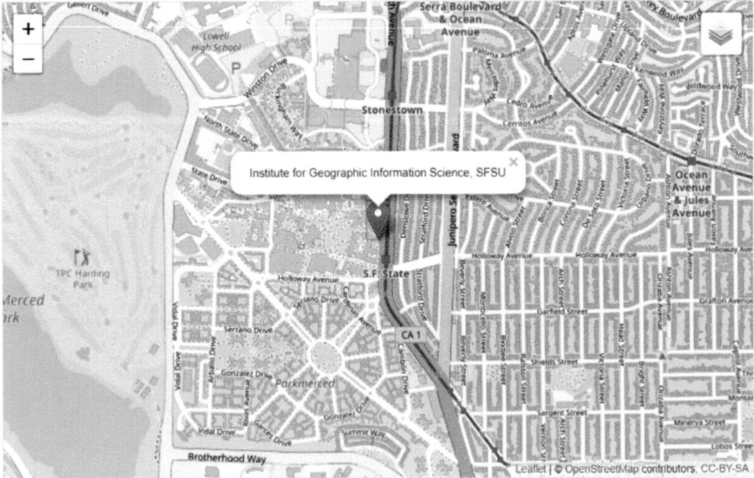

FIGURE 6.27 Leaflet map showing the location of the SFSU Institute for Geographic Information Science with choices of basemaps

With an interactive map, we do have the advantage of a good choice of base maps and the ability to resize and explore the map, but symbology is more limited, mostly just color and size, with only one variable in a legend. Note that interactive maps display in the Viewer window of RStudio or in R Markdown code output as you see in the html version of this book.

For a lot more information on the leaflet package in R, see:

- https://blog.rstudio.com/2015/06/24/leaflet-interactive-web-maps-with-r/
- https://github.com/rstudio/cheatsheets/blob/master/leaflet.pdf

6.8.2 Mapview

In the **mapview** package, either `mapview` or `mapView` produces perhaps the simplest interactive map display where the view provides the option of changing basemaps, and just a couple of options make it a pretty useful system for visualizing point data. So for environmental work involving samples, it is a pretty useful exploratory tool.

```
library(sf); library(mapview)
sierra <- st_as_sf(sierraFeb, coords = c("LONGITUDE", "LATITUDE"), crs = 4326)
mapview(sierra, zcol="TEMPERATURE", popup = TRUE)
```

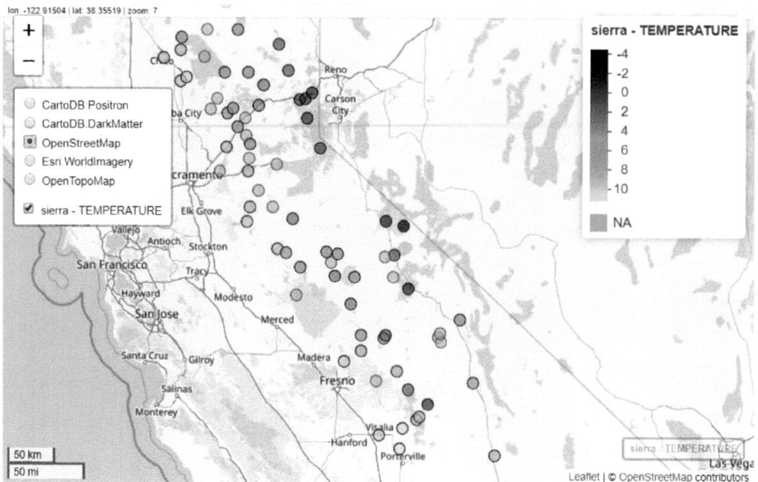

6.8.3 tmap (view mode)

You can change to an interactive mode with tmap by using `tmap_mode("view")` so you might think that you can do all the great things that `tmap` does in normal plot mode here, but as we've seen, interactive maps don't provide the same level of symbology. The view mode of tmap, like mapview, is just a wrapper around leaflet, so we'll focus on the latter. The key parameter needed is `tmap_mode`, which must be set to `"view"` to create an interactive map.

```
library(tmap); library(sf)
tmap_mode("view")
tmap_options(max.categories = 8)
sierra <- st_as_sf(sierraFeb, coords = c("LONGITUDE", "LATITUDE"), crs = 4326)
bounds <- st_bbox(sierra)
#sierraMap <-
tm_basemap(leaflet::providers$Esri.NatGeoWorldMap) +
  tm_shape(sierra) + tm_symbols(col="TEMPERATURE",
  palette=c("blue","red"), style="cont",n=8,size=0.2) +
  tm_legend() +
  tm_layout(legend.position=c("RIGHT","TOP"))
#sierraMap
#tmap_leaflet(sierraMap)
```

One nice feature of tmap view mode (and mapview) is the ability to select the basemap interactively using the layers symbol. There are lots of basemap available with leaflet (and thus with tmap view). To explore them, see http://leaflet-extras.github.io/leaflet-providers/preview/index.html but recognize that not all of these will work everywhere; many are highly localized or may only work at certain scales.

The following map using tmap could presumably be coded in leaflet, but tmap makes it a lot easier. We'll use it to demonstrate picking the basemap, which really can help in communicating our data with context (Figure 6.28).

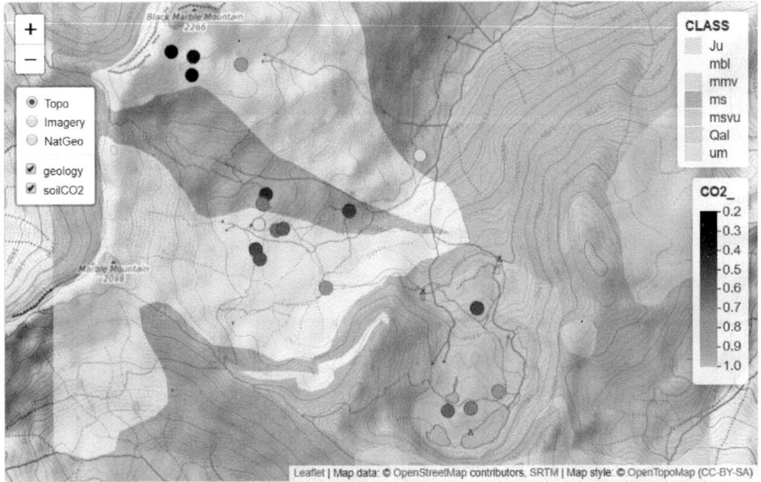

FIGURE 6.28 View (interactive) mode of tmap with selection of basemaps

```
library(sf); library(tmap); library(igisci)
tmap_mode = "view"
tmap_options(basemaps=c(Topo="OpenTopoMap", Imagery = "Esri.WorldImagery",
                        NatGeo="Esri.NatGeoWorldMap"))
soilCO2 <- st_read(ex("marbles/co2july95.shp"))
geology <- st_read(ex("marbles/geology.shp"))
```

```
mblCO2map <- tm_basemap() +
  tm_shape(geology) + tm_fill(col="CLASS", alpha=0.5) +
  tm_shape(soilCO2) + tm_symbols(col="CO2_",
  palette="viridis", style="cont",n=8,size=0.6) +
  tm_legend() +
  tm_layout(legend.position=c("RIGHT","TOP"))
mblCO2map
```

6.8.4 Interactive mapping of individual penguins abstracted from a big dataset

In a study of spatial foraging patterns by Adélie penguins (*Pygoscelis adeliae*), Ballard et al. (2019) collected data over five austral summer seasons (parts of December and January in 2005-06, 2006-07, 2007-08, 2008-09, 2012-13) period using SPLASH tags that combine Argos satellite tracking with a time-depth recorder, with 11,192 observations. The 24 SPLASH tags were re-used over the period of the project with a total of 162 individual breeding penguins, each active over a single season.

Our code takes advantage of the abstraction capabilities of base R and `dplyr` to select an individual penguin from the large data set and prepare its variables for useful display. Then the interactive mode of tmap works well for visualizing the penguin data – both the cloud of all observations and the focused individual – since we can zoom in to see the locations in context, and basemaps can be chosen. The tmap code colors the individual penguin's locations based on a decimal day derived from day and time. While interactive view is more limited in options, it does at least support a legend and title (Figure 6.29).

```
library(sf); library(tmap); library(tidyverse); library(igisci)
sat_table <- read_csv(ex("penguins/sat_table_splash.csv"))
obs <- st_as_sf(filter(sat_table,include==1), coords=c("lon","lat"), crs=4326)
uniq_ids <- unique(obs$uniq_id) # uniq_id identifies an individual penguin
onebird <- obs %>%
  filter(uniq_id==uniq_ids[sample.int(length(uniq_ids), size=1)]) %>%
  mutate(decimalDay = as.numeric(day) + as.numeric(hour)/24)
```

```
tmap_mode = "view"
tmap_options(basemaps=c(Terrain = "Esri.WorldTerrain",
                        Imagery = "Esri.WorldImagery",
                        OceanBasemap = "Esri.OceanBasemap",
                        Topo="OpenTopoMap",
                        Ortho="GeoportailFrance.orthos"))
penguinMap <- tm_basemap() +
  tm_shape(obs) + tm_symbols(col="gray", size=0.01, alpha=0.5) +
  tm_shape(onebird) + tm_symbols(col="decimalDay",
  palette="viridis", style="cont",n=8,size=0.6) +
  tm_legend() +
  tm_layout()
tmap_leaflet(penguinMap)
```

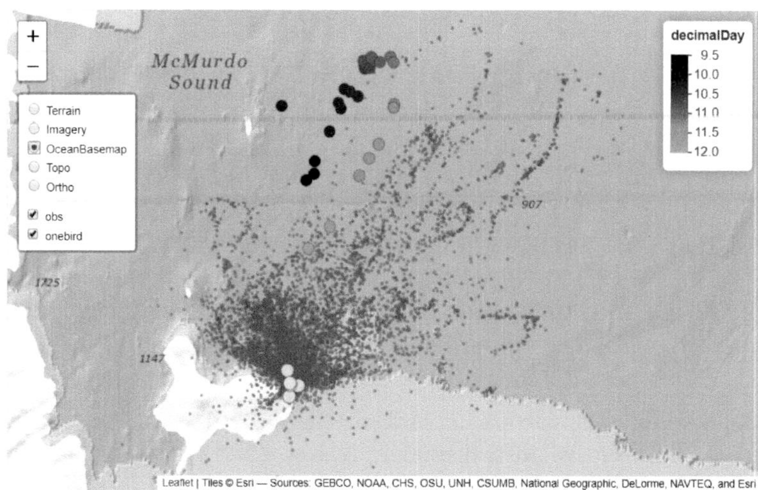

FIGURE 6.29 Observations of Adélie penguin migration from a 5-season study of a large colony at Ross Island in the SW Ross Sea, Antarctica; and an individual – H36CROZ0708 – from season 0708. Data source: Ballard et al. (2019). Fine-scale oceanographic features characterizing successful Adélie penguin foraging in the SW Ross Sea. Marine Ecology Progress Series 608:263-277.

Some final thoughts on maps and plotting/viewing packages: There's a lot more to creating maps from spatial data, but we need to look at spatial analysis to create some products that we might want to create maps from. In the next chapter, in addition to looking at spatial analysis methods we'll also continue our exploration of map design, ranging from very simple exploratory outputs to more carefully designed products for sharing with others.

6.9 Exercises: Spatial Data and Maps

6.9.1 Project preparation

Create a new RStudio project, and name it "Spatial". We're going to use this for data we create and new data we want to bring in. We'll still be reading in data from the data package, but working in this project we'll be getting used to (a) working with our own data and (b) storing data to be used for later projects. Once we've created the project, we'll want to create a `data` folder to store data in. This code should do this:

```
dirname <- "data"
if (!dir.exists(dirname)) {
  dir.create(dirname)
}
```

Exercise 6.1. Using the method of building simple sf geometries, build a simple 1×1 square object and plot it. Remember that you have to close the polygon, so the first vertex is the same as the last (of 5) vertices. Provide your code only.

Exercise 6.2. Build a map in ggplot of Colorado, Wyoming, and Utah with these boundary vertices in GCS. As with the square, remember to close each figure, and assign the crs to what is needed for GCS: 4326. Submit map as exported plot, and code in the submittal text block.

- Colorado: (-109,41),(-102,41),(-102,37),(-109,37)
- Wyoming: (-111,45),(-104,45),(-104,41),(-111,41)
- Utah: (-114,42),(-111,42),(-111,41),(-109,41),(-109,37),(-114,37)

Exercise 6.3. Add in the code for other western states and create kind of a western US map. Go ahead and use the following code:

```
AZ <- st_polygon(list(rbind(c(-114,37),c(-109,37),c(-109,31.3),c(-111,31.3),
  c(-114.8,32.5),c(-114.6,32.7),c(-114.1,34.3),c(-114.5,35),
  c(-114.5,36),c(-114,36),c(-114,37))))
NM <- st_polygon(list(rbind(c(-109,37),c(-103,37),c(-103,32),c(-106.6,32),
```

```
  c(-106.5,31.8),c(-108.2,31.8),c(-108.2,31.3),c(-109,31.3),c(-109,37))))
CA <- st_polygon(list(rbind(c(-124,42),c(-120,42),c(-120,39),c(-114.5,35),
  c(-114.1,34.3),c(-114.6,32.7),c(-117,32.5),c(-118.5,34),c(-120.5,34.5),
  c(-122,36.5),c(-121.8,36.8),c(-122,37),c(-122.4,37.3),c(-122.5,37.8),
  c(-123,38),c(-123.7,39),c(-124,40),c(-124.4,40.5),c(-124,41),c(-124,42))))
NV <- st_polygon(list(rbind(c(-120,42),c(-114,42),c(-114,36),c(-114.5,36),
  c(-114.5,35),c(-120,39),c(-120,42))))
OR <- st_polygon(list(rbind(c(-124,42),c(-124.5,43),
  c(-124,46),c(-123,46),c(-122.7,45.5),c(-119,46),c(-117,46),
  c(-116.5,45.5),c(-117.2,44.5),c(-117,44),
  c(-117,42),c(-120,42),c(-124,42))))
WA <- st_polygon(list(rbind(c(-124,46),c(-124.8,48.4),c(-123,48),
  c(-123,49),c(-117,49),
  c(-117,46),c(-119,46),c(-122.7,45.5),c(-123,46),c(-124,46))))
ID <- st_polygon(list(rbind(c(-117,49),
  c(-116,49),c(-116,48),c(-114.4,46.5),c(-114.4,45.5),
  c(-114,45.6),c(-113,44.5),c(-111,44.5),
  c(-111,42),c(-114,42),c(-117,42),
  c(-117,44),c(-117.2,44.5),c(-116.5,45.5),
  c(-117,46),c(-117,49))))
MT <- st_polygon(list(rbind(c(-116,49),c(-104,49),
  c(-104,45),c(-111,45),
  c(-111,44.5),c(-113,44.5),c(-114,45.6),c(-114.4,45.5),
  c(-114.4,46.5),c(-116,48),c(-116,49))))
```

Then build sfcWestStates from these geometries using the `st_sfc()` function, specifying GCS (4326) as the crs, and plot with either ggplot or tmap.

Exercise 6.4. Create an sf class from the western states adding the fields `name`, `abb`, `area_sqkm`, and `population`, and create a map labeling with the name. Note that the added fields need to be in the same order as the boundaries created above. Also remember to make the area and pop variables numeric, which will happen if they're brought in as a vector along with character variables.

- Colorado, CO, 269837, 5758736
- Wyoming, WY, 253600, 578759
- Utah, UT, 84899, 3205958
- Arizona, AZ, 295234, 7278717
- New Mexico, NM, 314917, 2096829
- California, CA, 423970, 39368078
- Nevada, NV, 286382, 3080156
- Oregon, OR, 254806, 4237256
- Washington, WA, 184827, 7705281
- Idaho, ID, 216443, 1839106
- Montana, MT, 380800, 1085407

Exercise 6.5. Store the W_States as a shape file in the data folder with st_write. (Look up how to use st_write with ?st_write – it's pretty simple.) Note that this will fail if it already exists, so include the parameter `delete_layer = TRUE`.

Exercise 6.6. Highest Peaks: Create a tibble for the highest peaks in the western states, with the following names, elevations in m, longitude and latitude, use st_as_sf to create an sf from it, and add them to that map. Then use st_write again to store these as "data/peaks.shp" again using `delete_layer = TRUE`:

- Wheeler Peak, 4011, -105.4, 36.5
- Mt. Whitney, 4421, -118.2, 36.5
- Boundary Peak, 4007, -118.35, 37.9
- Kings Peak, 4120, -110.3, 40.8
- Gannett Peak, 4209, -109, 43.2
- Mt. Elbert, 4401, -106.4, 39.1
- Humphreys Peak, 3852, -111.7, 35.4
- Mt. Hood, 3428, -121.7, 45.4
- Mt. Rainier, 4392, -121.8, 46.9
- Borah Peak, 3859, -113.8, 44.1
- Granite Peak, 3903, -109.8, 45.1

Note: the easiest way to do this is with the tribble function, starting with:

```
peaks <- tribble(
  ~peak, ~elev, ~longitude, ~latitude,
  "Wheeler Peak", 4011, -105.4, 36.5,
```

Exercise 6.7. California freeways. From the CA_counties and CAfreeways feature data in igisci, make a simple map in ggplot, with freeways colored red).

Exercise 6.8. After adding the terra library, create a raster from the built-in `volcano` matrix of elevations from Auckland's **Maunga Whau Volcano**, and use plot() to display it. We'd do more with that dataset, but we don't know what the cell size is.

Exercise 6.9. Western States tmap: Use tmap to create a map from the W_States (polygons) and peaksp (points) data we created earlier. Include a basemap using the maptiles package. Hints: you'll want to use tm_text with text set to "peak" to label the points, along with the parameter `auto.placement=TRUE`. Use this as an opportunity to test the shape files you've written earlier by using `st_read` with those shape files. Experiment with `tm_symbol` sizes, and `just`, `xmod`, and `ymod` settings with `tm_text`.

Exercise 6.10. tmap view mode. Also using the western states data, create a tmap in view mode, but don't use the state borders since the basemap will have them. Just before adding shapes, set the basemap to `leaflet::providers\$Esri.NatGeoWorldMapEsri.NatGeoWorldMap`, then continue to the peaks after the `+` to see the peaks on a National Geographic basemap (Figure 6.30)

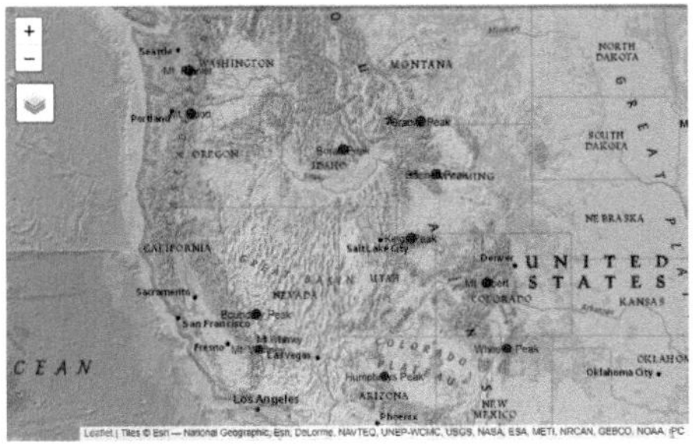

FIGURE 6.30 tmap View mode (goal)

7

Spatial Analysis

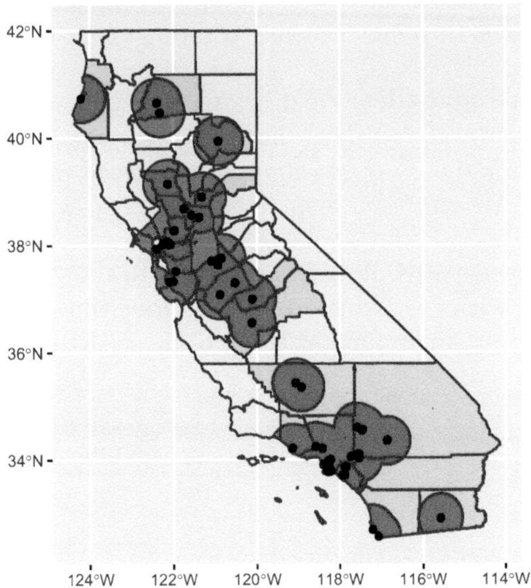

Spatial analysis provides an expansive analytical landscape with theoretical underpinnings dating to at least the 1950s (Berry and Marble (1968)), contributing to the development of geographic information systems as well as spatial tools in more statistically oriented programming environments like R. We won't attempt to approach any thorough coverage of these methods, and we would refer the reader for more focused consideration using R-based methods to sources such as Lovelace, Nowosad, and Muenchow (2019) and Hijmans (n.d.).

In this chapter, we'll continue working with spatial data, and explore spatial analysis methods. We'll look at a selection of useful geospatial abstraction and analysis methods, what are also called *geoprocessing* tools in a GIS. The R Spatial world has grown in recent years to include an increasingly good array of tools. This chapter will focus on *vector* GIS methods, and here we don't mean the same thing as the vector data objects in R nomenclature (Python's pandas package, which uses a similar data structure, calls these "series" instead of vectors), but instead on feature geometries and their spatial relationships based on the coordinate referencing system.

7.1 Data Frame Operations

But before we get into those specifically spatial operations, it's important to remember that feature data at least are also data frames, so we can use the methods we already know with them. For instance, we can look at properties of variables and then filter for features that meet a criterion, like all climate station points at greater than 2,000 m elevation, or all above 38°N latitude. To be able to work with latitude as a variable, we'll need to use `remove=FALSE` (the default is to remove them) to retain them when we use `st_as_sf`.

Adding a basemap with maptiles: We'd like to have a basemap, so we'll create one with the `maptiles` package (install if you don't have it.) *Warning: the `get_tiles` function goes online to get the basemap data, so if you don't have a good internet connection or the site goes down, this may fail.* We can then display the basemap with `tm_rgb`.

For temperature, we'll reverse a RColorBrewer palette to show a reasonable color scheme by reversing its `RdBu` palette with `rev` (which took me a *long* time to figure out – color schemes are much more challenging than you might think because there are *many* highly varied uses of color.)

```
library(tmap); library(RColorBrewer); library(sf); library(tidyverse);
library(maptiles); library(igisci)
tmap_mode("plot")
newname <- unique(str_sub(sierraFeb$STATION_NAME, 1,
                          str_locate(sierraFeb$STATION_NAME, ",")-1))
sierraFeb2000 <- st_as_sf(bind_cols(sierraFeb,STATION=newname) %>%
                          dplyr::select(-STATION_NAME) %>%
                          filter(ELEVATION >= 2000, !is.na(TEMPERATURE)),
                 coords=c("LONGITUDE","LATITUDE"), crs=4326)
sierraBase <- get_tiles(sierraFeb2000)
tm_shape(sierraBase) + tm_rgb() +
  tm_shape(sierraFeb2000) +
  tm_symbols(col = "TEMPERATURE", midpoint=NA, palette=rev(brewer.pal(8,"RdBu"))) +
  tm_text(text = "STATION", size=0.5, auto.placement=T, xmod=0.5, ymod=0.5) +
  tm_graticules(lines=F)
```

Now let's include LATITUDE. Let's see what we get by filtering for both `ELEVATION >= 2000` and `LATITUDE >= 38`:

```
sierraFeb %>%
  filter(ELEVATION >= 2000 & LATITUDE >= 38)
```

```
## # A tibble: 1 x 7
##   STATION_NAME                        COUNTY ELEVA~1 LATIT~2 LONGI~3 PRECI~4 TEMPE~5
##   <chr>                               <fct>    <dbl>   <dbl>   <dbl>   <dbl>   <dbl>
## 1 BODIE CALIFORNIA STATE HISTORI~ Mono      2551.    38.2   -119.    39.6    -4.4
## # ... with abbreviated variable names 1: ELEVATION, 2: LATITUDE, 3: LONGITUDE,
## #   4: PRECIPITATION, 5: TEMPERATURE
```

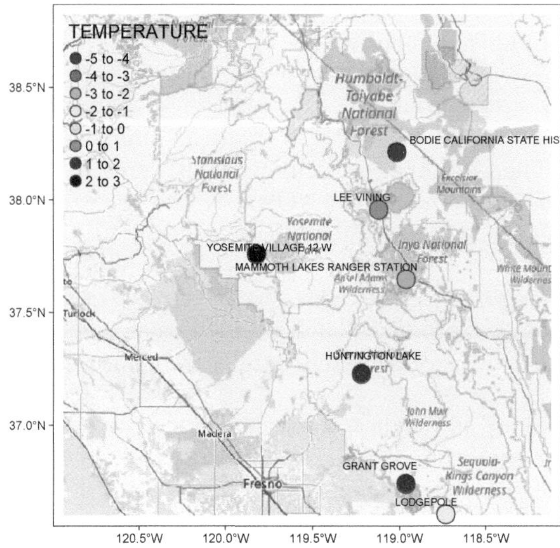

FIGURE 7.1 Plotting filtered data: above 2,000 m and 38°N latitude with a basemap

The only one left is Bodie (Figure 7.2). Maybe that's why Bodie, a ghost town now, has such a reputation for being so friggin' cold (at least for California). I've been snowed on there in summer.

7.1.1 Using grouped summaries, and filtering by a selection

We've been using February Sierra climate data for demonstrating various things, but this wasn't the original form of the data downloaded, so let's look at the original data and use

FIGURE 7.2 A Bodie scene, from Bodie State Historic Park (https://www.parks.ca.gov/)

some dplyr methods to restructure it, in this case to derive annual summaries. We'll use the monthly normals to derive annual values.

We looked at the very useful `group_by ... summarize()` group-summary process earlier 3.4.4. We'll use this and a selection process to create an annual Sierra climate dataframe we can map and analyze. The California monthly normals were downloaded from the National Centers for Environmental Information at NOAA, where you can select desired parameters, monthly normals as frequency, and limit to one state.

- https://www.ncei.noaa.gov/
- https://www.ncei.noaa.gov/products/land-based-station/us-climate-normals
- https://www.ncei.noaa.gov/access/search/data-search/normals-monthly-1991-2020

First we'll have a quick look at the data to see how it's structured. These were downloaded as monthly normals for the State of California, as of 2010, so there are 12 months coded `201001:201012`, with obviously the same `ELEVATION`, `LATITUDE`, and `LONGITUDE`, but monthly values for climate data like `MLY-PRCP-NORMAL`, etc.

```
head(read_csv(ex("sierra/908277.csv")),n=15)
```

```
## # A tibble: 15 x 11
##    STATION         STATI~1 ELEVA~2 LATIT~3 LONGI~4   DATE MLY-P~5 MLY-S~6 MLY-T~7
##    <chr>           <chr>   <chr>   <chr>   <chr>    <dbl>   <dbl>   <dbl>   <dbl>
##  1 GHCND:USC0004~ TWENTY~ 602     34.128~ -116.0~ 201001    13.2        0    10.6
##  2 GHCND:USC0004~ TWENTY~ 602     34.128~ -116.0~ 201002    15.0   -19754.   12.4
##  3 GHCND:USC0004~ TWENTY~ 602     34.128~ -116.0~ 201003    11.4        0    15.7
##  4 GHCND:USC0004~ TWENTY~ 602     34.128~ -116.0~ 201004     3.3        0    19.3
##  5 GHCND:USC0004~ TWENTY~ 602     34.128~ -116.0~ 201005     2.29       0    24.4
##  6 GHCND:USC0004~ TWENTY~ 602     34.128~ -116.0~ 201006     0.25       0    28.8
##  7 GHCND:USC0004~ TWENTY~ 602     34.128~ -116.0~ 201007    12.2        0    31.9
##  8 GHCND:USC0004~ TWENTY~ 602     34.128~ -116.0~ 201008    20.6        0    31.2
##  9 GHCND:USC0004~ TWENTY~ 602     34.128~ -116.0~ 201009    10.2        0    27.4
## 10 GHCND:USC0004~ TWENTY~ 602     34.128~ -116.0~ 201010     5.08       0    21
## 11 GHCND:USC0004~ TWENTY~ 602     34.128~ -116.0~ 201011     5.08       0    14.3
## 12 GHCND:USC0004~ TWENTY~ 602     34.128~ -116.0~ 201012    14.7        0     9.9
## 13 GHCND:USC0004~ BUTTON~ 82      35.4047 -119.4~ 201001    32.5        0     7.9
## 14 GHCND:USC0004~ BUTTON~ 82      35.4047 -119.4~ 201002    30.0        0    10.9
## 15 GHCND:USC0004~ BUTTON~ 82      35.4047 -119.4~ 201003    30.7        0    13.9
## # ... with 2 more variables: `MLY-TMAX-NORMAL` <dbl>, `MLY-TMIN-NORMAL` <dbl>,
## #   and abbreviated variable names 1: STATION_NAME, 2: ELEVATION, 3: LATITUDE,
## #   4: LONGITUDE, 5: `MLY-PRCP-NORMAL`, 6: `MLY-SNOW-NORMAL`,
## #   7: `MLY-TAVG-NORMAL`
```

To get just the Sierra data, it seemed easiest to just provide a list of relevant county names to filter the counties, do a bit of field renaming, then read in a previously created selection of Sierra weather/climate stations.

```
sierraCounties <- st_make_valid(CA_counties) %>%
  filter(NAME %in% c("Alpine","Amador","Butte","Calaveras","El Dorado",
                     "Fresno","Inyo","Kern","Lassen","Madera","Mariposa",
```

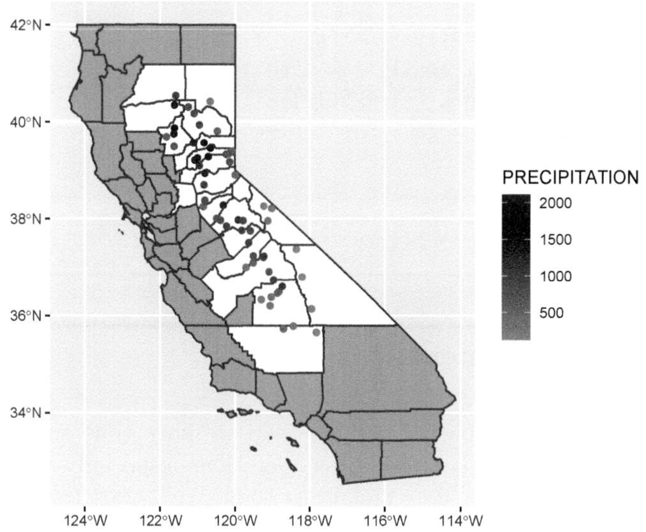

FIGURE 7.3 Sierra data

```
                    "Mono","Nevada","Placer","Plumas","Sacramento","Shasta",
                    "Sierra","Tehama","Tulare","Tuolumne","Yuba"))
normals <- read_csv(ex("sierra/908277.csv")) %>%
  mutate(STATION = str_sub(STATION,7,str_length(STATION)))
sierraStations <- read_csv(ex("sierra/sierraStations.csv"))
```

To get annual values, we'll want to use the stations as groups in a `group_by %>% summarize` process. For values that stay the same for a station (`LONGITUDE`, `LATITUDE`, `ELEVATION`), we'll use `first()` to get just one of the 12 identical monthly values. For values that vary monthly, we'll `sum()` the monthly precipitations and get the `mean()` of monthly temperatures to get appropriate annual values. We'll also use a `right_join` to keep only the stations that are in the Sierra. *Have a look at this script to make sure you understand what it's doing; it's got several elements you've been introduced to before, and that you should understand.* At the end, we'll use `st_as_sf` to make sf data out of the data frame, and retain the `LONGITUDE` and `LATITUDE` variables in case we want to use them as separate variables (Figure 7.3).

```
sierraAnnual <- right_join(sierraStations,normals,by="STATION") %>%
  filter(!is.na(STATION_NA)) %>%
    dplyr::select(-STATION_NA) %>%
    group_by(STATION_NAME) %>% summarize(LONGITUDE = first(LONGITUDE),
                                         LATITUDE = first(LATITUDE),
                                         ELEVATION = first(ELEVATION),
                                         PRECIPITATION = sum(`MLY-PRCP-NORMAL`),
                                         TEMPERATURE = mean(`MLY-TAVG-NORMAL`)) %>%
    mutate(STATION_NAME = str_sub(STATION_NAME,1,str_length(STATION_NAME)-6)) %>%
  filter(PRECIPITATION > 0) %>% filter(TEMPERATURE > -100) %>%
  st_as_sf(coords = c("LONGITUDE", "LATITUDE"), crs=4326, remove=F)
ggplot() +
```

```
  geom_sf(data=CA_counties, aes(), fill="gray") +
  geom_sf(data=sierraCounties, aes(), fill="white") +
  geom_sf(data=sierraAnnual, aes(col=PRECIPITATION)) +
  scale_color_gradient(low="orange", high="blue")
```

7.2 Spatial Analysis Operations

Again, there is a lot more to spatial analysis than we have time to cover. But we'll explore some particularly useful spatial analysis operations, especially those that contribute to statistical analysis methods we'll be looking at soon. We'll start by continuing to look at subsetting or filtering methods, but ones that use spatial relationships to identify what we want to retain or remove from consideration.

7.2.1 Using topology to subset

Using spatial relationships can be useful in filtering our data, and there are quite a few topological relationships that can be explored. See Lovelace, Nowosad, and Muenchow (2019) for a lot more about topological operations. We'll look at a relatively simple one that identifies whether a feature is within another one, and apply this method to filter for features within a selection of five counties, which we'll start by identifying by name.

```
nSierraCo <- CA_counties %>%
  filter(NAME %in% c("Plumas","Butte","Sierra","Nevada","Yuba"))
```

Then we'll use those counties to select towns (places) that occur within them ...

```
CA_places <- st_read(ex("sierra/CA_places.shp"))
nCAsel <- lengths(st_within(CA_places, nSierraCo)) > 0 # to get TRUE & FALSE
nSierraPlaces <- CA_places[nCAsel,]
```

... and do the same for the `sierraFeb` weather stations (Figure 7.4).

```
sierra <- st_as_sf(read_csv(ex("sierra/sierraFeb.csv")),
                   coords=c("LONGITUDE","LATITUDE"), crs=4326)
nCAselSta <- lengths(st_within(sierra, nSierraCo)) > 0 # to get TRUE & FALSE
nSierraStations <- sierra[nCAselSta,]
```

```
library(maptiles)
nsierraBase <- get_tiles(nSierraStations, provider="OpenTopoMap")
```

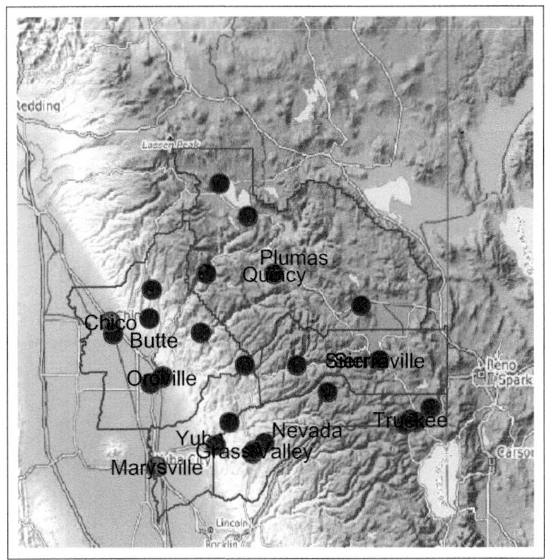

FIGURE 7.4 Northern Sierra stations and places

```
library(tmap)
tmap_mode("plot")
tm_shape(nsierraBase) + tm_rgb(alpha=0.5) +
  tm_shape(nSierraCo) + tm_borders() + tm_text("NAME") +
  tm_shape(nSierraStations) + tm_symbols(col="blue") +
  tm_shape(nSierraPlaces) + tm_symbols(col="red", alpha=0.5) + tm_text("AREANAME")
```

So far, the above is just subsetting for a map, which may be all we're wanting to do, but we'll apply this selection to a distance function in the next section to explore a method using a reduced data set.

7.2.2 Centroid

Related to the topological concept of relating features inside other features is creating a new feature in the middle of an existing one, specifically a point placed in the middle of a polygon: a *centroid*. Centroids are useful when you need point data for some kind of analysis, but that point needs to represent the polygon it's contained within. The methods we'll see in the code below include:

- `st_centroid()` : to derive a single point for each tract that should be approximately in its middle. It must be topologically contained within it, not fooled by an annulus ("doughnut") shape, for example.
- `st_make_valid()` : to make an invalid geometry valid (or just check for it). This function has just become essential now that `sf` supports spherical instead of just planar data, which ends up containing "slivers" where boundaries slightly under- or overlap since they were originally built from a planar projection. The `st_make_valid()` appears to make minor (and typically invisible) corrections to these slivers.

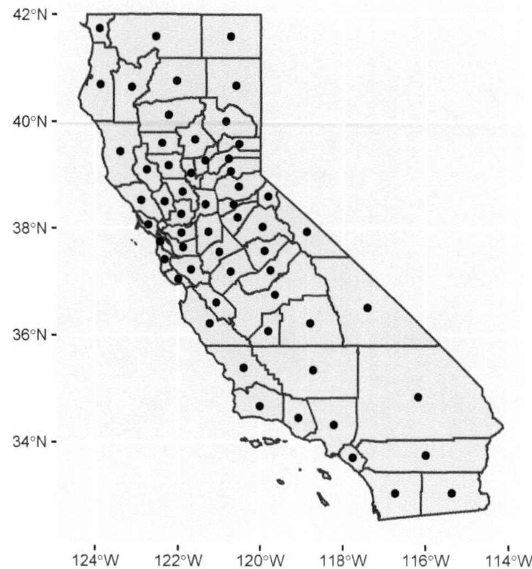

FIGURE 7.5 California county centroids

- `st_bbox` : reads the bounding box, or spatial extent, of any dataset, which we can then use to set the scale to focus on that area. In this case, we'll focus on the census centroids instead of the statewide `CAfreeways` for instance.

Let's see the effect of the centroids and bbox. We'll start with all county centroids (Figure 7.5).

```
library(tidyverse); library(sf); library(igisci)
CA_counties <- st_make_valid(CA_counties) # needed fix for data


ggplot() +
  geom_sf(data=CA_counties) +
  geom_sf(data=st_centroid(CA_counties))
```

Here's an example that also applies the bounding box to establish a mapping scale that covers the Bay Area while some of the data (TRI locations and CAfreeways) are state-wide (Figure 7.6).

```
library(tidyverse)
library(sf)
library(igisci)
BayAreaTracts <- st_make_valid(BayAreaTracts)
censusCentroids <- st_centroid(BayAreaTracts)
TRI_sp <- st_as_sf(read_csv(ex("TRI/TRI_2017_CA.csv")),
                   coords = c("LONGITUDE", "LATITUDE"),
                   crs=4326) # simple way to specify coordinate reference
bnd <- st_bbox(censusCentroids)
ggplot() +
```

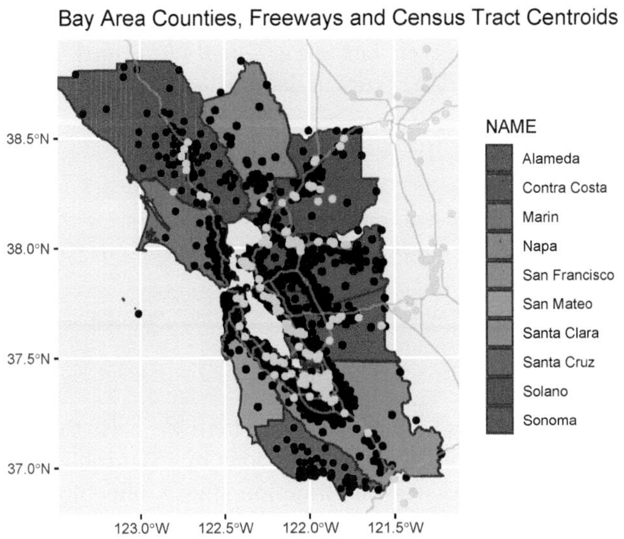

FIGURE 7.6 Map scaled to cover Bay Area tracts using a bbox

```
geom_sf(data = BayAreaCounties, aes(fill = NAME)) +
geom_sf(data = censusCentroids) +
geom_sf(data = CAfreeways, color = "grey", alpha=0.5) +
geom_sf(data = TRI_sp, color = "yellow") +
coord_sf(xlim = c(bnd[1], bnd[3]), ylim = c(bnd[2], bnd[4])) +
  labs(title="Bay Area Counties, Freeways and Census Tract Centroids")
```

7.2.3 Distance

Distance is a fundamental spatial measure, used to not only create spatial data (distance between points in surveying or distance to GNSS satellites) but also to analyze it. Note that we can either be referring to planar (from projected coordinate systems) or spherical (from latitude and longitude) great-circle distances. Two common spatial operations involving distance in vector spatial analysis are (a) deriving distances among features, and (b) creating buffer polygons of a specific distance away from features. In raster spatial analysis, we'll look at deriving distances from target features to each raster cell.

But first, let's take a brief excursion up the Nile River, using great circle distances to derive a longitudinal profile and channel slope...

7.2.3.1 Great circle distances

Earlier we created a simple river profile using Cartesian coordinates in metres for x, y, and z (elevation), but what if our xy locations are in geographic coordinates of latitude and longitude? Using the haversine method, we can derive great-circle ("as the crow flies") distances between latitude and longitude pairs. The following function uses a haversine algorithm described at https://www.movable-type.co.uk/scripts/latlong.html (*Calculate Distance, Bearing and More Between Latitutde/Longitude Points"* (n.d.)) to derive these distances in metres, provided lat/long pairs in degrees as `haversineD(lat1,lon1,lat2,lon2)`:

```
haversineD <- function(lat1deg,lon1deg,lat2deg,lon2deg){
  lat1 <- lat1deg/180*pi; lat2 <- lat2deg/180*pi # convert to radians
  lon1 <- lon1deg/180*pi; lon2 <- lon2deg/180*pi
  a <- sin((lat2-lat1)/2)^2 + cos(lat1)*cos(lat2)*sin((lon2-lon1)/2)^2
  c <- 2*atan2(sqrt(a),sqrt(1-a))
  c * 6.371e6 # mean earth radius is ~ 6.371 million metres
}
```

We'll use these to help us derived channel slope where we need the longitudinal distance in metres along the Nile River, but we have locations in geographic coordinates (crs 4326). But first, here's a few quick checks on the function since I remember that 1 minute (1/60 degree) of latitude is equal to 1 nautical mile (NM) or ~1.15 statute miles, and the same applies to longitude at the equator. Then longitude is half that at 60 degrees north or south.

```
paste("1'lat=",haversineD(30,120,30+1/60,120)/0.3048/5280,"miles at 30°N")
paste("1'lat=",haversineD(30,120,30+1/60,120)/0.3048/6076,"NM at 30°N")
paste("1'lon=",haversineD(0,0,0,1/60)/0.3048/6076,"NM at the equator")
paste("1'lon=",haversineD(60,0,60,1/60)/0.3048/6076,"NM at 60°N")
```

```
## [1] "1'lat= 1.15155540233128 miles at 30°N"
## [1] "1'lat= 1.000693305515 NM at 30°N"
## [1] "1'lon= 1.00069330551494 NM at the equator"
## [1] "1'lon= 0.500346651434429 NM at 60°N"
```

Thanks to Stephanie Kate in my Fall 2020 Environmental Data Science course at SFSU, we have geographic coordinates and elevations in metres for a series of points along the Nile River, including various tributaries. We'll focus on the main to Blue Nile and generate a map (Figure 7.7) and longitudinal profile (Figure 7.8), using great circle distances along the channel.

```
library(igisci); library(tidyverse)
bNile <- read_csv(ex("Nile/NilePoints.csv")) %>%
  filter(nile %in% c("main","blue")) %>%
  arrange(elev_meter) %>%
  mutate(d=0, longd=0, s=1e-6) # initiate distance, longitudinal distance & slope
```

```
for(i in 2:length(bNile$x)){
  bNile$d[i] <- haversineD(bNile$y[i],bNile$x[i],bNile$y[i-1],bNile$x[i-1])
  bNile$longd[i] <- bNile$longd[i-1] + bNile$d[i]
  bNile$s[i-1] <- (bNile$elev_meter[i]-bNile$elev_meter[i-1])/bNile$d[i]
}
```

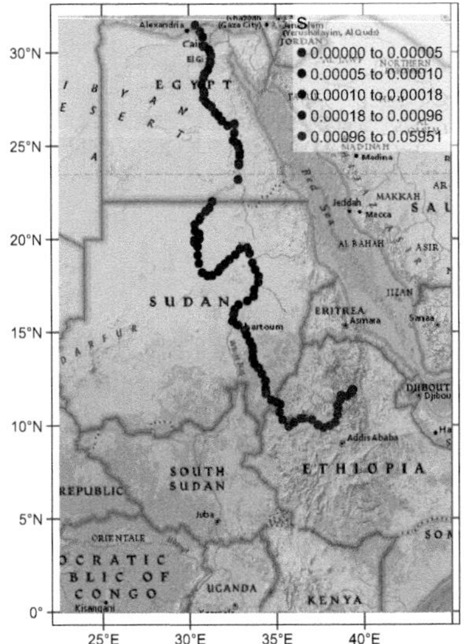

FIGURE 7.7 Nile River points, colored by channel slope

```
library(sf); library(tmap); library(maptiles)
Nile_sf <- st_as_sf(bNile,coords=c("x","y"),crs=4326)
nileBase <- get_tiles(Nile_sf, provider="Esri.NatGeoWorldMap")
tm_shape(nileBase) + tm_graticules() + tm_rgb()  +
  tm_shape(Nile_sf) + tm_dots(size=0.1,col="s",style="quantile",
                              palette=c("blue","red"))+
  tm_layout(legend.position=c("RIGHT","TOP"),
            legend.bg.color = "white",legend.bg.alpha=0.5)
```

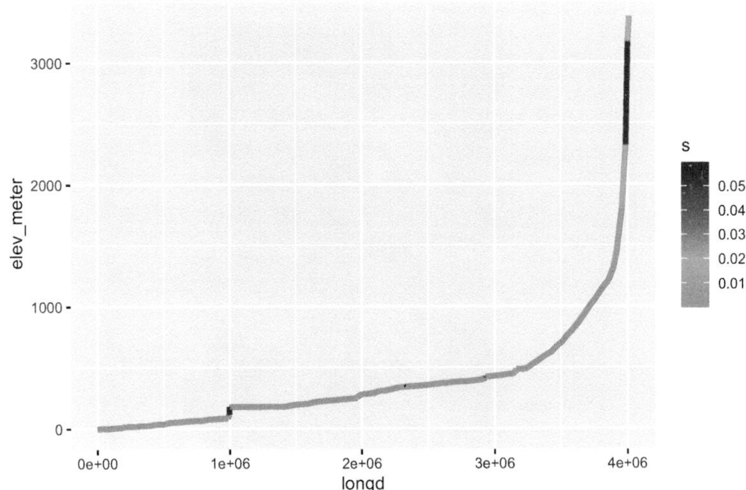

FIGURE 7.8 Nile River channel slope as range of colors from green to red, with great circle channel distances derived using the haversine method

```
ggplot(bNile, aes(longd,elev_meter)) + geom_line(aes(col=s), size=1.5) +
  scale_color_gradient(low="green", high="red") +
  ggtitle("Nile River to Blue Nile longitudinal profile")
```

7.2.3.2 Distances among features

The sf `st_distance` and terra `distance` functions derive distances between features, either among features of one data set object or between all features in one and all features in another data set.

To see how this works, we'll look at a purposefully small dataset: a selection of Marble Mountains soil CO_2 sampling sites and in-cave water quality sampling sites. We'll start by reading in the data, and filtering to just get cave water samples and a small set of nearby soil CO_2 sampling sites:

```
library(igisci); library(sf); library(tidyverse)
soilCO2all <- st_read(ex("marbles/co2july95.shp"))
cave_H2Osamples <- st_read(ex("marbles/samples.shp")) %>%
  filter((SAMPLES_ID >= 50) & (SAMPLES_ID < 60)) # these are the cave samples
soilCO2 <- soilCO2all %>% filter(LOC > 20) # soil CO2 samples in the area
```

Figure 7.9 shows the six selected soil CO_2 samples.

```
library(tmap); library(maptiles)
marblesTopoBase <- get_tiles(soilCO2, provider="OpenTopoMap")
tmap_mode("plot")
```

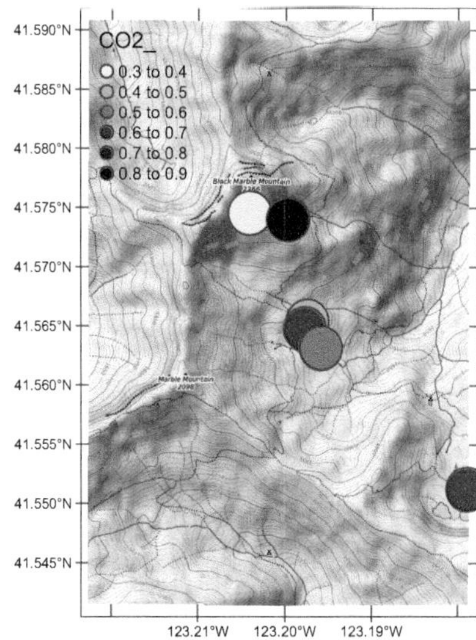

FIGURE 7.9 Selection of soil CO2 sampling sites, July 1995

```
tm_shape(marblesTopoBase) + tm_graticules() + tm_rgb() +
  tm_shape(soilCO2) + tm_symbols(col="CO2_", palette="Reds", size=4)# +


#  tm_shape(soilCO2) + tm_text("LOC")
```

If you just provide the six soil CO_2 sample points as a single feature data set input to the st_distance function, it returns a matrix of distances between each, with a diagonal of zeros where the distance would be to itself:

```
st_distance(soilCO2)
```

```
## Units: [m]
##           1         2          3          4        5         6
## 1    0.0000  370.5894 1155.64981 1207.62172 3333.529 1439.2366
## 2  370.5894    0.0000  973.73559 1039.45041 3067.106 1248.9347
## 3 1155.6498  973.7356    0.00000   75.14093 2207.019  283.6722
## 4 1207.6217 1039.4504   75.14093    0.00000 2174.493  237.7024
## 5 3333.5290 3067.1061 2207.01903 2174.49276    0.000 1938.2985
## 6 1439.2366 1248.9347  283.67215  237.70236 1938.299    0.0000
```

Then we'll look at distances between this same set of soil CO_2 samples with water samples collected in caves, where the effect of elevated soil CO_2 values might influence solution processes reflected in cave waters (Figure 7.10).

```
library(tmap)
tmap_mode("plot")
marblesTopoBase <- get_tiles(soilCO2, provider="OpenTopoMap")

tm_shape(marblesTopoBase) + tm_graticules() + tm_rgb() +
  tm_shape(cave_H2Osamples) + tm_symbols(col="CATOT", palette="Blues") +
  tm_shape(soilCO2) + tm_symbols(col="CO2_", palette="Reds")


soilwater <- st_distance(soilCO2, cave_H2Osamples)
soilwater
```

```
## Units: [m]
##          [,1]     [,2]     [,3]      [,4]      [,5]      [,6]      [,7]
## [1,] 2384.210 2271.521 2169.301 1985.6093 1980.0314 2004.1697 1905.9543
## [2,] 2029.748 1920.783 1826.579 1815.2653 1820.0496 1835.9465 1798.1216
## [3,] 1611.381 1480.688 1334.287  842.8228  846.3210  863.1026  849.7313
## [4,] 1637.998 1506.938 1357.582  781.9741  782.4215  801.6262  776.0732
## [5,] 1590.469 1578.730 1532.465 1526.3315 1567.5677 1520.1399 1820.3651
## [6,] 1506.358 1375.757 1220.000  567.8210  578.1919  589.1443  638.8615
```

In this case, the six soil CO_2 samples are the rows, and the seven cave water sample locations are the columns. We aren't really relating the values but just looking at distances. An analysis of this data might not be very informative because the caves aren't very near the

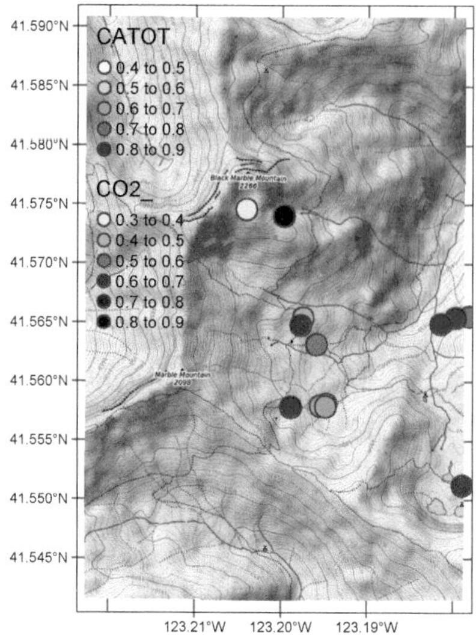

FIGURE 7.10 Selection of soil CO2 and in-cave water samples

soil samples, and conduit cave hydrology doesn't lend itself to looking at euclidean distance, but the purpose of this example is just to comprehend the results of the `sf::st_distance` or similarly the `terra::distance` function.

7.2.3.3 Distance to the nearest feature, abstracted from distance matrix

However, let's process the matrix a bit to find the distance from each soil CO_2 sample to the closest cave water sample:

```
soilCO2d <- soilCO2 %>% mutate(dist2cave = min(soilwater[1,]))
for (i in 1:length(soilwater[,1])) soilCO2d[i,"dist2cave"] <- min(soilwater[i,])
soilCO2d %>% dplyr::select(DESCRIPTIO, dist2cave) %>% st_set_geometry(NULL)
```

```
##                                         DESCRIPTIO     dist2cave
## 1             Super Sink moraine, low point W of 241 1905.9543 [m]
## 2      Small sink, first in string on hike to SuperSink 1798.1216 [m]
## 3       meadow on bench above Mbl Valley, 25' E of PCT  842.8228 [m]
## 4       red fir forest S of upper meadow, 30' W of PCT  776.0732 [m]
## 5          saddle betw Lower Skyhigh (E) & Frying Pan 1520.1399 [m]
## 6 E end of wide soil-filled 10m fissure N of lower camp  567.8210 [m]
```

... or since we can also look at distances to lines or polygons, we can find the distance from CO_2 sampling locations to the closest stream (Figure 7.11),

```
library(igisci); library(sf); library(tidyverse)
soilCO2all <- st_read(ex("marbles/co2july95.shp"))
streams <- st_read(ex("marbles/streams.shp"))
```

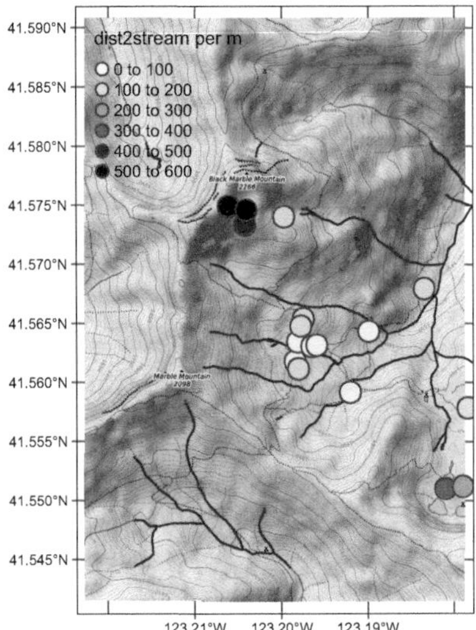

FIGURE 7.11 Distance from CO2 samples to closest streams (not including lakes)

```
strCO2 <- st_distance(soilCO2all, streams)
strCO2d <- soilCO2all %>% mutate(dist2stream = min(strCO2[1,]))
for (i in 1:length(strCO2[,1])) strCO2d[i,"dist2stream"] <- min(strCO2[i,])
```

```
marblesTopoBase <- get_tiles(soilCO2, provider="OpenTopoMap")
tm_shape(marblesTopoBase) + tm_graticules() + tm_rgb() +
  tm_shape(streams) + tm_lines(col="blue") +
  tm_shape(strCO2d) + tm_symbols(col="dist2stream")
```

7.2.3.4 Nearest feature detection

As we just saw, the matrix derived from `st_distance` when applied to feature data sets could be used for further analyses, such as this distance to the nearest place. But there's another approach to this using a specific function that identifies the nearest feature: `st_nearest_feature`, so we'll look at this with the previously filtered northern Sierra places and weather stations. We start by using `st_nearest_feature` to create an index vector of the nearest place to each station, and grab its location geometry:

```
nearest_Place <- st_nearest_feature(nSierraStations, nSierraPlaces)
near_Place_loc <- nSierraPlaces$geometry[nearest_Place]
```

Then add the distance to that nearest place to our `nSierraStations sf`. Note that we're using `st_distance` with two vector inputs as before, but ending up with just one distance per feature (instead of a matrix) with the setting `by_element=TRUE`...

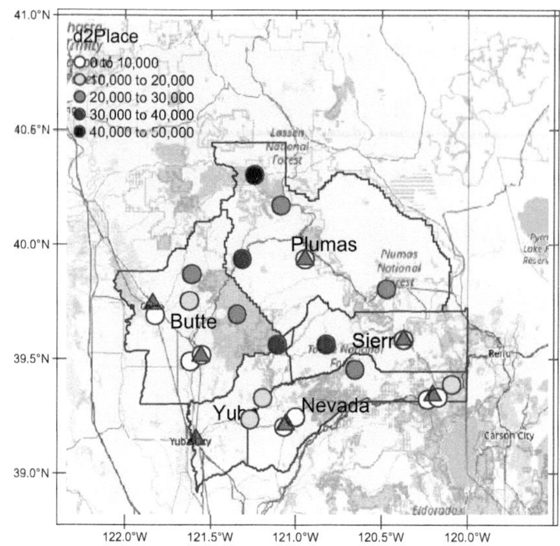

FIGURE 7.12 Distance to towns (places) from weather stations

```
nSierraStations <- nSierraStations %>%
  mutate(d2Place = st_distance(nSierraStations, near_Place_loc, by_element=TRUE),
         d2Place = units::drop_units(d2Place))
```

... and of course map the results (Figure 7.12):

```
sierraStreetBase <- get_tiles(nSierraStations, provider="OpenStreetMap")
tm_shape(sierraStreetBase) + tm_graticules() + tm_rgb() +
  tm_shape(nSierraCo) + tm_borders() + tm_text("NAME") +
  tm_shape(nSierraStations) + tm_symbols(col="d2Place") +
  tm_shape(nSierraPlaces) + tm_symbols(col="blue", alpha=0.5, size=0.5, shape=24)
```

That's probably enough examples of using distance to nearest feature so we can see how it works, but to see an example with a larger data set, an example in **Appendix A7.2** looks at the proximity of TRI sites to health data at census tracts.

7.2.4 Buffers

Creating buffers, or polygons defining the area within some distance of a feature, is commonly used in GIS. Since you need to specify that distance (or read it from a variable for each feature), you need to know what the horizontal distance units are for your data. If GCS, these will be decimal degrees and 1 degree is a long way, about 111 km (or 69 miles), though that differs for longitude where 1 degree of longitude is 111 km * cos(latitude). If in UTM, the horizontal distance will be in metres, but in the US, state plane coordinates are typically in feet. So let's read the trails shapefile from the Marble Mountains and look for its units:

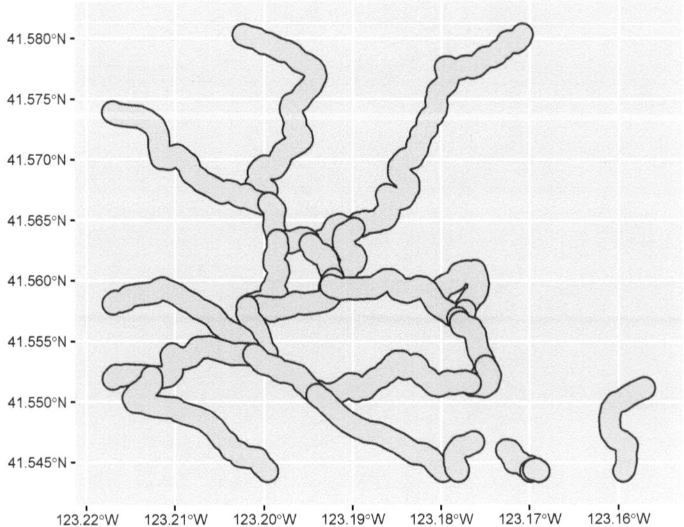

FIGURE 7.13 100 m trail buffer, Marble Mountains

```
library(igisci)
library(sf); library(tidyverse)
trails <- st_read(ex("marbles/trails.shp"))
```

Then we know we're in metres, so we'll create a 100 m buffer this way (Figure 7.13).

```
trail_buff0 <- st_buffer(trails,100)
ggplot(trail_buff0) + geom_sf()
```

7.2.4.1 The `units` package

Since the spatial data are in UTM, we can just specify the distance in metres. However if the data were in decimal degrees of latitude and longitude (GCS), we would need to let the function know that we're providing it metres so it can transform that to work with the GCS, using `units::set_units(100, "m")` instead of just `100`, for the above example where we are creating a 100 m buffer.

7.2.5 Spatial overlay: union and intersection

Overlay operations are described in the `sf` cheat sheet under "Geometry operations". These are useful to explore, but a couple we'll look at are union and intersection.

Normally these methods have multiple inputs, but we'll start with one that can also be used to dissolve boundaries, **`st_union`** – if only one input is provided it appears to do a dissolve (Figure 7.14):

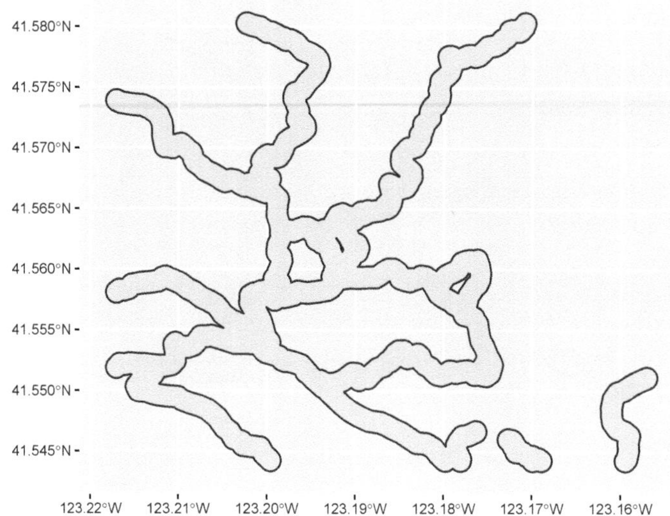

FIGURE 7.14 Unioned trail buffer, dissolving boundaries

```
trail_buff <- st_union(trail_buff0)
ggplot(trail_buff) + geom_sf()
```

For a clearer example with the normal multiple input, we'll **intersect** a 100 m buffer around streams and a 100 m buffer around trails ...

```
streams <- st_read(ex("marbles/streams.shp"))
trail_buff <- st_buffer(trails, 100)
str_buff <- st_buffer(streams,100)
strtrInt <- st_intersection(trail_buff,str_buff)
```

...to show areas that are close to streams and trails (Figure 7.15):

```
ggplot(strtrInt) + geom_sf(fill="green") +
  geom_sf(data=trails, col="red") +
  geom_sf(data=streams, col="blue")
```

Or how about a union of these two buffers? We'll also dissolve the boundaries using union with a single input (the first union) to dissolve those internal overlays (Figure 7.16):

```
strtrUnion <- st_union(st_union(trail_buff,str_buff))
ggplot(strtrUnion) + geom_sf(fill="green") +
  geom_sf(data=trails, col="red") +
  geom_sf(data=streams, col="blue")
```

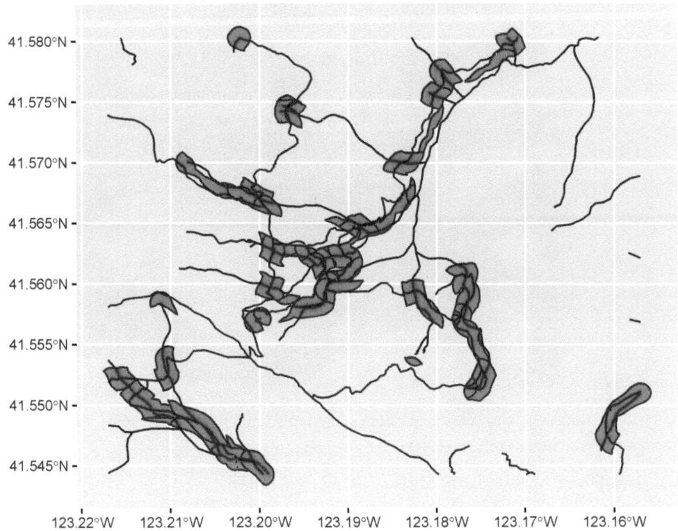

FIGURE 7.15 Intersection of trail and stream buffers

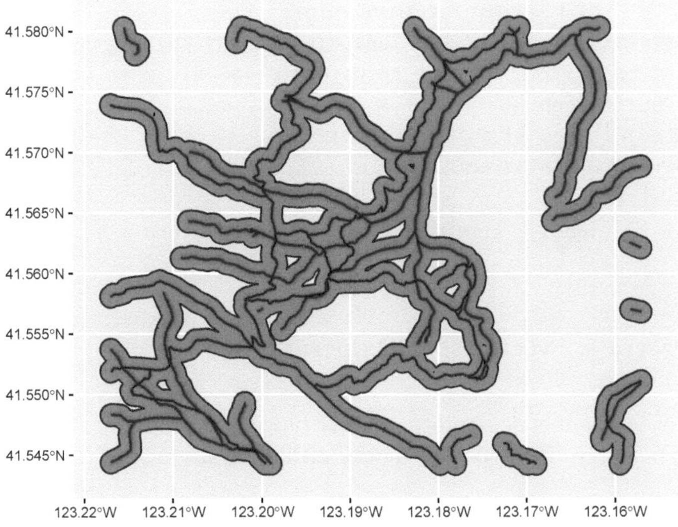

FIGURE 7.16 Union of two sets of buffer polygons

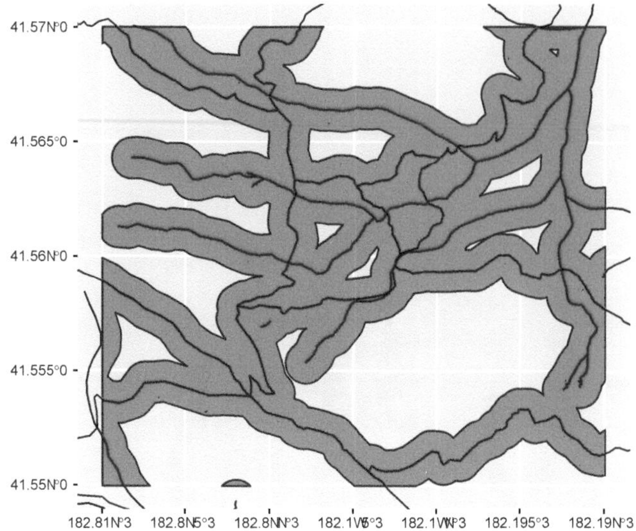

FIGURE 7.17 Cropping with specified x and y limits

7.2.6 Clip with st_crop

Clipping a GIS layer to a rectangle (or to a polygon) is often useful. We'll clip to a rectangle based on latitude and longitude limits we'll specify, however since our data is in UTM, we'll need to use **st_transform** to get it to the right coordinates (Figure 7.17).

```
xmin=-123.21; xmax=-123.18; ymin=41.55; ymax=41.57
clipMatrix <- rbind(c(xmin,ymin),c(xmin,ymax),c(xmax,ymax),
                    c(xmax,ymin),c(xmin,ymin))
clipGCS <- st_sfc(st_polygon(list(clipMatrix)),crs=4326)
bufferCrop <- st_crop(strtrUnion,st_transform(clipGCS,crs=st_crs(strtrUnion)))
bnd <- st_bbox(bufferCrop)
ggplot(bufferCrop) + geom_sf(fill="green") +
  geom_sf(data=trails, col="red") +
  geom_sf(data=streams, col="blue") +
  coord_sf(xlim = c(bnd[1], bnd[3]), ylim = c(bnd[2], bnd[4]))
```

7.2.6.1 sf or terra for vector spatial operations?

Note: there are other geometric operations in `sf` beyond what we've looked at. See the cheat sheet and other resources at https://r-spatial.github.io/sf .

The `terra` system also has many vector tools (such as `terra::union` and `terra::buffer`), that work with and create SpatVector spatial data objects. As noted earlier, these S4 spatial data can be created from (S3) sf data with `terra::vect` and vice versa with `sf::st_as_sf`. We'll mostly focus on sf for vector and terra for raster, but to learn more about SpatVector data and operations in terra, see https://rspatial.org.

For instance, there's a similar `terra` operation for one of the things we just did, but instead of using `union` to dissolve the internal boundaries like `st_union` did, uses `aggregate` to remove boundaries between polygons with the same codes:

```
trails <- vect(ex("marbles/trails.shp"))
trailbuff <- aggregate(buffer(trails,100))
plot(trailbuff)
```

7.2.7 Spatial join with `st_join`

A spatial join can do many things, but we'll just look at its simplest application – connecting a point with the polygon that it's in. In this case, we want to join the attribute data in `BayAreaTracts` with EPA Toxic Release Inventory (TRI) point data (at factories and other point sources) and then display a few ethnicity variables from the census, associated spatially with the TRI point locations. We'll be making better maps later; this is a quick `plot()` display to show that the spatial join worked (Figure 7.18).

```
library(igisci); library(tidyverse); library(sf)
TRI_sp <- st_as_sf(read_csv(ex("TRI/TRI_2017_CA.csv")),
                   coords = c("LONGITUDE", "LATITUDE"), crs=4326) %>%
  st_join(st_make_valid(BayAreaTracts)) %>%
  filter(!is.na(TRACT)) %>%  # removes points not in BayAreaTracts
  dplyr::select(POP2012:HISPANIC)
plot(TRI_sp)
```

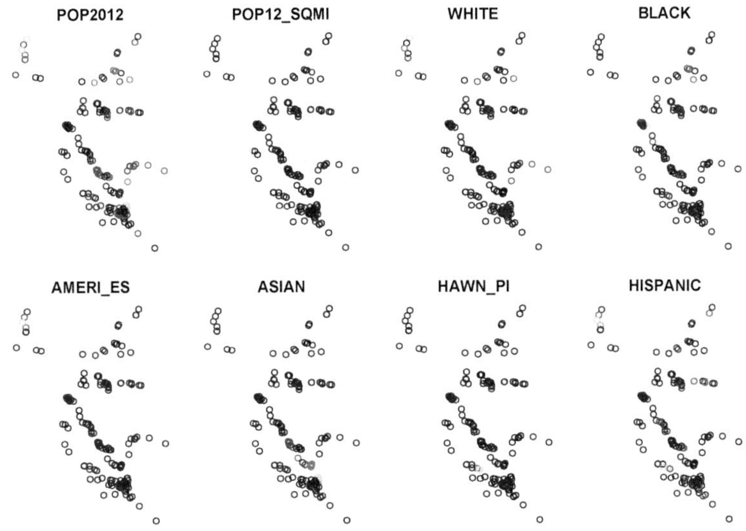

FIGURE 7.18 TRI points with census variables added via a spatial join

7.2.8 Further exploration of spatial analysis

We've looked at a small selection of spatial analysis functions; there's a lot more we can do, and the R spatial world is rapidly growing. There are many other methods discussed in the "Spatial data operations" and "Geometry operations" chapters of *Geocomputation* (https://geocompr.robinlovelace.net/, Lovelace, Nowosad, and Muenchow (2019)) that are worth exploring, and still others in the `terra` package described at https://rspatial.org, Hijmans (n.d.).

For example, getting a good handle on coordinate systems is certainly an important area to learn more about, and you will find coverage of transformation and reprojection methods in the "Reprojecting geographic data" and data access methods in the "Geographic data I/O" chapter of *Geocomputation*, and in the "Coordinate Reference Systems" section of https://rspatial.org. The Simple Features for R reference (*Simple Features for r* (n.d.)) site includes a complete reference for all of its functions, as does https://rspatial.org for `terra`.

And then there's analysis of raster data, but we'll leave that for the next chapter.

7.3 Exercises: Spatial Analysis

Exercise 7.1. Maximum Elevation Assuming you wrote them in the last chapter, read in your western states "data/W_States.shp" and peaks "data/peaks.shp" data, then use a spatial join to add the peak points to the states to provide a new attribute maximum elevation, and display that using geom_sf_text() with the state polygons. Note that joining from the states to the peaks works because there's a one-to-one correspondence between states and maximum peaks. (If you didn't write them, you'll need to repeat your code that built them from the previous chapter.)

Exercise 7.2. Using two shape files of your own choice, not from igisci, read them in as sf data, and perform a spatial join to join the polygons to the point data. Then display the point feature data frame, and create a ggplot scatter plot or boxplot graph where the two variables are related, preferably with a separate categorical variable to provide color.

Exercise 7.3. From the above spatially joined data create a map where the points are colored by data brought in from the spatial join; can be either a categorical or continuous variable.

Exercise 7.4. Transect Buffers: Using data from the `SFmarine` folder in igisci extdata, `transects.shp` (`ex("SFmarine/transects.shp")`) and `mainland.shp`(to make a nicer map showing land for reference), use `st_buffer` to create 1000 m buffers around the transect points, merged to remove boundaries (Figure 7.19).

Exercise 7.5. Create an sf that represents all areas within 50 km of a TRI facility in California that has >10,000 pounds of total stack air release for all gases, clipped (intersected) with the state of California as provided in the CA counties data set. Then use this to create a map of California showing the clipped buffered areas in red, with California counties and those selected large-release TRI facilities in black dots, as shown here. To create a variable

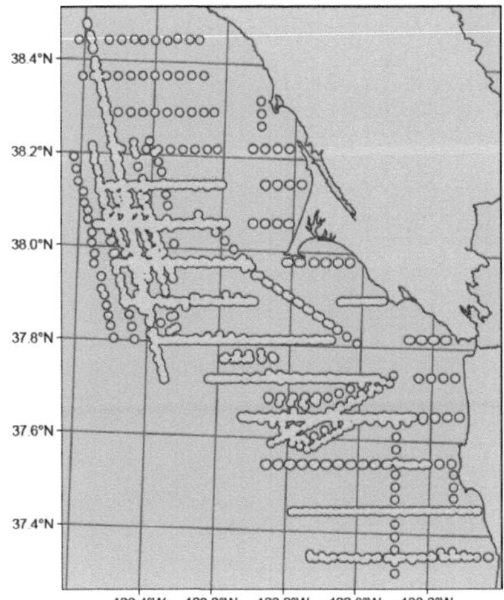

FIGURE 7.19 Transect Buffers (goal)

representing 50 km to provide to the dist parameter in st_buffer [this is needed since the spatial data are in GCS], use the units::set_units function, something like this: `D50km <- units::set_units(50, "km")`. See the first figure in this chapter for your goal.

8

Raster Spatial Analysis

Raster spatial analysis is particularly important in environmental analysis, since much environmental data are continuous in nature, based on continuous measurements from instruments (like temperature, pH, air pressure, water depth, elevation), and raster models work well with continuous data. In the Spatial Data and Maps chapter, we looked at creating rasters from scratch, or converted from features, and visualizing them. Here we'll explore raster analytical methods, commonly working from existing information-rich rasters like elevation data, where we'll start by looking at terrain functions.

We'll make a lot of use of `terra` functions in this chapter, as this package is replacing the `raster` package which has been widely used. One raster package that's probably also worth considering is the `stars` package.

8.1 Terrain functions

Elevation data are particularly information-rich, and a lot can be derived from them that informs us about the nature of landscapes and what drives surface hydrologic and geomorphic processes as well as biotic habitat (some slopes are drier than others, for instance). We'll start by reading in some elevation data from the Marble Mountains of California (Figure 8.1) and use terra's terrain function to derive slope (Figure 8.2), aspect (Figure 8.3), and hillshade rasters.

```
library(terra); library(igisci)
elev <- rast(ex("marbles/elev.tif"))
plot(elev)
```

```
slope <- terrain(elev, v="slope")
plot(slope)
```

```
aspect <- terrain(elev, v="aspect")
plot(aspect)
```

... then we'll use terra::classify to make six discrete categorical slope classes (though the legend suggests it's continuous) (Figure 8.4)...

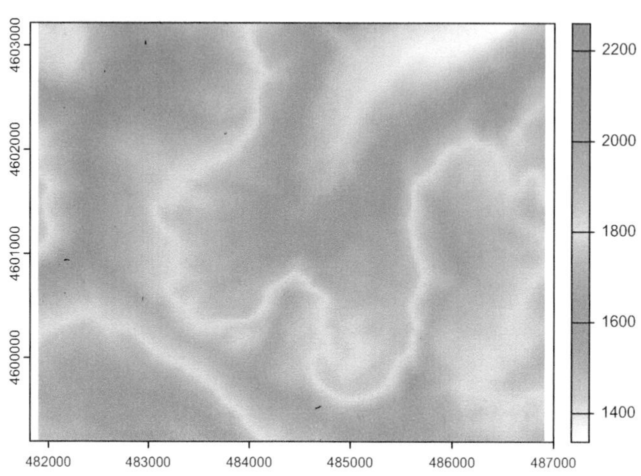

FIGURE 8.1 Marble Mountains (California) elevation

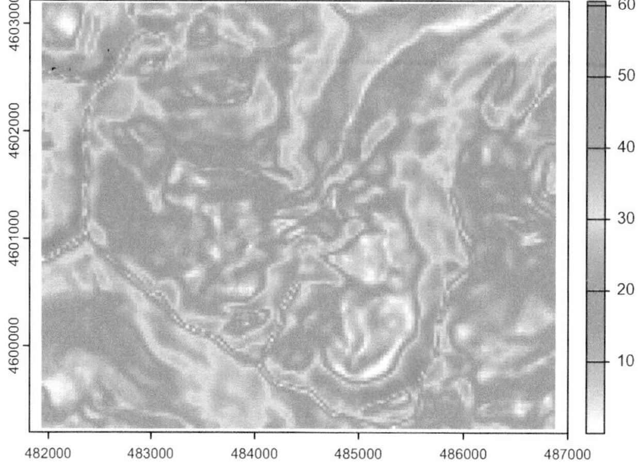

FIGURE 8.2 Slope

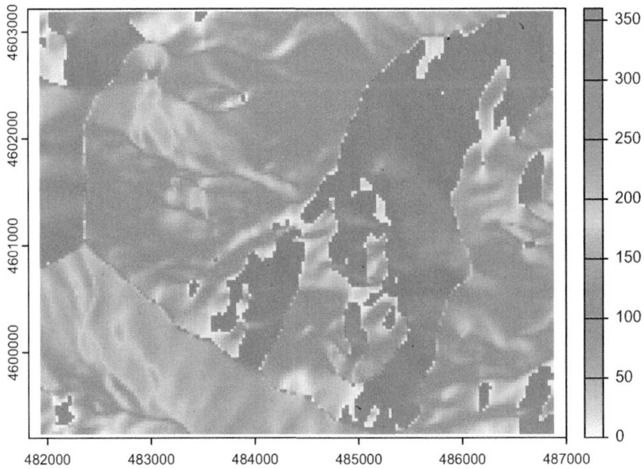

FIGURE 8.3 Aspect

```
slopeclasses <-matrix(c(00,10,1, 10,20,2, 20,30,3,
                        30,40,4, 40,50,5, 50,90,6), ncol=3, byrow=TRUE)
slopeclass <- classify(slope, rcl = slopeclasses)
plot(slopeclass)
```

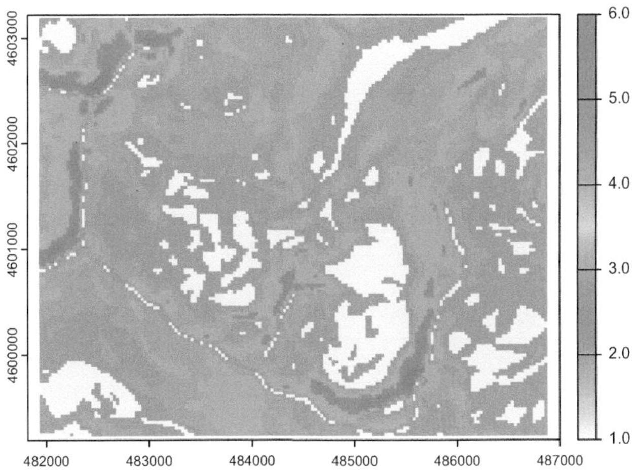

FIGURE 8.4 Classified slopes

Then a hillshade effect raster with slope and aspect as inputs after converting to radians (Figure 8.5):

```
hillsh <- shade(slope/180*pi, aspect/180*pi, angle=40, direction=330)
plot(hillsh, col=gray(0:100/100))
```

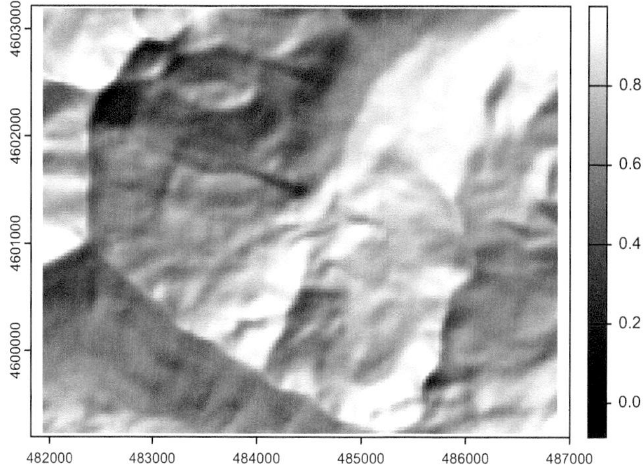

FIGURE 8.5 Hillshade

8.2 Map Algebra in terra

Map algebra was originally developed by Dana Tomlin in the 1970s and 1980s (Tomlin (1990)), and was the basis for his Map Analysis Package. It works by assigning raster outputs from an algebraic expression of raster inputs. Map algebra was later incorporated in Esri's Grid and Spatial Analyst subsystems of ArcInfo and ArcGIS. Its simple and elegant syntax makes it still one of the best ways to manipulate raster data.

Let's look at a couple of simple map algebra statements to derive some new rasters, such as converting elevation in metres to feet (Figure 8.6).

```
elevft <- elev / 0.3048
plot(elevft)
```

... including some that create and use Boolean (true-false) values, where 1 is true and 0 is false, so might answer the question "Is it steep?" (as long as we understand 1 means Yes or true) (Figure 8.7)...

```
steep <- slope > 20
plot(steep)
```

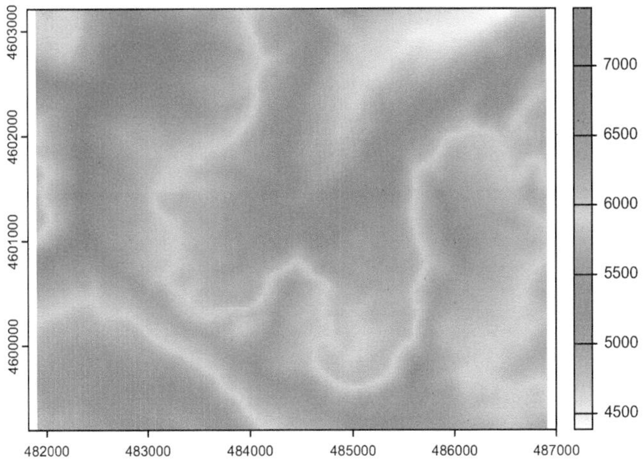

FIGURE 8.6 Map algebra conversion of elevations from metres to feet

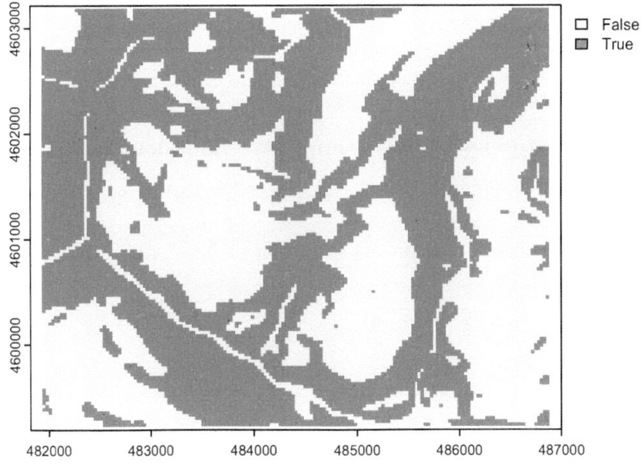

FIGURE 8.7 Boolean: slope > 20

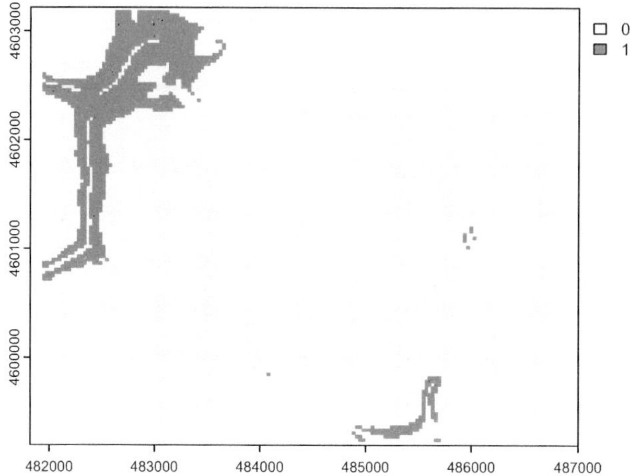

FIGURE 8.8 Boolean intersection: (slope > 20) * (elev > 2000)

... or Figure 8.8, which shows all areas that are steeper than 20 degrees *and* above 2,000 m elevation.

```
plot(steep * (elev > 2000))
```

You should be able to imagine that map algebra is particularly useful when applying a model equation to data to create a prediction map. For instance, later we'll use `lm()` to derive linear regression parameters for predicting February temperature from elevation in the Sierra ...

$$Temperature_{prediction} = 11.88 - 0.006elevation$$

... which can be coded in map algebra something like the following if we have an elevation raster, to create a tempPred raster:

```
tempPred <- 11.88 - 0.006 * elevation
```

8.3 Distance

A continuous raster of distances from significant features can be very informative in environmental analysis. For instance, distance from the coast or a body of water may be an important variable for habitat analysis and ecological niche modeling. The goal of the following is to derive distances from streams as a continuous raster.

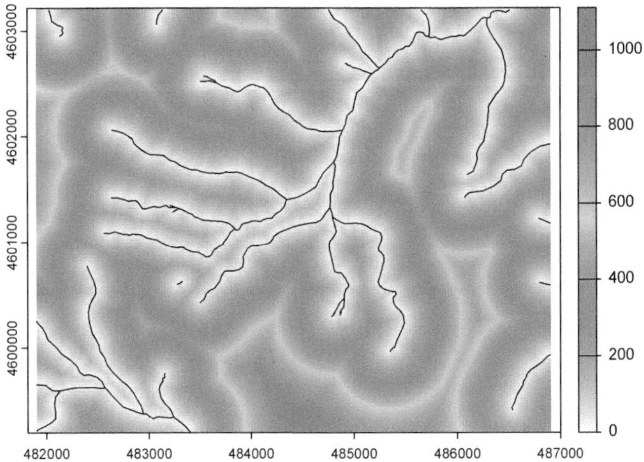

FIGURE 8.9 Stream distance raster

We'll need to know what cell size to use and how far to extend our raster. If we have an existing study area raster, this process is simple. We start by converting streams from a SpatVector to a SpatRaster, as we did a couple of chapters ago. The `terra::distance()` function then uses this structure to provide the cells that we're deriving distance from and then uses that same cell size and extent for the output raster. If we instead used vector features, the `distance` function would return point-to-point distances, very different from deriving continuous rasters of Euclidean distance (Figure 8.9).

```
streams <- vect(ex("marbles/streams.shp"))
elev <- rast(ex("marbles/elev.tif"))
stdist <- terra::distance(rasterize(streams,elev))
plot(stdist)
lines(streams)
```

If we didn't have an elevation raster, we could use the process we employed while converting features to rasters in the Spatial Data and Maps chapter, where we derived a raster template from the extent of streams, as shown here:

```
streams <- vect(ex("marbles/streams.shp"))
XMIN <- ext(streams)$xmin
XMAX <- ext(streams)$xmax
YMIN <- ext(streams)$ymin
YMAX <- ext(streams)$ymax
aspectRatio <- (YMAX-YMIN)/(XMAX-XMIN)
cellSize <- 30
NCOLS <- as.integer((XMAX-XMIN)/cellSize)
NROWS <- as.integer(NCOLS * aspectRatio)
templateRas <- rast(ncol=NCOLS, nrow=NROWS,
                    xmin=XMIN, xmax=XMAX, ymin=YMIN, ymax=YMAX,
```

```
                    vals=1, crs=crs(streams))
strms <- rasterize(streams,templateRas)
stdist <- terra::distance(strms)
plot(stdist)
lines(streams)
```

In deriving distances, it's useful to remember that distances can go on forever (well, on the planet they may go around and around, if we were using spherical coordinates) so that's another reason we have to specify the raster structure we want to populate.

8.4 Extracting Values

A very useful method or environmental analysis and modeling is to extract values from rasters at specific point locations. The point locations might be observations of species, soil samples, or even random points, and getting continuous (or discrete) raster observations can be very useful in a statistical analysis associated with those points. The distance from streams raster we derived earlier, or elevation, or terrain derivatives like slope and aspect might be very useful in a ecological niche model, for instance. We'll start by using random points and use these to extract values from four rasters:

- **elev**: read in from elev.tif
- **slope**: created from elev with **terrain**
- **str_dist**: euclidean distance to streams
- **geol**: rasterized from geology polygon features

```
library(igisci); library(terra)
geolshp <- vect(ex("marbles/geology.shp"))
streams <- vect(ex("marbles/streams.shp"))
elev <- rast(ex("marbles/elev.tif"))
slope <- terrain(elev, v="slope")
str_dist <- terra::distance(rasterize(streams,elev))
geol <- rasterize(geolshp,elev,field="CLASS")
```

Note that in contrast to the other rasters, the stream distance raster ends up with no name, so we should give it a name:

```
names(slope)
```

```
## [1] "slope"
```

```
names(str_dist)
```

```
## [1] "layer"
```

```
names(str_dist) <- "str_dist"
```

On the raster `names` property: You'll find that many terra functions may not assign the `names` property you'd expect, so it's a good idea to check with `names()` and maybe set it to what we want, as we've just done for `str_dist`. As we'll see later with the focal statistics function, the name of the input is used even though we've modified it in the function result, and that may create confusion when we use it. We just saw that the `distance()` function produced an empty name, and there may be others you'll run into. For many downstream uses, the names property may not matter, but it will be important when we extract values from rasters into points where the `names` property is assigned to the variable created for the points.

Then we'll create 200 **random xy points** 10.2.3 within the extent of `elev`, and assign it the same `crs`.

```
library(sf)
x <- runif(200, min=xmin(elev), max=xmax(elev))
y <- runif(200, min=ymin(elev), max=ymax(elev))
rsamps <- st_as_sf(data.frame(x,y), coords = c("x","y"), crs=crs(elev))
```

To visualize where the random points land, we'll map them on the geology sf, streams, and contours created from elev using default settings. The `terra::as.contour` function will create these as SpatVector data, which along with `streams` we'll convert with `sf::st_as_sf` to display in ggplot (Figure 8.10).

```
library(tidyverse)
cont <- st_as_sf(as.contour(elev, nlevels=30))
geology <- st_read(ex("marbles/geology.shp"))
```

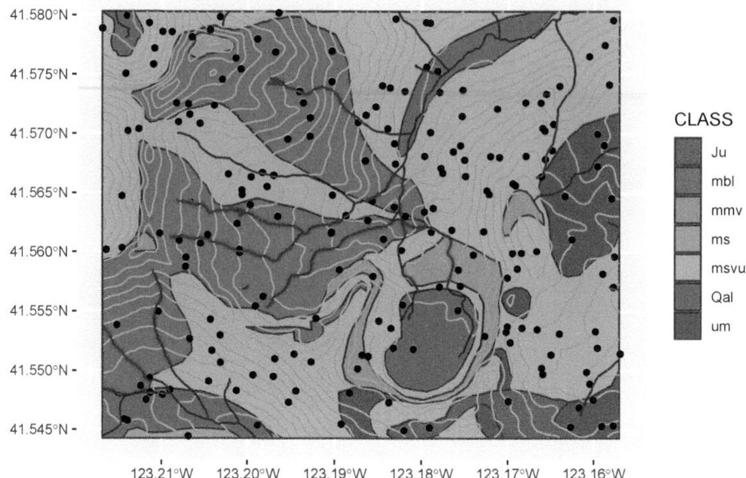

FIGURE 8.10 Random points in the Marble Valley area, Marble Mountains, California

```
ggplot() +
  geom_sf(data=geology, aes(fill=CLASS)) +
  geom_sf(data=cont, col="gray") +
  geom_sf(data=rsamps) +
  geom_sf(data=st_as_sf(streams), col="blue")
```

Now we'll extract data from each of the rasters, using an S4 version of `rsamps`, and then bind them together with the `rsamps` simple features. We'll have to use `terra::vect` and `sf::st_as_sf` to convert feature data to the type required by specific tools, and due to a function naming issue, we'll need to use the package prefix with `terra::extract`, but otherwise the code is pretty straightforward.

```
rsampS4 <- vect(rsamps)
elev_ex <- terra::extract(elev, rsampS4) %>% dplyr::select(-ID)
slope_ex <- terra::extract(slope, rsampS4) %>% dplyr::select(-ID)
geol_ex <- terra::extract(geol, rsampS4) %>%
  dplyr::rename(geology = CLASS) %>% dplyr::select(-ID)
strD_ex <- terra::extract(str_dist, rsampS4)  %>% dplyr::select(-ID)
rsampsData <- bind_cols(rsamps, elev_ex, slope_ex, geol_ex, strD_ex)
```

Then plot the map with the points colored by geology (Figure 8.11)...

```
ggplot() +
  geom_sf(data=cont, col="gray") +
  geom_sf(data=rsampsData, aes(col=geology)) +
  geom_sf(data=st_as_sf(streams), col="blue")
```

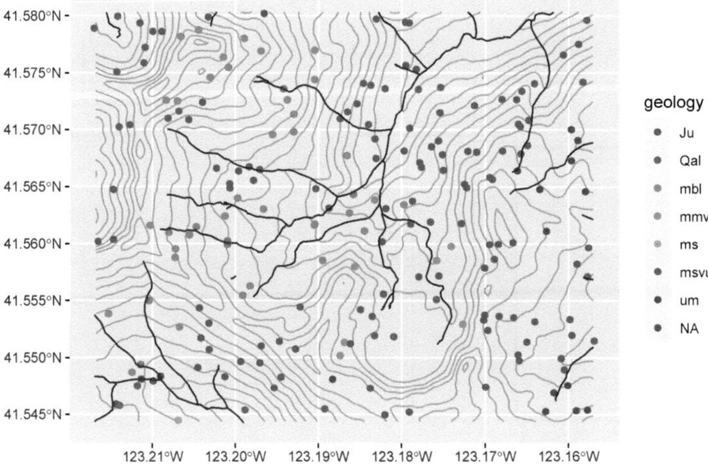

FIGURE 8.11 Points colored by geology extracted from raster

... and finally `str_dist` by `elev`, colored by `geology`, derived by extracting. We'll filter out the NAs along the edge (Figure 8.12). Of course other analyses and visualizations are possible.

```
rsampsData %>%
  filter(!is.na(geology)) %>%
  ggplot(aes(x=str_dist,y=elev,col=geology)) +
  geom_point() + geom_smooth(method = "lm", se=F)
```

Here's a similar example, but using water sample data, which we can then use to relate to extracted raster values to look at relationships such as in Figures 8.13, 8.14, and 8.15. It's

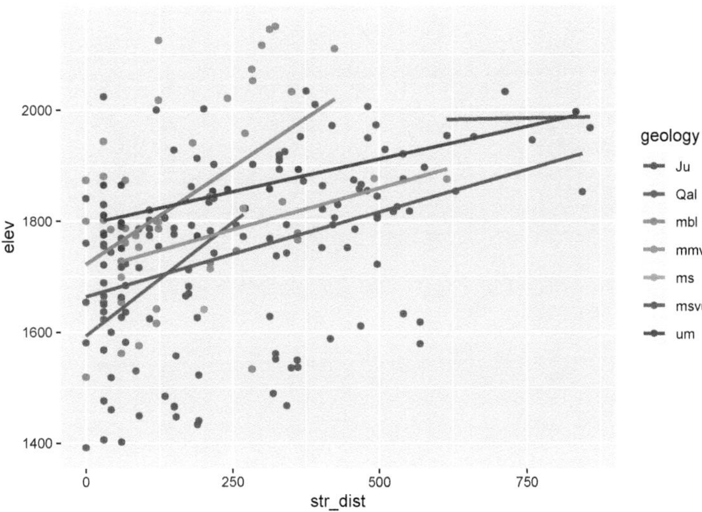

FIGURE 8.12 Elevation by stream distance, colored by geology, random point extraction

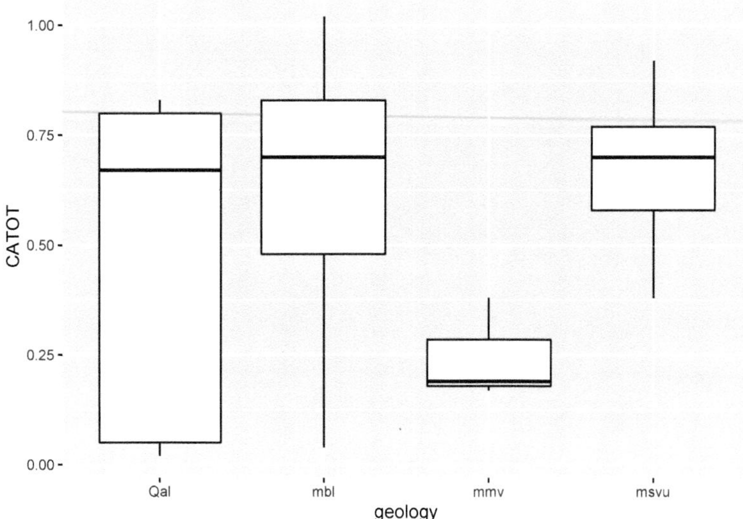

FIGURE 8.13 Dissolved calcium carbonate grouped by geology extracted at water sample points

worthwhile to check various results along the way, as we did above. Most of the code is very similar to what we used above, including dealing with naming the distance rasters.

```
streams <- vect(ex("marbles/streams.shp"))
trails <- vect(ex("marbles/trails.shp"))
elev <- rast(ex("marbles/elev.tif"))
geolshp <- vect(ex("marbles/geology.shp"))
sampsf <- st_read(ex("marbles/samples.shp")) %>%
```

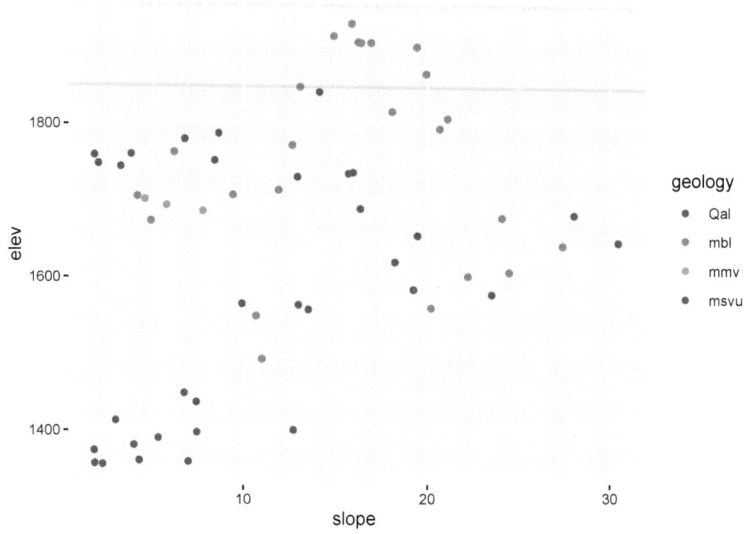

FIGURE 8.14 Slope by elevation colored by extracted geology

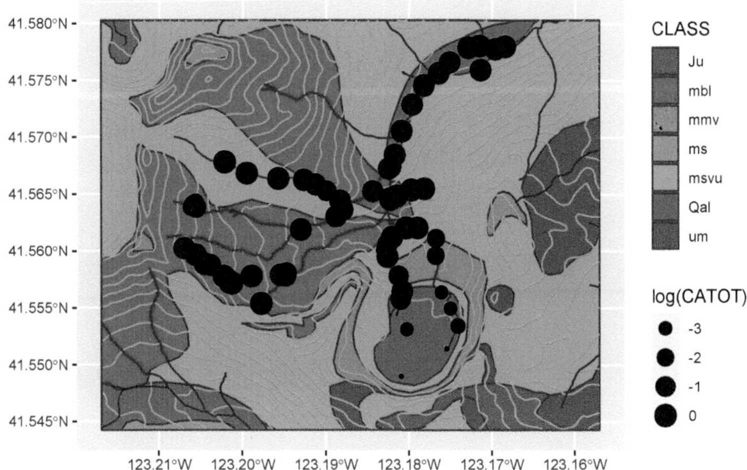

FIGURE 8.15 Logarithm of calcium carbonate total hardness at sample points, showing geologic units

```
  dplyr::select(CATOT, MGTOT, PH, TEMP, TDS)
samples <- vect(sampsf)
strms <- rasterize(streams,elev)
tr <- rasterize(trails,elev)
geol <- rasterize(geolshp,elev,field="CLASS")
stdist <- terra::distance(strms); names(stdist) <- "stDist"
trdist <- terra::distance(tr); names(trdist) = "trDist"
slope <- terrain(elev, v="slope")
aspect <- terrain(elev, v="aspect")
elev_ex <- terra::extract(elev, samples) %>% dplyr::select(-ID)
slope_ex <- terra::extract(slope, samples) %>% dplyr::select(-ID)
aspect_ex <- terra::extract(aspect, samples) %>% dplyr::select(-ID)
geol_ex <- terra::extract(geol, samples) %>%
  dplyr::rename(geology = CLASS) %>% dplyr::select(-ID)
strD_ex <- terra::extract(stdist, samples) %>% dplyr::select(-ID)
trailD_ex <- terra::extract(trdist, samples) %>% dplyr::select(-ID)
samplePts <- cbind(samples,elev_ex,slope_ex,aspect_ex,geol_ex,strD_ex,trailD_ex)
samplePtsDF <- as.data.frame(samplePts)
```

```
head(samplePtsDF)
```

```
##   CATOT MGTOT   PH TEMP  TDS elev    slope     aspect geology  stDist
## 1  0.80  0.02 7.74 15.3 0.16 1357 1.721006 123.690068     Qal  0.0000
## 2  0.83  0.04 7.88 14.8 0.16 1359 6.926249 337.833654     Qal 30.0000
## 3  0.83  0.04 7.47 15.1 0.12 1361 4.222687 106.389540     Qal  0.0000
## 4  0.63  0.03 7.54 15.8 0.17 1356 2.160789 353.659808     Qal  0.0000
## 5  0.67  0.06 7.67 14.0 0.14 1374 1.687605  81.869898     Qal  0.0000
```

```
## 6   0.70   0.04 7.35   7.0 0.16 1399 12.713997    4.236395     msvu 192.0937
##      trDist
## 1 169.7056
## 2 318.9044
## 3 108.1665
## 4 241.8677
## 5 150.0000
## 6 342.0526
```

```
ggplot(data=samplePtsDF, aes(x=geology,y=CATOT)) + geom_boxplot()
```

```
ggplot(data=samplePtsDF, aes(x=slope,y=elev,col=geology)) + geom_point()
```

```
cont <- st_as_sf(as.contour(elev, nlevels=30))
```

```
ggplot() +
  geom_sf(data=st_as_sf(geolshp), aes(fill=CLASS)) +
  geom_sf(data=cont, col="gray") +
  geom_sf(data=st_as_sf(streams), col="blue") +
  geom_sf(data=sampsf, aes(size=log(CATOT)))
```

8.5 Focal Statistics

Focal (or neighborhood) statistics work with a continuous or categorical raster to pass a moving window through it, assigning the central cell with summary statistic applied to the neighborhood, which by default is a 3×3 neighborhood (`w=3`) centered on the cell. One of the simplest is a low-pass filter where `fun="mean"`. This applied to a continuous raster like elevation will look very similar to the original, so we'll apply a larger 9×9 (`w=9`) window so we can see the effect (Figure 8.16), which you can compare with the earlier plots of raw elevation.

```
elevLowPass9 <- terra::focal(elev,w=9,fun="mean")
names(elevLowPass9) <- "elevLowPass9" # otherwise gets "elev"
plot(elevLowPass9)
```

The effect is probably much more apparent in a hillshade, where the very smooth 9x9 low-pass filtered elevation will seem to create an out-of-focus hillshade (Figure 8.17).

```
slope9 <- terrain(elevLowPass9, v="slope")
aspect9 <- terrain(elevLowPass9, v="aspect")
```

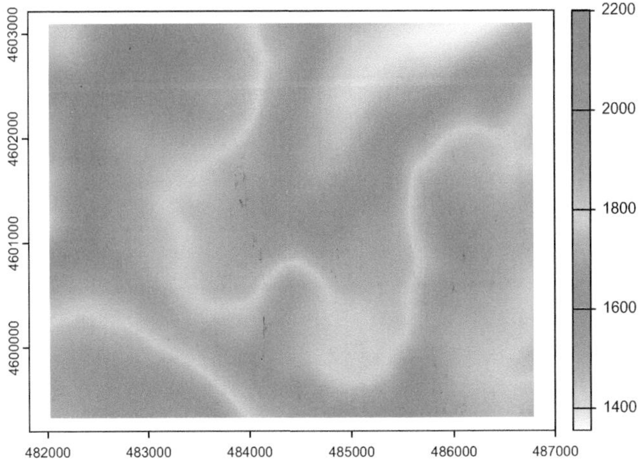

FIGURE 8.16 9x9 focal mean of elevation

```
plot(shade(slope9/180*pi, aspect9/180*pi, angle=40, direction=330),
     col=gray(0:100/100))
```

For categorical/factor data such as geology (Figure 8.18), the modal class in the neighborhood can be defined (Figure 8.19).

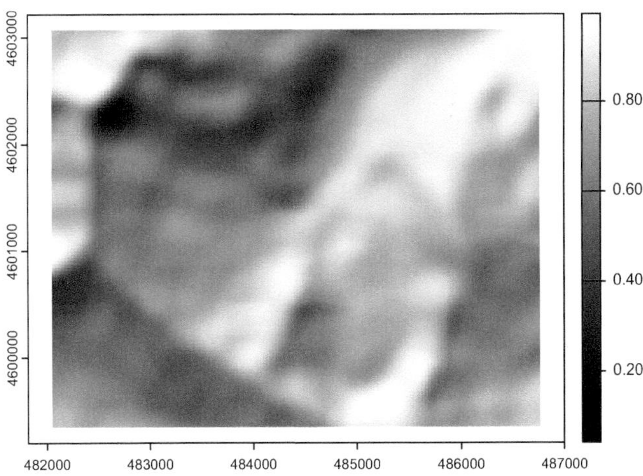

FIGURE 8.17 Hillshade of 9x9 focal mean of elevation

```
plot(geol)
```

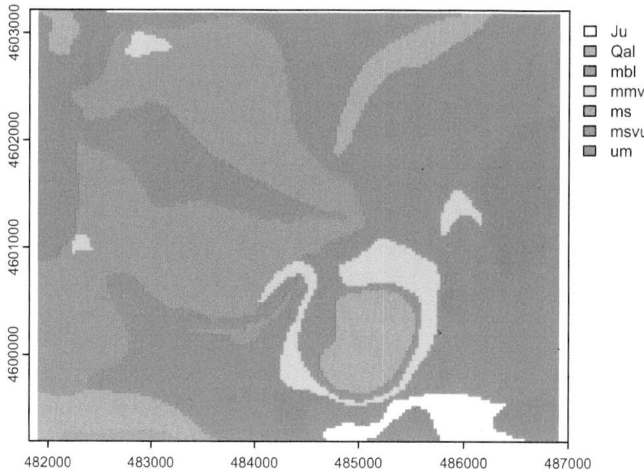

FIGURE 8.18 Marble Mountains geology raster

```
plot(terra::focal(geol,w=9,fun="modal"))
```

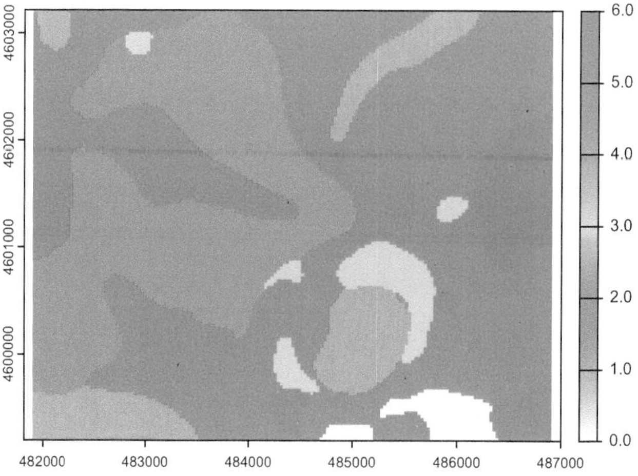

FIGURE 8.19 Modal geology in 9 by 9 neighborhoods

Note that while plot displayed these with a continuous legend, the modal result is going to be an integer value representing the modal class, the most common rock type in the neighborhood. This is sometimes called a *majority filter*. *Challenge*: how could we link the modes to the original character CLASS value, and produce a more useful map?

8.6 Zonal Statistics

Zonal statistics let you stratify by zone, and is a lot like the grouped summary (3.4.4) we've done before, but in this case the groups are connected to the input raster values by location. There's probably a more elegant way of doing this, but here are a few that are then joined together.

```
meanElev <- zonal(elev,geol,"mean") %>% rename(mean=elev)
maxElev <- zonal(elev,geol,"max") %>% rename(max=elev)
minElev <- zonal(elev,geol,"min") %>% rename(min=elev)
```

```
left_join(left_join(meanElev,maxElev,by="CLASS"),minElev,by="CLASS")
```

```
##   CLASS     mean  max  min
## 1    Ju 1940.050 2049 1815
## 2   Qal 1622.944 1825 1350
## 3   mbl 1837.615 2223 1463
## 4   mmv 1846.186 2260 1683
## 5    ms 1636.802 1724 1535
## 6  msvu 1750.205 2136 1336
## 7    um 1860.758 2049 1719
```

8.7 Exercises: Raster Spatial Analysis

Exercise 8.1. You can get the four values that define the extent of a raster with terra functions `xmin`, `xmax`, `ymin`, and `ymax`. Use these with the raster `elev` created from `"marbles/elev.tif"`, then derive 100 uniform random x and y values with those min and max values. Use cbind to display a matrix of 100 coordinate pairs.

Exercise 8.2. Create sf points from these 100 uniform random coordinate pairs. Use tmap to display them on a base of the elevation raster.

Exercise 8.3. Geology and elevation by stream and trail distance. Now use those points to extract values from stream distance, trail distance, geology, slope, aspect, and elevation, and display that sf data frame as a table, then plot trail distance (x) vs stream distance (y) colored by geology and sized by elevation (Figure 8.20).

Exercise 8.4. Create a slope raster from "SanPedro/dem.tif" then a "steep" raster of all slopes > 26 degrees, determined by a study of landslides to be a common landslide threshold, then display them using (palette="Reds", alpha=0.5, legend.show=F) along with roads "SanPedro/roads.shp" in "black", streams "SanPedro/streams.shp" in "blue", and watershed borders "SanPedro/SPCWatershed.shp" in "darkgreen" with lwd=2.

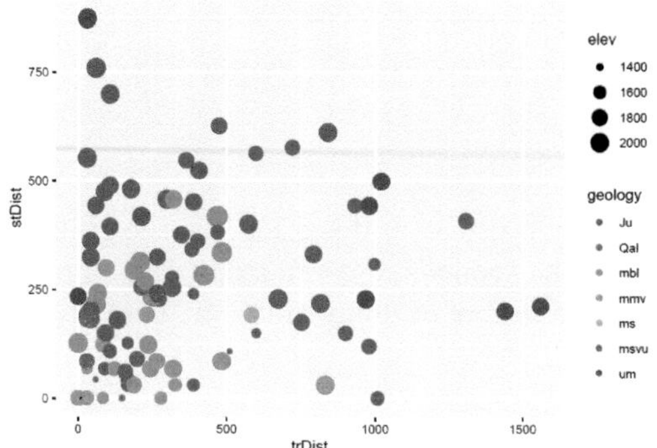

FIGURE 8.20 Geology and elevation by stream and trail distance (goal)

Exercise 8.5. Add a hillshade to that map.

9

Spatial Interpolation

A lot of environmental data is captured in point samples, which might be from soil properties, the atmosphere (like weather stations that capture temperature, precipitation, pressure, and winds), properties in surface water bodies or aquifers, etc. While it's often useful to just use those point observations and evaluate spatial patterns of those measurements, we also may want to predict a "surface" of those variables over the map, and that's where interpolation comes in.

In this section, we'll look at a few interpolators – nearest neighbor, inverse distance weighting (IDW) and kriging – but the reader should refer to more thorough coverage in Jakub Nowosad's *Geostatistics in R* (Nowosad (n.d.)) or at https://rspatial.org. For data, we'll use the Sierra climate data and derive the annual summaries that we looked at in the last chapter.

We'll be working in the Teale Albers projection, which is good for California, in metres (Figure 9.1), so we'll start by reprojecting Sierra data we created earlier (7.1.1) using `st_transform`.

```
Teale<-paste0("+proj=aea +lat_1=34 +lat_2=40.5 +lat_0=0 +lon_0=-120 ",
              "+x_0=0 +y_0=-4000000 +datum=WGS84 +units=m")
sierra <- st_transform(sierraAnnual, crs=Teale)
counties <- st_transform(CA_counties, crs=Teale)
sierraCounties <- st_transform(sierraCounties, crs=Teale)
ggplot() +
  geom_sf(data=counties) +
  geom_sf(data=sierraCounties, fill="grey", col="black") +
  geom_sf(data=sierra, aes(col=PRECIPITATION))
```

9.1 Null Model of the Original Data

We'll start by creating a "null model" of the z intercept, z being precipitation, as the root mean squared error (RMSE) of individual precipitation values as compared to the mean. (This is the same as the standard deviation if we could assume that were dealing with a population instead of a sample, so we'll borrow that symbol σ.) We'll then compare this with the RMSE of our model predictions to derive *relative performance*, since the models should have lower RMSE than the overall standard deviation.

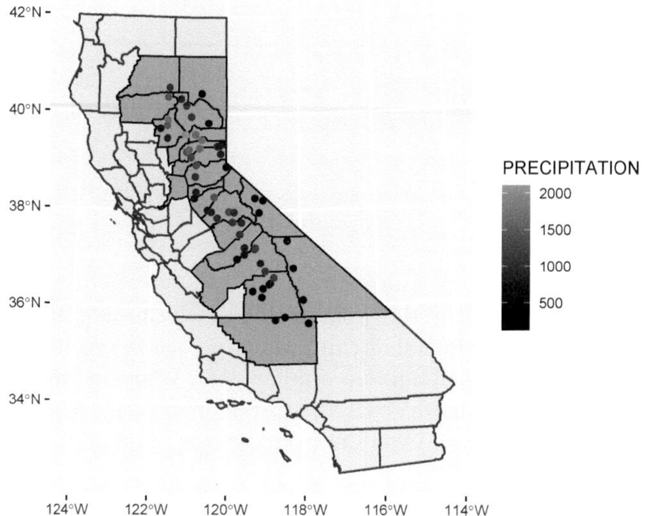

FIGURE 9.1 Precipitation map in Teale Albers in Sierra counties

$$\sigma = \sqrt{\frac{\sum_{i=1}^{n} (z_i - \bar{z})^2}{n}}$$

```
RMSE <- function(obs, pred) {
  sqrt(mean((obs-pred)**2, na.rm=TRUE))
}
null <- RMSE(mean(sierra$PRECIPITATION), sierra$PRECIPITATION)
null
```

```
## [1] 460.6205
```

The term "null model" can be confusing unless you consider that it's related to a null hypothesis of random variation about the mean. Note that it's not exactly the same as R's `sd()` function, which assumes you have a sample, not the population, and uses $n-1$ instead of n in the equation, as we can see here ...

```
sd(sierra$PRECIPITATION)
```

```
## [1] 464.6435
```

... and resolve `null` back to the sample standard deviation by multiplying by $\sqrt{\frac{n}{n-1}}$:

```
n <- length(sierra$PRECIPITATION)
null*sqrt(n/(n-1))
```

```
## [1] 464.6435
```

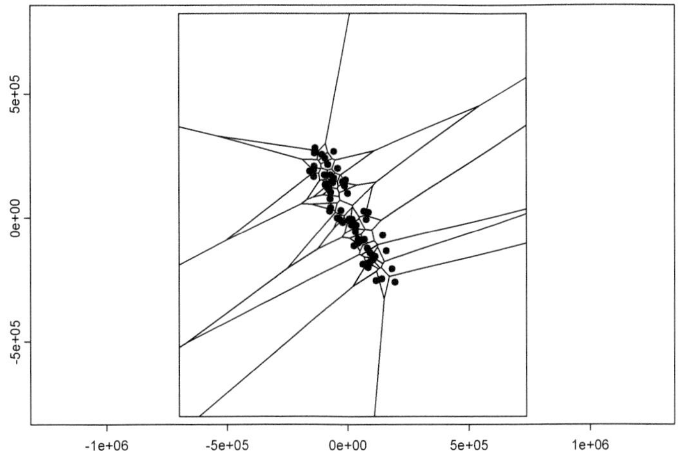

FIGURE 9.2 Voronoi polygons around Sierra stations

9.2 Voronoi Polygon

We're going to develop a simple prediction using Voronoi polygons (Voronoi (1908)), so we'll create these and see what they look like with the original data points also displayed (Figure 9.2). Note that we'll need to convert to S4 objects using `vect()`:

```
library(terra)
# proximity (Voronoi/Thiessen) polygons
sierraV <- vect(sierra)
v <- voronoi(sierraV)
plot(v)
points(sierraV)
```

Voronoi polygons keep going outward to some arbitrary limit, so for a more realistic tesselation, we'll crop out the Sierra counties we selected earlier that contain weather stations. Note that the Voronoi polygons contain all of the attribute data from the original points, just associated with polygon geometry now.

Now we can see a map that displays the original intent envisioned by the meteorologist Thiessen (Thiessen (1911)) to apply observations to a polygon area surrounding the station (Figure 9.3). Voronoi polygons or the raster determined from them (next) will also work with categorical data.

```
vSierra <- crop(v, vect(st_union(sierraCounties)))
plot(vSierra, "PRECIPITATION")
```

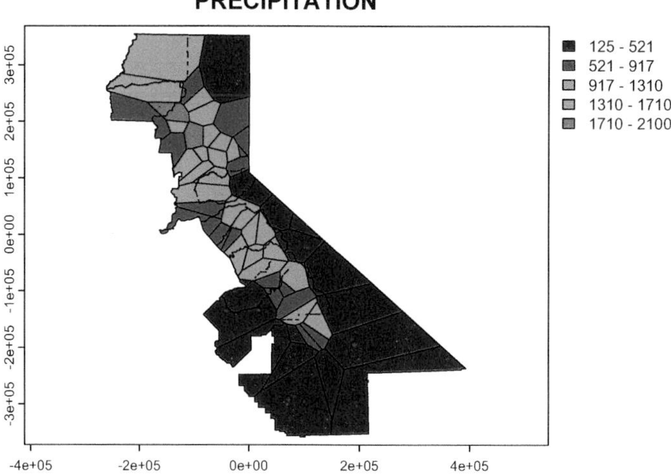

FIGURE 9.3 Precipitation mapped by Voronoi polygon

Now we'll rasterize those polygons (Figure 9.4). We'll start by building a blank raster of given dimensions of our selected counties, with a resolution of 1000 m.

```
# rasterize
r <- rast(vSierra, res=1000)  # Builds a blank raster
vr <- rasterize(vSierra, r, "PRECIPITATION")
plot(vr)
```

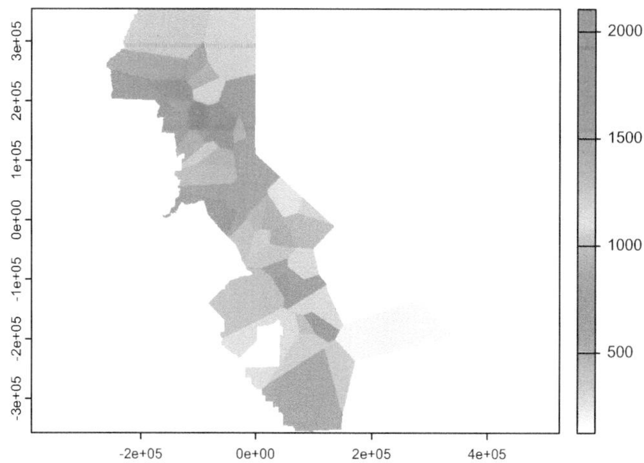

FIGURE 9.4 Rasterized Voronoi polygons

9.2.1 Cross-validation and relative performance

Cross-validation is used to see how well the model works by sampling it a defined number of times (folds) to pull out sets of training and testing samples, and then comparing the model's predictions with the (unused) testing data. Each time through, the model is built out of that fold's training data. For each fold, a random selection 10.2.3 is created by the sample function. The mean RMSE of all of the folds is the overall result, and provides an idea on how well the interpolation worked. **Relative performance** is a normalized measure of how well the model compares with the null model, and ranges from 0 (same RMSE as the null model) to 1 (zero error in our model). We'll use σ to represent the null model, since it's similar to the standard deviation:

$$perf = 1 - \frac{\overline{RMSE}}{\sigma}$$

The coding approach for the cross-validation process we'll use generally follows that documented at https://rspatial.org for use with `terra` data (Hijmans (n.d.)) and `gstat` (Pebesma (n.d.)) geostatistical methods.

```
# k-fold cross-validation
k <- 5
set.seed(42)
kf <- sample(1:k, size = nrow(sierraV), replace=TRUE) # get random numbers
   # between 1 and 5 to same size as input data points SpatVector sierraV
rmse <- rep(NA, k) # blank set of 5 NAs
for (i in 1:k) {
  test <- sierraV[kf == i, ]  # split into k sets of points,
                              # (k-1)/k going to training set,
  train <- sierraV[kf != i, ] # and 1/k to test, including all of the variables
  v <- voronoi(train)
  v
  test
  p <- terra::extract(v, test) # extract values from training data at test locations
  rmse[i] <- RMSE(test$PRECIPITATION, p$PRECIPITATION)
}
rmse
```

```
## [1] 470.1564 468.7459 468.6382 203.1462 301.4839
```

```
mean(rmse)
```

```
## [1] 382.4341
```

```
# relative model performance
perf <- 1 - (mean(rmse) / null)
round(perf, 3)
```

```
## [1] 0.17
```

9.3 Nearest Neighbor Interpolation

The previous assignment of data to Voronoi polygons can be considered to be a nearest neighbor interpolation where only one neighbor is used, but we can instead use multiple neighbors (Figure 9.5). In this case we'll use up to 5 (`nmax=5`). Presumably setting `idp` to zero makes it a nearest neighbor, along with the `nmax` setting .

```
library(gstat)
d <- data.frame(geom(sierraV)[,c("x", "y")], as.data.frame(sierraV))
head(d)
```

```
##           x          y        STATION_NAME  LONGITUDE LATITUDE ELEVATION
## 1 105093.88 -168859.68        ASH MOUNTAIN  -118.8253  36.4914     520.6
## 2  43245.02 -102656.13        AUBERRY 2 NW  -119.5128  37.0919       637
## 3  81116.40 -122688.39   BALCH POWER HOUSE -119.08833 36.90917     524.3
## 4  67185.02  -89772.17      BIG CREEK PH 1 -119.24194 37.20639    1486.8
## 5 145198.34  -70472.98      BISHOP AIRPORT -118.35806 37.37111    1250.3
## 6 -61218.92  140398.66 BLUE CANYON AIRPORT  -120.7102  39.2774    1608.1
##   PRECIPITATION TEMPERATURE
## 1        688.59    17.02500
## 2        666.76    16.05833
## 3        794.02    16.70000
## 4        860.05    11.73333
## 5        131.58    13.61667
## 6       1641.34    10.95000
```

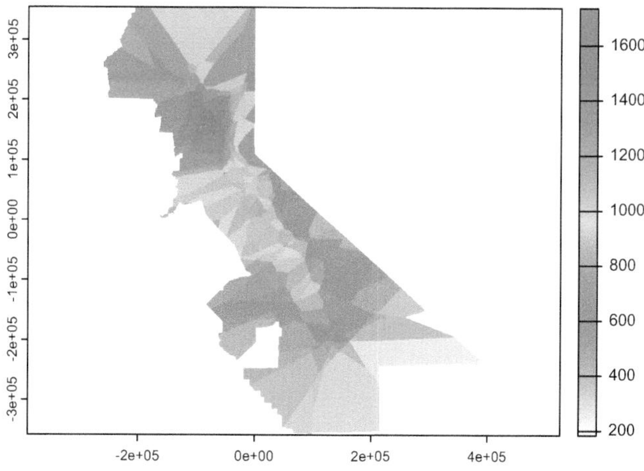

FIGURE 9.5 Nearest neighbor interpolation of precipitation

```
gs <- gstat(formula=PRECIPITATION~1, locations=~x+y, data=d, nmax=5,
            set=list(idp = 0))
nn <- interpolate(r, gs, debug.level=0)
nnmsk <- mask(nn, vr)
plot(nnmsk, 1)
```

9.3.1 Cross-validation and relative performance of the nearest neighbor model

Again using cross-validation and deriving relative performance for our nearest neighbor model, the higher RMSE and lower relative performance compared with the Voronoi model suggests that it's not any better than that single-neighbor model, so we might just stick with that.

```
rmsenn <- rep(NA, k)
for (i in 1:k) {
  test <- d[kf == i, ]
  train <- d[kf != i, ]
  gscv <- gstat(formula=PRECIPITATION~1, locations=~x+y, data=train,
                nmax=5, set=list(idp = 0))
  p <- predict(gscv, test, debug.level=0)$var1.pred
  rmsenn[i] <- RMSE(test$PRECIPITATION, p)
}
rmsenn
```

```
## [1] 315.4844 493.9716 583.9471 316.5636 365.5138
```

```
mean(rmsenn)
```

```
## [1] 415.0961
```

```
1 - (mean(rmsenn) / null)
```

```
## [1] 0.09883281
```

9.4 Inverse Distance Weighted (IDW)

The inverse distance weighted (IDW) interpolator is popular due to its ease of use and few statistical assumptions. Surrounding points influence the predicted value of a cell based on the inverse of their distance. The inverse distance weight has an an exponent set with `set=list(idp=2)`, and if that is 2 (so inverse distance squared), it is sometimes referred

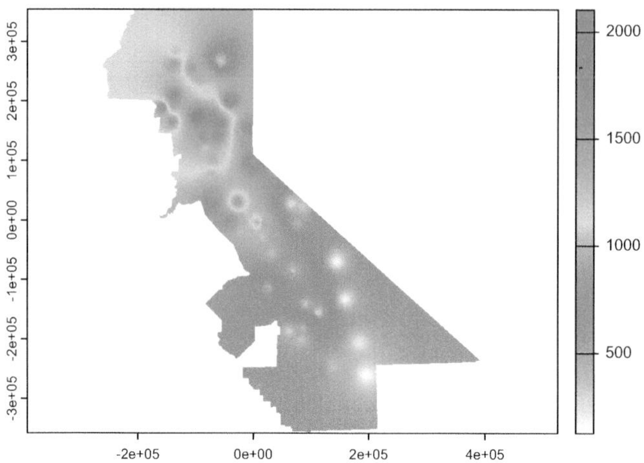

FIGURE 9.6 IDW interpolation, power = 2

to as a "gravity model" since it's the same as the way gravity works. For environmental variables, however, sampling locations are not like objects that create an effect but rather *observations of a continuous surface*. But we'll go ahead and start with the gravity model, inverse distance power of 2 (Figure 9.6).

```
library(gstat)
gs <- gstat(formula=PRECIPITATION~1, locations=~x+y, data=d, nmax=Inf,
            set=list(idp=2))
idw <- interpolate(r, gs, debug.level=0)
idwr <- mask(idw, vr)
plot(idwr, 1)
```

9.4.1 Using cross-validation and relative performance to guide inverse-distance weight choice

```
rmse <- rep(NA, k)
for (i in 1:k) {
  test <- d[kf == i, ]
  train <- d[kf != i, ]
  gs <- gstat(formula=PRECIPITATION~1, locations=~x+y, data=train,
              set=list(idp=2))
  p <- predict(gs, test, debug.level=0)
  rmse[i] <- RMSE(test$PRECIPITATION, p$var1.pred)
}
rmse
```

```
## [1] 366.3199 393.9248 512.9897 240.9113 360.9724
```

```
mean(rmse)
```

```
## [1] 375.0236
```

```
1 - (mean(rmse) / null)
```

```
## [1] 0.1858295
```

So what we can see from the above is the IDW has a bit better relative performance (from a lower mean RMSE). The IDW does come with an artifact of pits and peaks around the data points, which you can see in the map, and this effect can be decreased with a lower idp.

9.4.2 IDW: trying other inverse distance powers

A power of 1 might reduce the gravity-like emphasis on the sampling point, so we'll try that `(idp=1)` (Figure 9.7):

```
library(gstat)
gs <- gstat(formula=PRECIPITATION~1, locations=~x+y, data=d, nmax=Inf,
            set=list(idp=1))
idw <- interpolate(r, gs, debug.level=0)
idwr <- mask(idw, vr)
plot(idwr, 1)
```

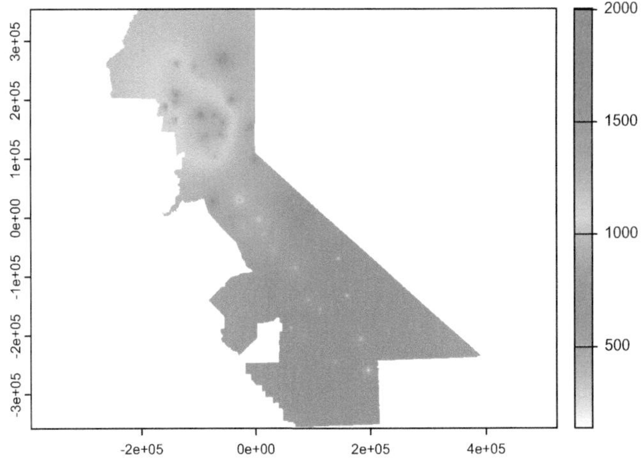

FIGURE 9.7 IDW interpolation, power = 1

```
rmse <- rep(NA, k)
for (i in 1:k) {
  test <- d[kf == i, ]
  train <- d[kf != i, ]
  gs <- gstat(formula=PRECIPITATION~1, locations=~x+y, data=train, set=list(idp=1))
  p <- predict(gs, test, debug.level=0)
  rmse[i] <- RMSE(test$PRECIPITATION, p$var1.pred)
}
rmse
```

```
## [1] 401.7832 412.2137 468.5602 355.0603 410.7361
```

```
mean(rmse)
```

```
## [1] 409.6707
```

```
1 - (mean(rmse) / null)
```

```
## [1] 0.1106113
```

... but the relative performance is no better. You should continue this experiment by trying other weights (remember to set them the same for the model you're mapping and for the cross-validation), or you could even loop it to optimize the setting for the highest relative performance, but I'll leave you to figure that out. Note that the number of folds may also have an impact, especially for small data sets like this.

9.5 Polynomials and Trend Surfaces

In the Voronoi, nearest-neighbor, and IDW methods above, predicted cell values were based on values of nearby input points, in the case of the Voronoi creating a polygon surrounding the input point to assign its value to (and the raster simply being those polygons assigned to raster cells.) Another approach is to create a polynomial model where values are assigned based on their location defined as x and y values. The simplest is first-order linear trend surface model, as the equation of a plane:

$$z = b_0 + b_1x + b_2y$$

where z is a variable like `PRECIPITATION`. Higher-order polynomials can model more complex surfaces, such as a fold using a second-order polynomial, which would look like:

$$z = b_0 + b_1x + b_2y + b_3x^2 + b_4xy + b_5y^2$$

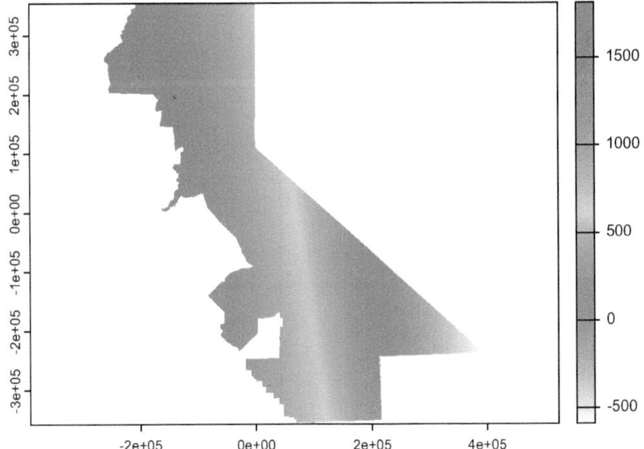

FIGURE 9.8 Linear trend

... or a third-order polynomial which might model a dome or basin:

$$z = b_0 + b_1x + b_2y + b_3x^2 + b_4xy + b_5y^2 + b_6x^3 + b_7x^2y + b_8xy^2 + b_9y^3$$

To enter the first order, your formula can look like `z~x+y` but anything beyond this pretty much requires using the `degree` setting. For the first order this would include `z~1` and `degree=1`; the second order just requires changing the degree to 2, etc.

We'll use gstat with a global linear trend surface by providing this formula as `TEMPERATURE~x+y`, and use the `degree` setting (Figure 9.8).

```
library(gstat)
gs <- gstat(formula=PRECIPITATION~1, locations=~x+y, degree=1, data=d)
interp <- interpolate(r, gs, debug.level=0)
prc1 <- mask(interp, vr)
plot(prc1, 1)
```

What we're seeing is a map of a sloping plane showing the general trend. Not surprisingly there are areas with values lower than the minimum input temperature and higher than the maximum input temperature, which we can check with:

```
min(d$PRECIPITATION)
```

```
## [1] 125.48
```

```
max(d$PRECIPITATION)
```

```
## [1] 2103.6
```

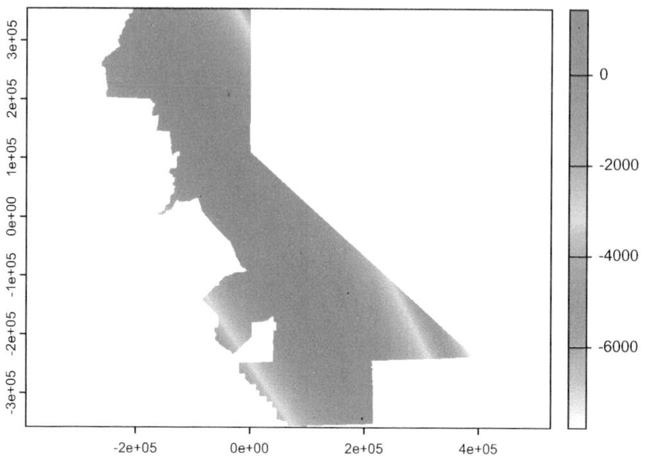

FIGURE 9.9 2nd order polynomial, precipitation

... but that's what we should expect when defining a trend that extends beyond our input point locations.

Then we'll try a second order (Figure 9.9):

```
gs <- gstat(formula=PRECIPITATION~1, locations=~x+y, degree=2, data=d)
interp2 <- interpolate(r, gs, debug.level=0)
prc2 <- mask(interp2, vr)
plot(prc2, 1)
```

... which exhibits this phenomenon all the more, though only away from our input data. Then a third order (Figure 9.10)...

```
gs <- gstat(formula=PRECIPITATION~1, locations=~x+y, degree=3, data=d)
interp3 <- interpolate(r, gs, debug.level=0)
prc3 <- mask(interp3, vr)
plot(prc3, 1)
```

... extends this to the extreme, though the effect is only away from our data points so it's just a cartographic problem we could address by controlling the range of values over which to apply the color range. Or we could use map algebra to assign all cells higher than the maximum inputs to the maximum and lower than the minimum inputs to the minimum (Figure 9.11).

```
prc3$var1.pred[prc3$var1.pred > max(d$PRECIPITATION)] <- max(d$PRECIPITATION)
prc3$var1.pred[prc3$var1.pred < min(d$PRECIPITATION)] <- min(d$PRECIPITATION)
plot(prc3$var1.pred)
```

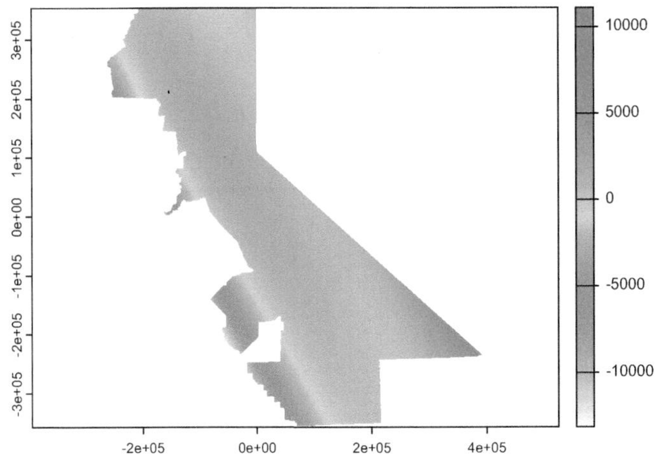

FIGURE 9.10 Third order polynomial, temperature

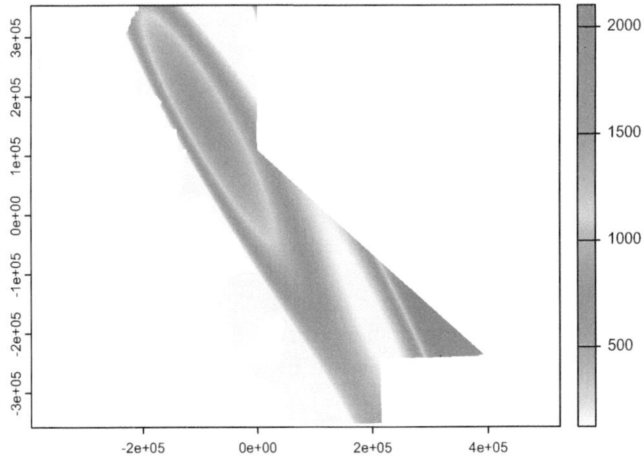

FIGURE 9.11 Third order polynomial with extremes flattened

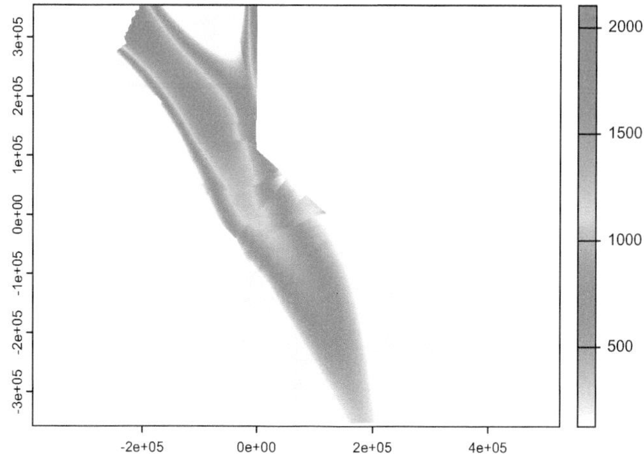

FIGURE 9.12 Third order local polynomial, precipitation

Then a third order *local* polynomial (setting nmax to less than Inf), where we'll similarly set the values outside the range to the min or max value. The nmax setting can be seen to have a significant effect on the smoothness of the result (Figure 9.12).

```
gs <- gstat(formula=PRECIPITATION~1, locations=~x+y, degree=3, nmax=30, data=d)
interp3loc <- interpolate(r, gs, debug.level=0)
prc3loc <- mask(interp3loc, vr)
prc3loc$var1.pred[prc3loc$var1.pred > max(d$PRECIPITATION)] <- max(d$PRECIPITATION)
prc3loc$var1.pred[prc3loc$var1.pred < min(d$PRECIPITATION)] <- min(d$PRECIPITATION)
plot(prc3loc$var1.pred)
```

9.6 Kriging

Kriging is a *stochastic* geostatistical interpolator that provides advantages for considering the statistical probability of results and characteristics of the original data such as spatial autocorrelation, and has been widely applied in fields such as mining (Krige's application) and considering environmental risk. There's a lot more to it than we can cover here, and the reader is encouraged to see Nowosad (n.d.) and other works to learn more. Kriging commonly requires a lot of exploratory work with visual interpretation of data to derive the right model parameters. It starts by looking at the variogram and creating a model with parameters to see which might best approximates the distribution.

9.6.1 Create a variogram.

Compare every point to every other point. The width setting (20,000 m) is the subsequent distance intervals into which data point pairs are grouped for semivariance estimates (Figure 9.13).

```
#p <- data.frame(geom(dtakm)[, c("x","y")], as.data.frame(dta))
gs <- gstat(formula=PRECIPITATION~1, locations=~x+y, data=d)
v <- variogram(gs, width=20e3)
v
```

```
##      np      dist     gamma dir.hor dir.ver   id
## 1    23  14256.47  45143.82       0       0 var1
## 2    60  30035.37  61133.79       0       0 var1
## 3    93  50919.48 163622.47       0       0 var1
## 4    97  71033.67 192267.26       0       0 var1
## 5   126  89759.82 186047.23       0       0 var1
## 6   109 109139.47 192847.68       0       0 var1
## 7   109 130125.18 139241.93       0       0 var1
## 8    91 149596.56 146620.61       0       0 var1
## 9    94 169606.82 150855.26       0       0 var1
## 10   90 188727.78 203683.08       0       0 var1
## 11   57 208856.39 220665.36       0       0 var1
```

```
plot(v)
```

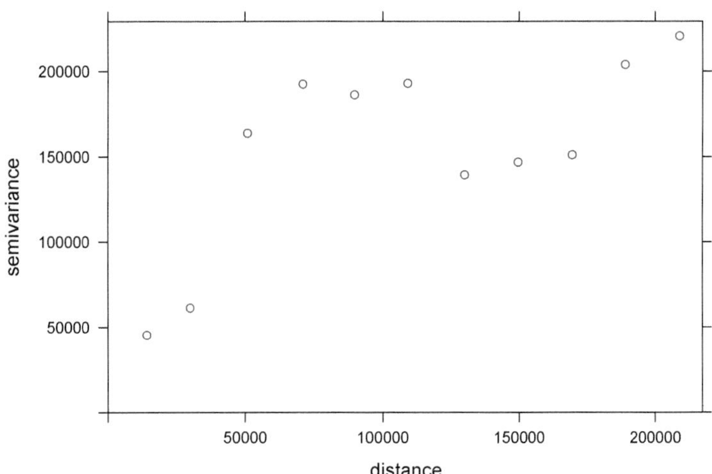

FIGURE 9.13 Variogram of precipitation at Sierra weather stations

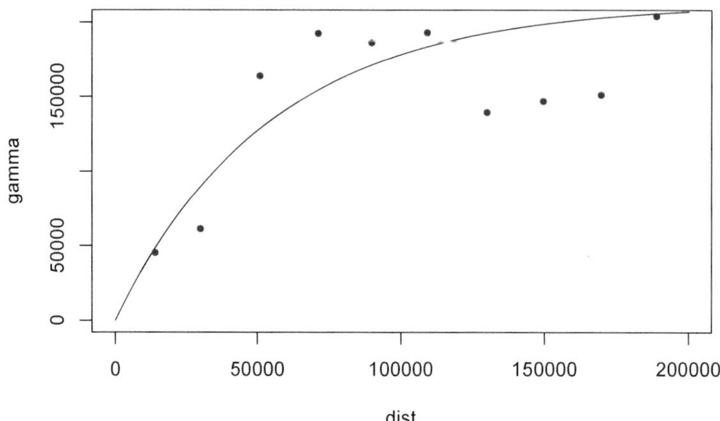

FIGURE 9.14 Fitted variogram

9.6.2 Fit the variogram based on visual interpretation

For geostatistical analysis leading to Kriging, we often use visual interpretation of the variogram to find the best fit, employing various models, such as exponential "Exp" (Figure 9.14).

```
fve <- fit.variogram(v, vgm(psill=2e5, model="Exp", range=200e3, nugget=0))
plot(variogramLine(fve, maxdist=2e5), type='l', ylim=c(0,2e5))
points(v[,2:3], pch=20, col='red')
```

The list of twenty variogram models can be provided with `vgm()`. We'll try spherical with "Sph" (Figure 9.15).

```
vgm()
```

```
##    short                                      long
## 1    Nug                              Nug (nugget)
## 2    Exp                         Exp (exponential)
## 3    Sph                           Sph (spherical)
## 4    Gau                            Gau (gaussian)
## 5    Exc        Exclass (Exponential class/stable)
## 6    Mat                              Mat (Matern)
## 7    Ste Mat (Matern, M. Stein's parameterization)
## 8    Cir                            Cir (circular)
## 9    Lin                              Lin (linear)
## 10   Bes                              Bes (bessel)
## 11   Pen                      Pen (pentaspherical)
## 12   Per                            Per (periodic)
```

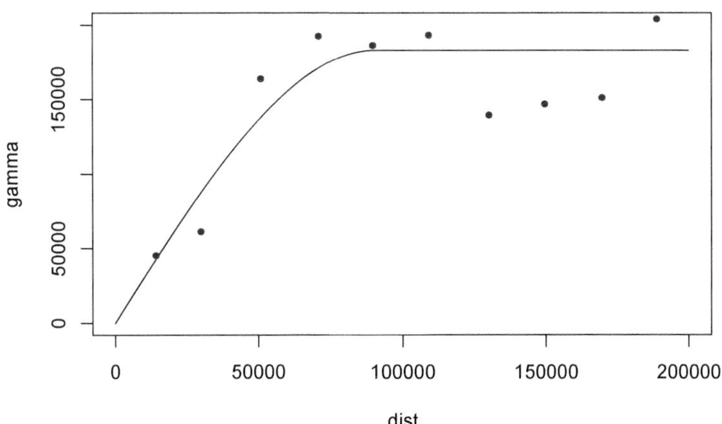

FIGURE 9.15 Spherical fit

```
## 13    Wav                                    Wav (wave)
## 14    Hol                                    Hol (hole)
## 15    Log                             Log (logarithmic)
## 16    Pow                                   Pow (power)
## 17    Spl                                  Spl (spline)
## 18    Leg                                Leg (Legendre)
## 19    Err                         Err (Measurement error)
## 20    Int                               Int (Intercept)
```

```
fvs <- fit.variogram(v, vgm(psill=2e5, model="Sph", range=20e3, nugget=0))
fvs
```

```
##   model    psill    range
## 1   Nug      0.0     0.00
## 2   Sph 183011.2 90689.21
```

```
plot(variogramLine(fvs, maxdist=2e5), type='l', ylim=c(0,2e5))
points(v[,2:3], pch=20, col='red')
```

The exponential model at least visually looks like the best fit, so we'll use the `fse` model we created first (Figure 9.16).

```
plot(v, fve)
```

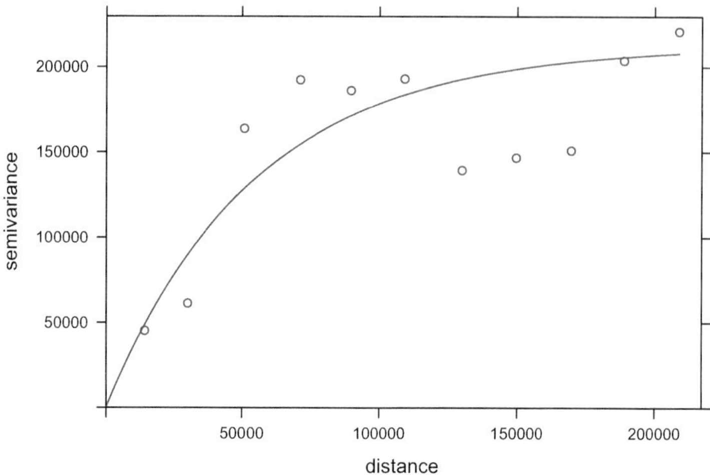

FIGURE 9.16 Exponential model

9.6.3 Ordinary Kriging

So we'll go with the `fve` model derived using an exponential form to develop an ordinary Kriging prediction and variance result. The prediction is the interpolation, and the variance shows the areas near the data points where we should have more confidence in the prediction; this ability is one of the hallmarks of the Kriging method (Figure 9.17).

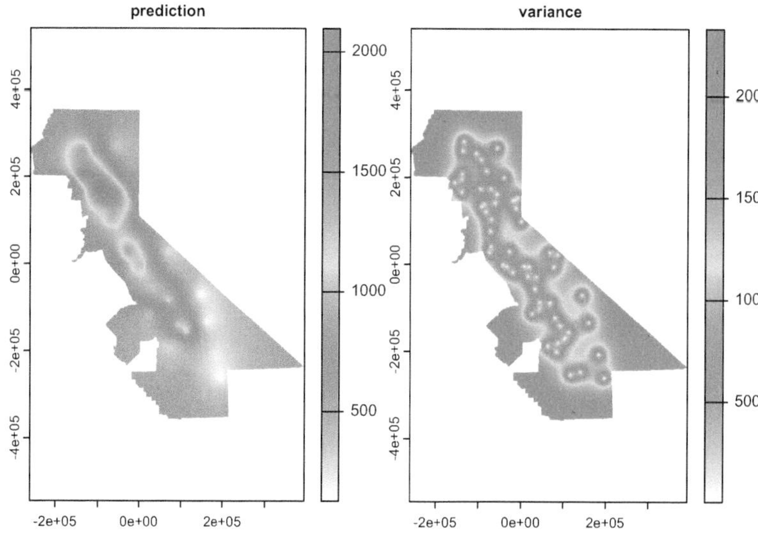

FIGURE 9.17 Ordinary Kriging

```
kr <- gstat(formula=PRECIPITATION~1, locations=~x+y, data=d, model=fve)
kp <- interpolate(r, kr, debug.level=0)
ok <- mask(kp, vr)
names(ok) <- c('prediction', 'variance')
plot(ok)
```

Clearly the prediction provides a clear pattern of precipitation, at least for the areas with low variance. The next step would be to do a cross-validation to compare its RMSE with IDW and other models, but we'll leave that for later.

To learn more about these interpolation methods and related geostatistical methods, see http://rspatial.org/terra/analysis/4-interpolation.html, Nowosad (n.d.), and the gstat user manual at http://www.gstat.org/gstat.pdf .

9.7 Exercises: Spatial Interpolation

Exercise 9.1. We're going to create a couple of interpolations of California climate variables, so start by building `CAclimate sf` from the `908277.csv` file in the `sierra` folder; this csv has stations from all of California. Note that you should use `na.rm=T` for summarizations of both `MLY-PRCP-NORMAL` and `MLY-TAVG-NORMAL`, and since there are some weird extreme unrealistic negative numbers for these, also filter for `PRECIPITATION > 0` and `TEMPERATURE > -100` – other values don't exist on our planet, much less California. There are also some "unknown" values in `LATITUDE`, `LONGITUDE`, and `ELEVATION`, so also filter for `LATITUDE != "unknown"`. (These low negative values plus codes like "unknown" are just other ways of saying `NA`). Create a plot of your points colored by `PRECIPITATION`. You can either use ggplot2 or tmap. Note that `LATITUDE`, `LONGITUDE` and `ELEVATION` get read in as `chr`, so you'll want to use `as.numeric` with them in the summarize section, so code like `LATITUDE = first(as.numeric(LATITUDE))`.

Exercise 9.2. Create **Voronoi polygons of precipitation** (Figure 9.18) Note that the voronoi tool is in `terra` so the data need to be converted to S4 objects with `vect()`. Then rasterize the resulting polygons, remembering that the coordinates are in decimal degrees so `res=0.01` would probably be a better setting. (Extra credit: what shape will the cells be on the map? Remember that we're working in decimal degrees, but the map is visually projected by the plotting system since it understands geospatial coordinate systems. Experiment with the res setting to visualize this.)

Exercise 9.3. Use `gstat` to create an IDW interpolation of the same data, as we did earlier for the `sierra` data, creating a data frame `d`, but now from `CAclimateT` (Figure 9.19).

Exercise 9.4. Use the IDW interpolator from the same data, but use `ELEVATION` instead of `PRECIPITATION`. Then create a hillshade from it using the `terra::shade` function (since the interpolator will return an `SpatRaster`. Then compare it with `"ca_hillsh_WGS84.tif"` from the external data.

Exercise 9.5. Hillshade of Voronoi station elevations. Optionally, how about a hillshade created from a voronoi of elevation? Remember that the hillshade above is created

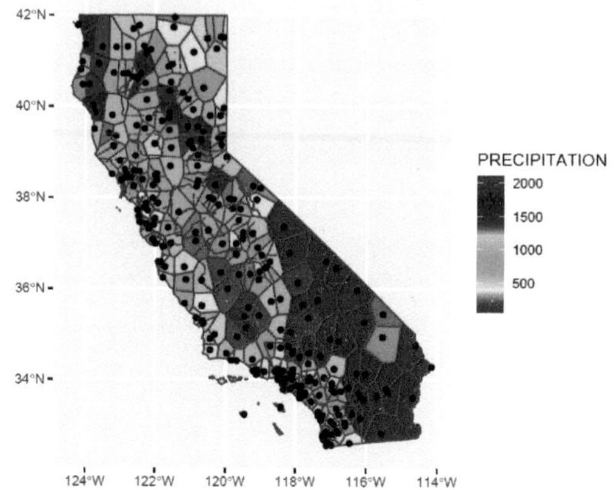

FIGURE 9.18 Voronoi polygons of precipitation (goal)

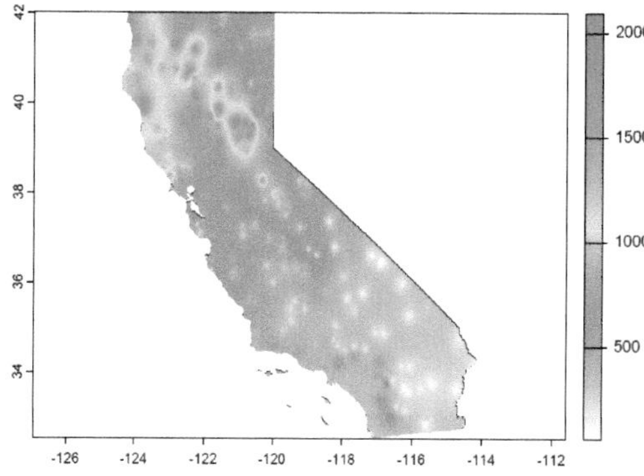

FIGURE 9.19 IDW (goal)

from point data that happens to have elevation data at the stations, and the Voronoi polygons are simply flat "plateaus" surrounding the weather stations.

Exercise 9.6. Create a linear trend surface of `PRECIPITATION` in California from the same data.

Part III

Statistics and Modeling

10

Statistical Summaries and Tests

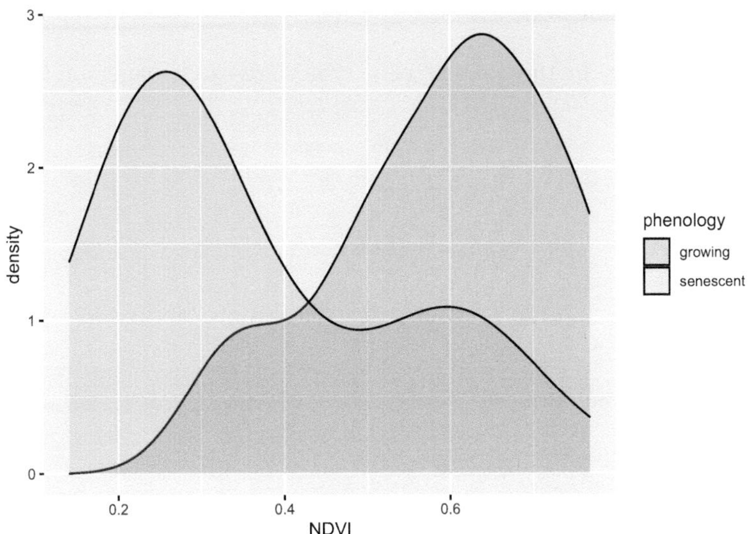

10.1 Goals of Statistical Analysis

To frame how we might approach statistical analysis and modeling, there are various goals that are commonly involved:

- To understand our data
  - nature of our data, through summary statistics and various graphics like histograms
  - spatial statistical analysis
  - time series analysis
- To *group* or *classify* things based on their properties
  - using factors to define groups, and deriving grouped summaries
  - comparing *observed* vs *expected* counts or probabilities
- To understand how variables relate to one another
  - or maybe even explain variations in other variables, through correlation analysis

- To *model* behavior and maybe *predict* it
  - various linear models
- To *confirm* our observations from exploration (field/lab/vis)
  - inferential statistics e.g. difference of means tests, ANOVA, χ^2
- To have the confidence to draw conclusions, make informed decisions
- To help *communicate* our work

These goals can be seen in the context of a typical research paper or thesis outline in environmental science:

- Introduction
- Literature Review
- Methodology
- Results
  - field, lab, geospatial data
- Analysis
  - statistical analysis
  - qualitative analysis
  - visualization
- Discussion
  - making sense of analysis
  - possibly recursive, with visualization
- Conclusion
  - conclusion about what the above shows
  - new questions for further research
  - possible policy recommendation

The scope and theory of statistical analysis and models is extensive, and there are many good books on the subject that employ the R language, including sources that focus on environmental topics like water resources (e.g. Helsel et al. (2020)). This chapter is a short review of some of these methods and how they apply to environmental data science.

10.2 Summary Statistics

Summary statistics such as mean, standard deviation, variance, minimum, maximum, and range are derived in quite a few R functions, commonly as a parameter or a sub-function (see `mutate`). An overall simple statistical summary is very easy to do in base R:

```
summary(tidy_eucoak)
```

```
##      site              site #          tree                 Date
##  Length:180         Min.   :1.000   Length:180         Min.   :2006-11-08
##  Class :character   1st Qu.:2.000   Class :character   1st Qu.:2006-12-07
##  Mode  :character   Median :4.000   Mode  :character   Median :2007-01-30
##                     Mean   :4.422                      Mean   :2007-01-29
##                     3rd Qu.:6.000                      3rd Qu.:2007-03-22
##                     Max.   :8.000                      Max.   :2007-05-07
##
##     month              rain_mm       rain_subcanopy      slope
##  Length:180         Min.   : 1.00   Min.   : 1.00   Min.   : 9.00
##  Class :character   1st Qu.:16.00   1st Qu.:16.00   1st Qu.:12.00
##  Mode  :character   Median :28.50   Median :30.00   Median :24.00
##                     Mean   :37.99   Mean   :34.84   Mean   :20.48
##                     3rd Qu.:63.25   3rd Qu.:50.00   3rd Qu.:27.00
##                     Max.   :99.00   Max.   :98.00   Max.   :32.00
##                     NA's   :36      NA's   :4
##      aspect         runoff_L       surface_tension runoff_rainfall_ratio
##  Min.   :100.0   Min.   : 0.000   Min.   :28.51   Min.   :0.00000
##  1st Qu.:143.0   1st Qu.: 0.000   1st Qu.:37.40   1st Qu.:0.00000
##  Median :196.0   Median : 0.825   Median :62.60   Median :0.03347
##  Mean   :186.6   Mean   : 2.244   Mean   :55.73   Mean   :0.05981
##  3rd Qu.:221.8   3rd Qu.: 3.200   3rd Qu.:72.75   3rd Qu.:0.08474
##  Max.   :296.0   Max.   :16.000   Max.   :72.75   Max.   :0.42000
##                  NA's   :8        NA's   :44      NA's   :8
```

10.2.1 Summarize by group: *stratifying a summary*

In the visualization chapter and elsewhere, we've seen the value of adding symbolization based on a categorical variable or factor. Summarizing by group has a similar benefit, and provides a tabular output in the form of a data frame, and the tidyverse makes it easy to extract several summary statistics at once. For instance, for the euc/oak study, we can create variables of the mean and maximum runoff, the mean and standard deviation of rainfall, for each of the sites. This table alone provides a useful output, but we can also use it in further analyses.

```
eucoakrainfallrunoffTDR %>%
  group_by(site) %>%
  summarize(
    rain = mean(rain_mm, na.rm = TRUE),
    rainSD = sd(rain_mm, na.rm = TRUE),
    runoffL_oak = mean(runoffL_oak, na.rm = TRUE),
    runoffL_euc = mean(runoffL_euc, na.rm = TRUE),
    runoffL_oakMax = max(runoffL_oak, na.rm = TRUE),
    runoffL_eucMax = max(runoffL_euc, na.rm = TRUE),
  )
```

```
## # A tibble: 8 x 7
##   site   rain rainSD runoffL_oak runoffL_euc runoffL_oakMax runoffL_eucMax
##   <chr> <dbl>  <dbl>       <dbl>       <dbl>          <dbl>          <dbl>
## 1 AB1    48.4   28.2      6.80        6.03           6.80           6.03
## 2 AB2    34.1   27.9      4.91        3.65           4.91           3.65
## 3 KM1    48     32.0      1.94        0.592          1.94           0.592
## 4 PR1    56.5   19.1      0.459       2.31           0.459          2.31
## 5 TP1    38.4   29.5      0.877       1.66           0.877          1.66
## 6 TP2    34.3   29.2      0.0955      1.53           0.0955         1.53
## 7 TP3    32.1   28.4      0.381       0.815          0.381          0.815
## 8 TP4    32.5   28.2      0.231       2.83           0.231          2.83
```

10.2.2 Boxplot for visualizing distributions by group

We've looked at this already in the visualization chapter, but a Tukey boxplot is a good way to visualize distributions by group. In this soil CO_2 study of the Marble Mountains (JD Davis, Amato, and Kiefer 2001) (Figure 10.1), some sites had much greater variance, and some sites tended to be low vs high (Figure 10.2).

```
soilCO2_97$SITE <- factor(soilCO2_97$SITE)
ggplot(data = soilCO2_97, mapping = aes(x = SITE, y = CO2pct)) +
  geom_boxplot()
```

10.2.3 Generating pseudorandom numbers

Functions commonly used in R books for quickly creating a lot of numbers to display (often with a histogram 4.3.1) are those that generate pseudorandom numbers. These are also useful in statistical methods that need a lot of these, such as in Monte Carlo simulation. The two most commonly used are:

- **`runif()`** generates a vector of `n` pseudorandom numbers ranging by default from `min=0` to `max=1` (Figure 10.3).
- **`rnorm()`** generates a vector of `n` normally distributed pseudorandom numbers with a default `mean=0` and `sd=0` (Figures 10.4 and 10.5)

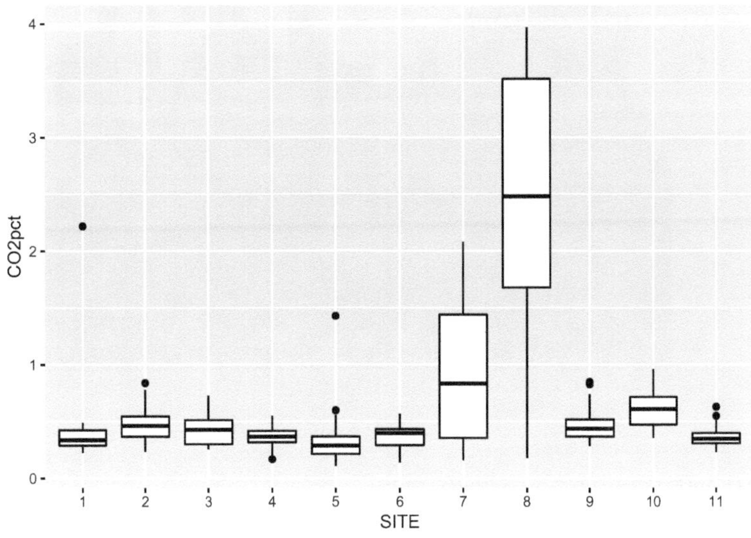

FIGURE 10.1 Tukey boxplot by group

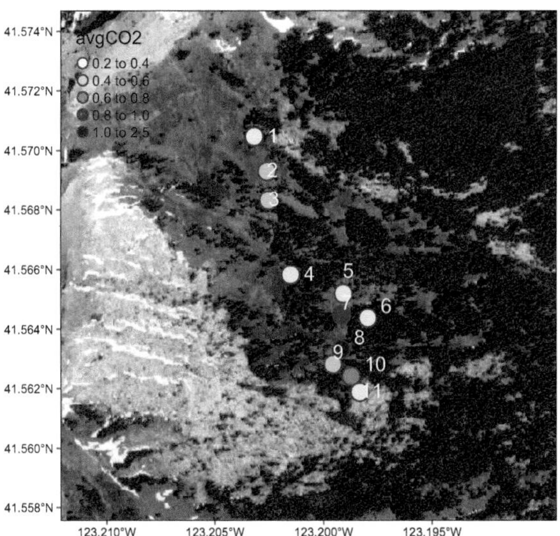

FIGURE 10.2 Marble Mountains average soil carbon dioxide per site

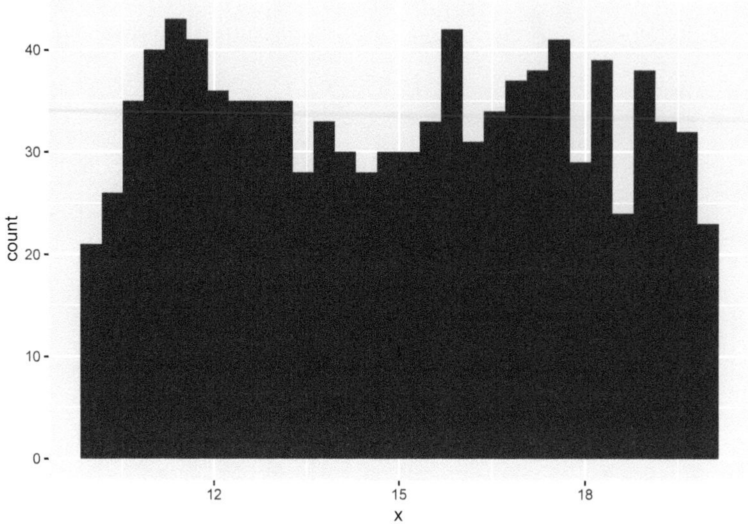

FIGURE 10.3 Random uniform histogram

Figure 10.6 shows both in action as x and y.

```
x <- as_tibble(runif(n=1000, min=10, max=20))
names(x) <- 'x'
ggplot(x, aes(x=x)) + geom_histogram()
```

```
y <- as_tibble(rnorm(n=1000, mean=100, sd=10))
```

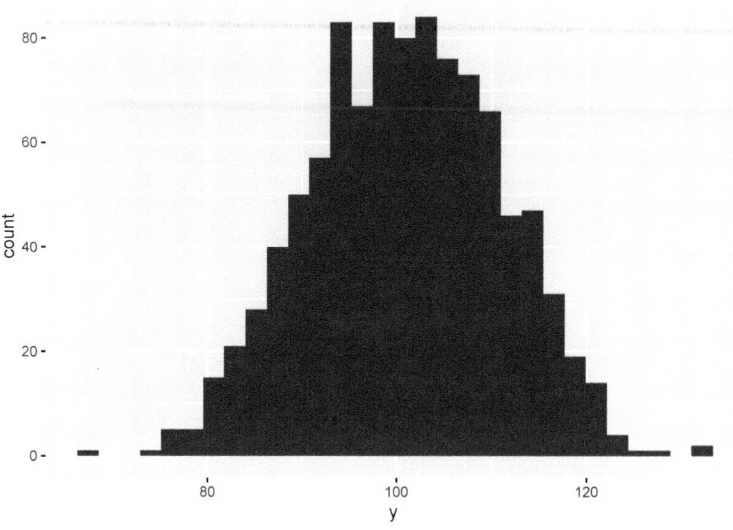

FIGURE 10.4 Random normal histogram

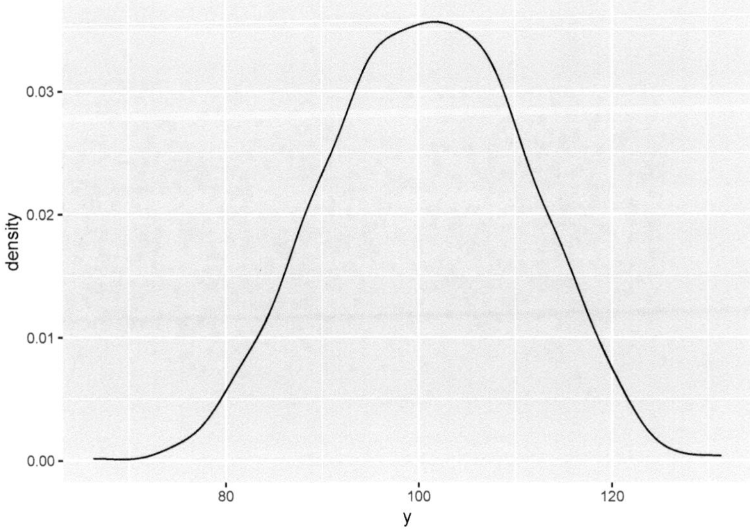

FIGURE 10.5 Random normal density plot

```
names(y) <- 'y'
ggplot(y, aes(x=y)) + geom_histogram()
```

```
ggplot(y, aes(x=y)) + geom_density()
```

```
xy <- bind_cols(x,y)
```

```
ggplot(xy, aes(x=x,y=y)) + geom_point()
```

10.3 Correlation r and Coefficient of Determination r^2

In the visualization chapter, we looked at creating scatter plots and also arrays of scatter plots where we could compare variables to visually see if they might be positively or negatively correlated. A statistic that is commonly used for this is the *Pearson product-moment correlation coefficient* or r statistic.

We'll look at the formula for r below, but it's easier to just use the `cor` function. You just need the two variables as vector inputs in R to return the r statistic. Squaring r to r^2 is the *coefficient of determination* and can be interpreted as the amount of the variation in the dependent variable y that that is "explained" by variation in the independent variable x. This coefficient will always be positive, with a maximum of 1 or 100%.

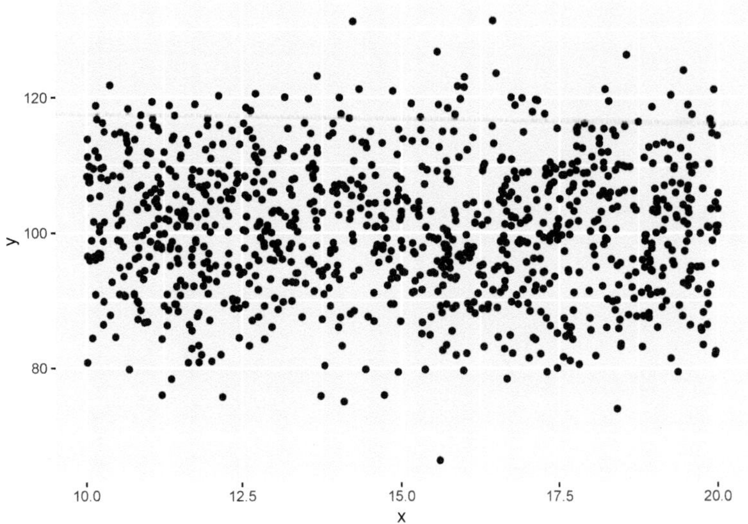

FIGURE 10.6 Random normal plotted against random uniform

If we create two random variables, uniform or normal, there shouldn't be any correlation. We'll do this five times each set:

```
for (i in 1:5) {print(paste(i, cor(rnorm(100), rnorm(100))))}
```

```
## [1] "1 0.0861276591535613"
## [1] "2 -0.0530030635693458"
## [1] "3 -0.00258999661653757"
## [1] "4 0.151902670698442"
## [1] "5 0.0299297305024441"
```

```
for (i in 1:5) {print(paste(i, cor(runif(100), runif(100))))}
```

```
## [1] "1 -0.0282873634049497"
## [1] "2 -0.137429308622517"
## [1] "3 0.0389749859037954"
## [1] "4 -0.00863501006288174"
## [1] "5 -0.00347423499468112"
```

```
for (i in 1:5) {print(paste(i, cor(rnorm(100), runif(100))))}
```

```
## [1] "1 0.112869627470136"
## [1] "2 0.0444231135099521"
## [1] "3 -0.0907625260655007"
## [1] "4 0.0485073708764654"
## [1] "5 -0.0297300718464998"
```

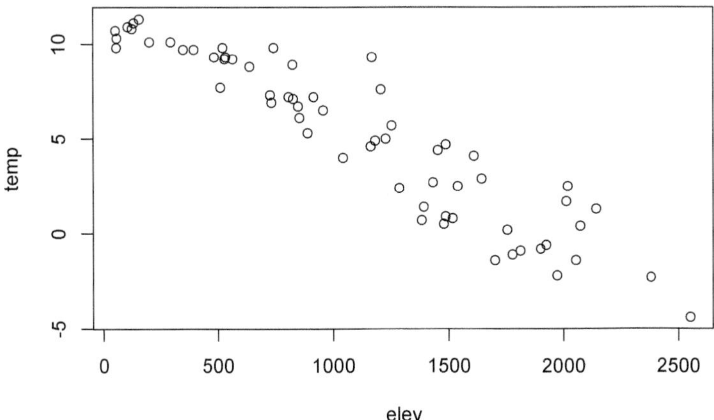

FIGURE 10.7 Scatter plot illustrating negative correlation

But variables such as temperature and elevation in the Sierra will be strongly negatively correlated, as seen in the scatter plot and the r value close to -1, and have a high r^2 (Figure 10.7).

```
library(igisci); library(tidyverse)
sierra <- sierraFeb %>% filter(!is.na(TEMPERATURE))
elev <- sierra$ELEVATION; temp <- sierra$TEMPERATURE
plot(elev, temp)
```

```
cor(elev, temp)
```

```
## [1] -0.9357801
```

```
cor(elev, temp)^2
```

```
## [1] 0.8756845
```

While you don't need to use this, since the `cor` function is easier to type, it's interesting to know that the formula for Pearson's correlation coefficient is something that you can actually code in R, taking advantage of its vectorization methods:

$$r = \frac{\sum(x_i - \overline{x})(y_i - \overline{y})}{\sqrt{\sum(x_i - \overline{x})^2 \sum(y_i - \overline{y})^2}}$$

```
r <- sum((elev-mean(elev))*(temp-mean(temp)))/
     sqrt(sum((elev-mean(elev))^2*sum((temp-mean(temp))^2)))
r
```

```
## [1] -0.9357801
```

```
r^2
```

```
## [1] 0.8756845
```

Another version of the formula runs faster, so might be what R uses, but you'll never notice the time difference:

$$r = \frac{n(\sum xy) - (\sum x)(\sum y)}{\sqrt{(n\sum x^2 - (\sum x)^2)(n\sum y^2 - (\sum y)^2)}}$$

```
n <- length(elev)
r <- (n*sum(elev*temp)-sum(elev)*sum(temp))/
  sqrt((n*sum(elev^2)-sum(elev)^2)*(n*sum(temp^2)-sum(temp)^2))
r
```

```
## [1] -0.9357801
```

```
r^2
```

```
## [1] 0.8756845
```

... and as you can see, all three methods give the same results.

10.3.1 Displaying correlation in a pairs plot

We can use another type of pairs plot from the **psych** package to look at the correlation coefficient in the upper-right part of the pairs plot, since correlation can be determined between x and y or y and x; the result is the same. In contrast to what we'll see in regression models, there doesn't have to be one *explanatory* (or independent) variable and one *response* (or dependent) variable; either one will do. The r value shows both the direction (positive or negative) and the magnitude of the correlation, with values closer to 1 or -1 being more correlated (Figure 10.8).

```
library(psych)
sierraFeb %>%
     dplyr::select(ELEVATION:TEMPERATURE) %>%
     pairs.panels(method = "pearson", # correlation method
             hist.col = "#00AFBB",
             density = TRUE, # show density plots
             ellipses = F, smooth = F) # unneeded
```

We can clearly see in the graphs the negative relationships between elevation and temperature and between latitude and longitude (what is that telling us?), and these correspond to strongly negative correlation coefficients.

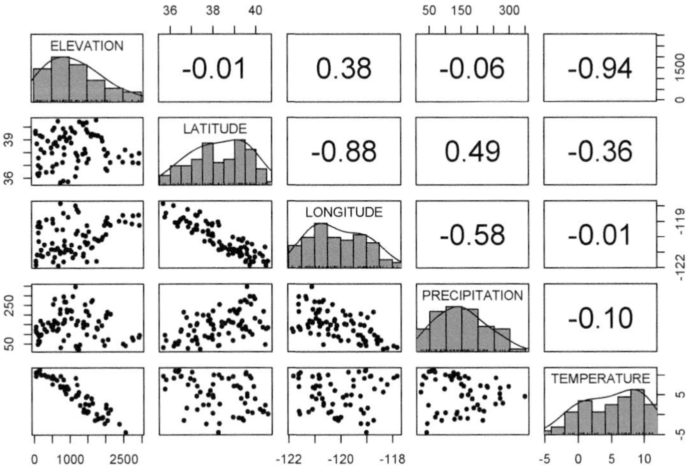

FIGURE 10.8 Pairs plot with r values

One problem with Pearson: While there's nothing wrong with the correlation coefficient, Pearson's name, along with some other pioneers of statistics like Fisher, is tainted by an association with the racist field of *eugenics*. But the mathematics is not to blame, and the correlation coefficient is still as useful as ever. Maybe we can just say that Pearson *discovered* the correlation coefficient...

10.4 Statistical Tests

Tests that compare our data to other data or look at relationships among variables are important statistical methods, and you should refer to statistical references to best understand how to apply the appropriate methods for your research.

10.4.1 Comparing samples and groupings with a t test and a non-parametric Kruskal-Wallis Rank Sum test

A common need in environmental research is to compare samples of a phenomenon (e.g. Figure 10.9) or compare samples with an assumed standard population. The simplest application of this is the t-test, which can only involve comparing two samples or one sample with a population. After this, we'll look at analysis of variance, extending this to allow for more than two groups. Start by building `XSptsPheno` from code in the Histogram section (4.3.1) of the Visualization chapter, then proceed:

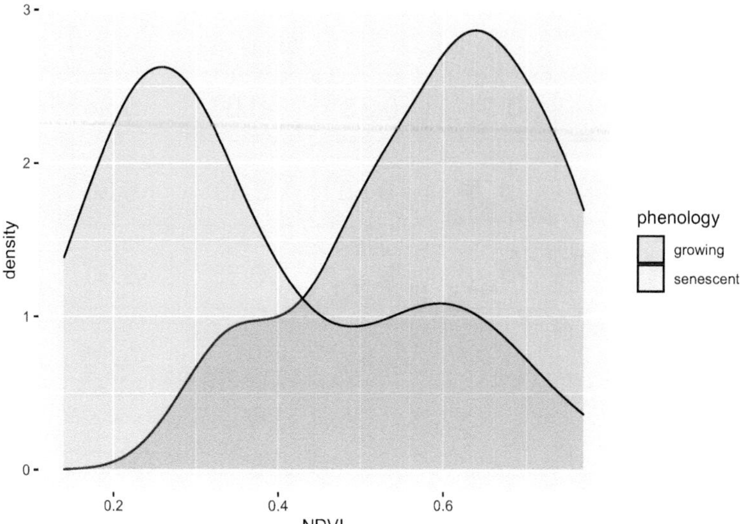

FIGURE 10.9 NDVI by phenology

```
XSptsPheno %>%
  ggplot(aes(NDVI, fill=phenology)) +
  geom_density(alpha=0.2)
```

```
t.test(NDVI~phenology, data=XSptsPheno)
```

```
##
##  Welch Two Sample t-test
##
## data:  NDVI by phenology
## t = 5.4952, df = 52.03, p-value = 1.19e-06
## alternative hypothesis: true difference in means between group growing and group senescent is not equal to 0
## 95 percent confidence interval:
##  0.1421785 0.3057412
## sample estimates:
##   mean in group growing mean in group senescent
##               0.5901186               0.3661588
```

One condition for using a t test is that our data are normally distributed. While these data sets *appear* reasonably normal, though with a bit of bimodality especially for the senescent group, the Shapiro-Wilk test (which uses a null hypothesis of normal) has a p value < 0.05 for the senescent group, so the data can't be assumed to be normal.

```
shapiro.test(XSptsPheno$NDVI[XSptsPheno$phenology=="growing"])
```

```
##
##  Shapiro-Wilk normality test
##
## data:  XSptsPheno$NDVI[XSptsPheno$phenology == "growing"]
## W = 0.93608, p-value = 0.07918
```

```
shapiro.test(XSptsPheno$NDVI[XSptsPheno$phenology=="senescent"])
```

```
##
##  Shapiro-Wilk normality test
##
## data:  XSptsPheno$NDVI[XSptsPheno$phenology == "senescent"]
## W = 0.88728, p-value = 0.004925
```

Therefore we should use a non-parametric alternative such as the Kruskal-Wallis Rank Sum test:

```
kruskal.test(NDVI~phenology, data=XSptsPheno)
```

```
##
##  Kruskal-Wallis rank sum test
##
## data:  NDVI by phenology
## Kruskal-Wallis chi-squared = 19.164, df = 1, p-value = 1.199e-05
```

A bit of a review on significance tests and p values

First, each type of test will be testing a particular summary statistic, like the t test is comparing means and the analysis of variance will be comparing variances. In confirmatory statistical tests, you're always seeing if you can reject the null hypothesis that there's no difference, so in the t test or the rank sum test above that compares two samples, it's that there's no difference between two samples.

There will of course nearly always be some difference, so you might think that you'd always reject the null hypothesis that there's no difference. That's where random error and probability comes in. You can accept a certain amount of error – say 5% – to be able to say that the difference in values could have occurred by chance. That 5% or 0.05 is often called the "significance level" and is the probability of the two values being the same that you're willing to accept. (The remainder 0.95 is often called the confidence level, the extent to which you're confident that rejecting the null hypothesis and accepting the working hypothesis is correct.)

When R reports the p (or Pr) probability value, you compare that value to that significance level to see if it's lower, and then use that to possibly reject the null hypothesis and accept the working hypothesis. R will simply report the p value and commonly show asterisks along with it to indicate if it's lower than various common significance levels, like 0.1, 0.05, 0.01, and 0.001.

10.4.1.1 Runoff and Sediment Yield under Eucalyptus vs Oaks – is there a difference?

Starting with the Data Abstraction chapter, one of the data sets we've been looking at is from a study of runoff and sediment yield under paired eucalyptus and coast live oak sites, and we might want to analyze these data statistically to consider some basic research questions. These are discussed at greater length in Thompson, Davis, and Oliphant (2016), but the key questions are:

- *Is the runoff under eucalyptus canopy significantly different from that under oaks?*
- *Is the sediment yield under eucalyptus canopy significantly different from that under oaks?*

We'll start with the first, since this was the focus on the first part of the study where multiple variables that influence runoff were measured, such as soil hydrophobicity resulting from the chemical effects of eucalyptus, and any rainfall contrasts at each site and between sites. For runoff, we'll then start by test for normality of each of the two samples (euc and oak) which shows clearly that both samples are non-normal.

```
shapiro.test(tidy_eucoak$runoff_L[tidy_eucoak$tree == "euc"])
```

```
##
##  Shapiro-Wilk normality test
##
## data:  tidy_eucoak$runoff_L[tidy_eucoak$tree == "euc"]
## W = 0.74241, p-value = 4.724e-11
```

```
shapiro.test(tidy_eucoak$runoff_L[tidy_eucoak$tree == "oak"])
```

```
##
##  Shapiro-Wilk normality test
##
## data:  tidy_eucoak$runoff_L[tidy_eucoak$tree == "oak"]
## W = 0.71744, p-value = 1.698e-11
```

So we might apply the non-parametric Kruskal-Wallis test ...

```
kruskal.test(runoff_L~tree, data=tidy_eucoak)
```

```
##
##  Kruskal-Wallis rank sum test
##
## data:  runoff_L by tree
## Kruskal-Wallis chi-squared = 2.2991, df = 1, p-value = 0.1294
```

... and no significant difference can be seen. If we look at the data graphically, this makes sense, since the distributions are not dissimilar (Figure 10.10).

```
tidy_eucoak %>%
  ggplot(aes(log(runoff_L),fill=tree)) +
  geom_density(alpha=0.2)
```

However, some of this may result from major variations among sites, which is apparent in a site-grouped boxplot (Figure 10.11).

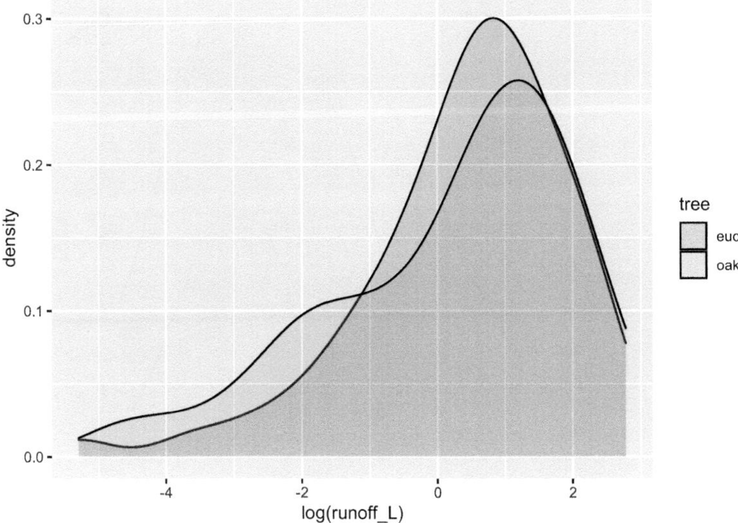

FIGURE 10.10 Runoff under eucalyptus and oak in Bay Area sites

```
ggplot(data = tidy_eucoak) +
  geom_boxplot(aes(x=site, y=runoff_L, color=tree))
```

We might restrict our analysis to Tilden Park sites in the East Bay, where there's more of a difference (Figure 10.12), but the sample size is very small.

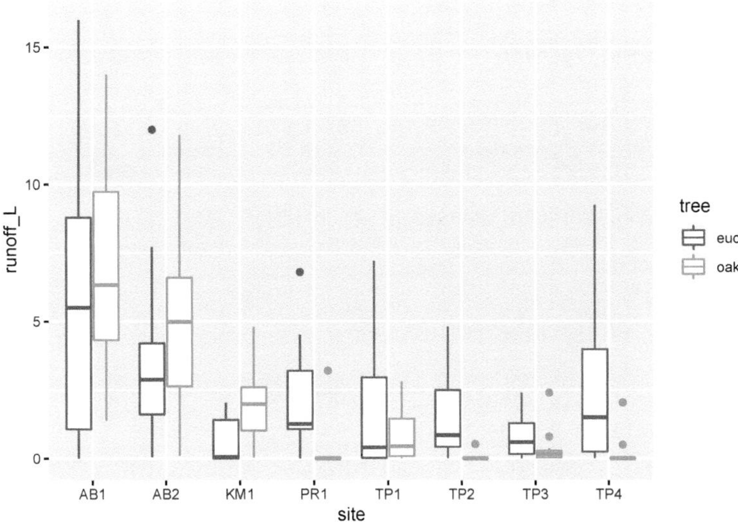

FIGURE 10.11 Runoff at various sites contrasting euc and oak

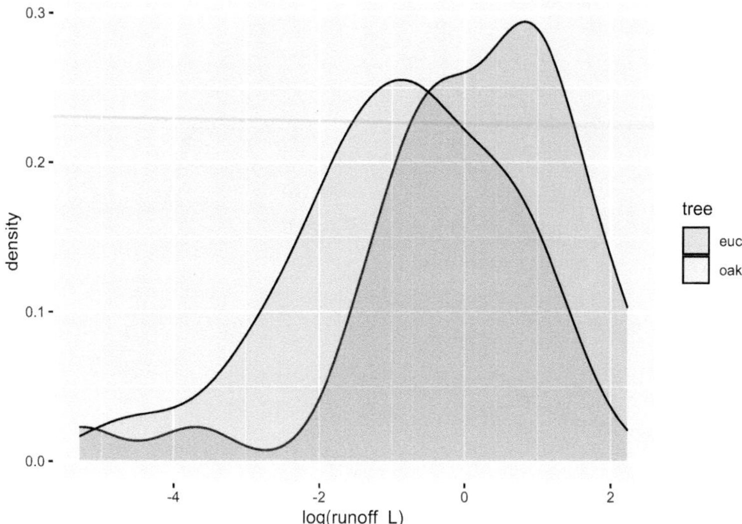

FIGURE 10.12 East Bay sites

```
tilden <- tidy_eucoak %>% filter(str_detect(tidy_eucoak$site,"TP"))
tilden %>%
  ggplot(aes(log(runoff_L),fill=tree)) +
  geom_density(alpha=0.2)
```

```
shapiro.test(tilden$runoff_L[tilden$tree == "euc"])
```

```
##
##  Shapiro-Wilk normality test
##
## data:  tilden$runoff_L[tilden$tree == "euc"]
## W = 0.73933, p-value = 1.764e-07
```

```
shapiro.test(tilden$runoff_L[tilden$tree == "oak"])
```

```
##
##  Shapiro-Wilk normality test
##
## data:  tilden$runoff_L[tilden$tree == "oak"]
## W = 0.59535, p-value = 8.529e-10
```

So once again, as is common with small sample sets, we need a non-parametric test.

```
kruskal.test(runoff_L~tree, data=tilden)
```

```
##
##  Kruskal-Wallis rank sum test
```

```
##
## data:  runoff_L by tree
## Kruskal-Wallis chi-squared = 14.527, df = 1, p-value = 0.0001382
```

Sediment Yield

In the year runoff was studied, there were no runoff events sufficient to mobilize sediments. The next year, January had a big event, so we collected sediments and processed them in the lab.

From the basic sediment yield question listed above we can consider two variants:

- Is there a difference between eucs and oaks in terms of fine sediment yield?
- Is there a difference between eucs and oaks in terms of total sediment yield? (includes litter)

So here, we will need to extract fine and total sediment yield from the data and derive group statistics by site (Figure 10.13). As usual, we'll use a faceted density plot to visualize the distributions (Figure 10.14). Then we'll run the test.

```
eucoaksed <- read_csv(ex("eucoak/eucoaksediment.csv"))
summary(eucoaksed)
```

```
##       id                 site               trtype               slope
##  Length:14          Length:14          Length:14          Min.   : 9.00
##  Class :character   Class :character   Class :character   1st Qu.:12.00
##  Mode  :character   Mode  :character   Mode  :character   Median :21.00
##                                                           Mean   :20.04
##                                                           3rd Qu.:25.00
##                                                           Max.   :32.00
```

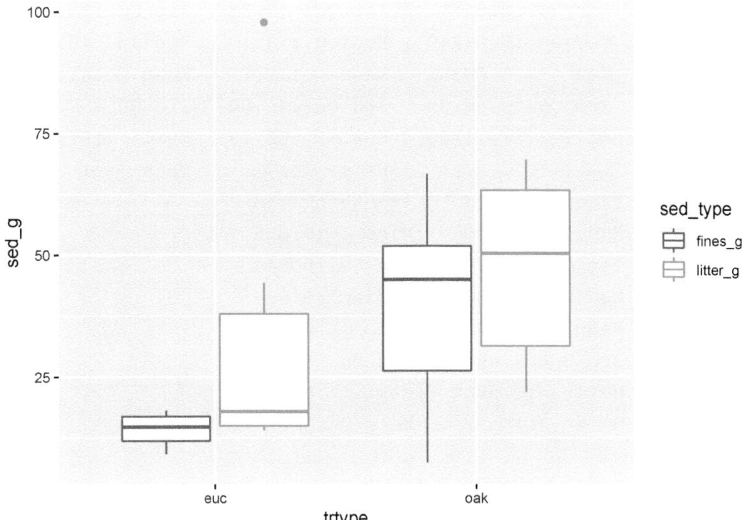

FIGURE 10.13 Eucalyptus and oak sediment runoff box plots

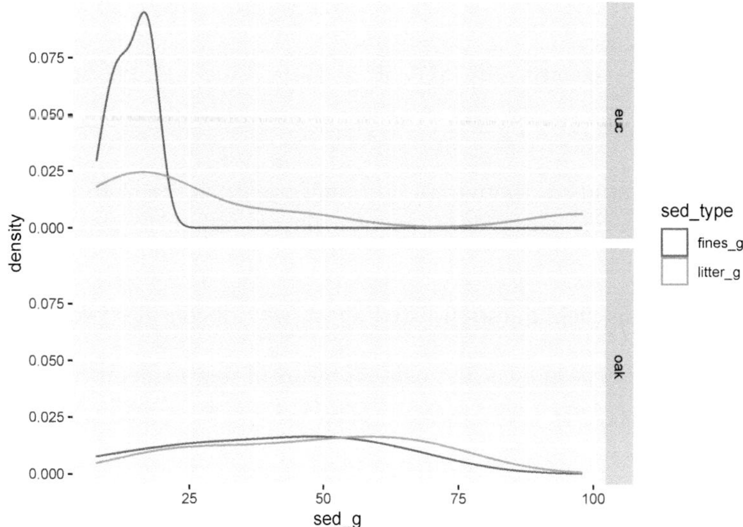

FIGURE 10.14 Facet density plot of eucalyptus and oak sediment runoff

```
##
##   bulkDensity        litter          Jan08rain      mean_runoff_ratio
##  Min.   :0.960   Min.   : 25.00   Min.   :228.1   Min.   :0.00450
##  1st Qu.:1.060   1st Qu.: 51.25   1st Qu.:290.8   1st Qu.:0.02285
##  Median :1.125   Median : 77.00   Median :301.1   Median :0.04835
##  Mean   :1.156   Mean   : 76.64   Mean   :298.5   Mean   :0.06679
##  3rd Qu.:1.245   3rd Qu.: 95.75   3rd Qu.:317.0   3rd Qu.:0.12172
##  Max.   :1.490   Max.   :135.00   Max.   :328.5   Max.   :0.16480
##
##  med_runoff_ratio  std_runoff_ratio     fines_g          litter_g
##  Min.   :0.00000   Min.   :0.01070   Min.   : 7.50   Min.   :14.00
##  1st Qu.:0.01105   1st Qu.:0.01642   1st Qu.:13.30   1st Qu.:18.80
##  Median :0.04950   Median :0.02740   Median :18.10   Median :40.40
##  Mean   :0.06179   Mean   :0.04355   Mean   :27.77   Mean   :41.32
##  3rd Qu.:0.09492   3rd Qu.:0.05735   3rd Qu.:45.00   3rd Qu.:57.50
##  Max.   :0.16430   Max.   :0.11480   Max.   :66.70   Max.   :97.80
##                                      NA's   :1       NA's   :1
##     total_g        fineTotalRatio    fineRainRatio
##  Min.   : 23.50   Min.   :0.1300   Min.   :0.025
##  1st Qu.: 35.00   1st Qu.:0.2700   1st Qu.:0.044
##  Median : 60.50   Median :0.3900   Median :0.064
##  Mean   : 69.12   Mean   :0.3715   Mean   :0.097
##  3rd Qu.: 95.50   3rd Qu.:0.4900   3rd Qu.:0.141
##  Max.   :125.90   Max.   :0.5500   Max.   :0.293
##  NA's   :1        NA's   :1        NA's   :1
```

```
eucoaksed %>%
  group_by(trtype) %>%
```

```
  summarize(meanfines = mean(fines_g, na.rm=T), sdfines = sd(fines_g, na.rm=T),
            meantotal = mean(total_g, na.rm=T), sdtotal = sd(total_g, na.rm=T))
```

```
## # A tibble: 2 x 5
##   trtype meanfines sdfines meantotal sdtotal
##   <chr>      <dbl>   <dbl>     <dbl>   <dbl>
## 1 euc         14.2    3.50      48.6    35.0
## 2 oak         39.4   20.4       86.7    26.2
```

```
eucoakLong <- eucoaksed %>%
  pivot_longer(col=c(fines_g,litter_g),
               names_to = "sed_type",
               values_to = "sed_g")
eucoakLong %>%
  ggplot(aes(trtype, sed_g, col=sed_type)) +
  geom_boxplot()
```

```
eucoakLong %>%
  ggplot(aes(sed_g, col=sed_type)) +
  geom_density() +
  facet_grid(trtype ~ .)
```

Tests of euc vs oak based on fine sediments:

```
shapiro.test(eucoaksed$fines_g[eucoaksed$trtype == "euc"])
shapiro.test(eucoaksed$fines_g[eucoaksed$trtype == "oak"])
t.test(fines_g~trtype, data=eucoaksed)
```

```
##
##  Shapiro-Wilk normality test
##
## data:  eucoaksed$fines_g[eucoaksed$trtype == "euc"]
## W = 0.9374, p-value = 0.6383
```

```
##
##  Shapiro-Wilk normality test
##
## data:  eucoaksed$fines_g[eucoaksed$trtype == "oak"]
## W = 0.96659, p-value = 0.8729
```

```
##
##  Welch Two Sample t-test
##
## data:  fines_g by trtype
## t = -3.2102, df = 6.4104, p-value = 0.01675
## alternative hypothesis: true difference in means between group euc and group oak is not equal to 0
## 95 percent confidence interval:
##  -44.059797  -6.278299
## sample estimates:
## mean in group euc mean in group oak
##          14.21667          39.38571
```

Tests of euc vs oak based on total sediments:

```
shapiro.test(eucoaksed$total_g[eucoaksed$trtype == "euc"])
shapiro.test(eucoaksed$total_g[eucoaksed$trtype == "oak"])
kruskal.test(total_g~trtype, data=eucoaksed)
```

```
##
##  Shapiro-Wilk normality test
##
## data:  eucoaksed$total_g[eucoaksed$trtype == "euc"]
## W = 0.76405, p-value = 0.02725

##
##  Shapiro-Wilk normality test
##
## data:  eucoaksed$total_g[eucoaksed$trtype == "oak"]
## W = 0.94988, p-value = 0.7286

##
##  Kruskal-Wallis rank sum test
##
## data:  total_g by trtype
## Kruskal-Wallis chi-squared = 3.449, df = 1, p-value = 0.06329
```

So we used a t test for the `fines_g`, and the test suggests that there's a significant difference in sediment yield for fines, but the Kruskal-Wallis test on total sediment (including litter) did not show a significant difference. Both results support the conclusion that oaks in this study produced more soil erosion, largely because the Eucalyptus stands generate so much litter cover, and that litter also made the total sediment yield not significantly different. See Thompson, Davis, and Oliphant (2016) for more information on this study and its conclusions.

10.4.2 Analysis of variance

The purpose of analysis of variance (ANOVA) is to compare groups based upon continuous variables. It can be thought of as an extension of a t test where you have more than two groups, or as a linear model where one variable is a factor. In a confirmatory statistical test, you'll want to see if you can reject the null hypothesis that there's no difference between the within-sample variances and the between-sample variances.

- The response variable is a continuous variable
- The explanatory variable is the grouping – categorical (a factor in R)

From a study of a karst system in Tennessee (J. D. Davis and Brook 1993), we might ask the question:

Are water samples from streams draining sandstone, limestone, and shale (Figure 10.15) different based on solutes measured as total hardness?

We can look at this spatially (Figure 10.16) as well as by variables graphically (Figure 10.17).

FIGURE 10.15 Water sampling in varying lithologies in a karst area

```
wChemData <- read_excel(ex("SinkingCove/SinkingCoveWaterChem.xlsx")) %>%
  mutate(siteLoc = str_sub(Site,start=1L, end=1L))
wChemTrunk <- wChemData %>% filter(siteLoc == "T") %>%
  mutate(siteType = "trunk")
wChemDrip <- wChemData %>% filter(siteLoc %in% c("D","S")) %>%
  mutate(siteType = "dripwater")
wChemTrib <- wChemData %>% filter(siteLoc %in% c("B", "F", "K", "W", "P")) %>%
  mutate(siteType = "tributary")
wChemData <- bind_rows(wChemTrunk, wChemDrip, wChemTrib)
sites <- read_csv(ex("SinkingCove/SinkingCoveSites.csv"))
wChem <- wChemData %>%
```

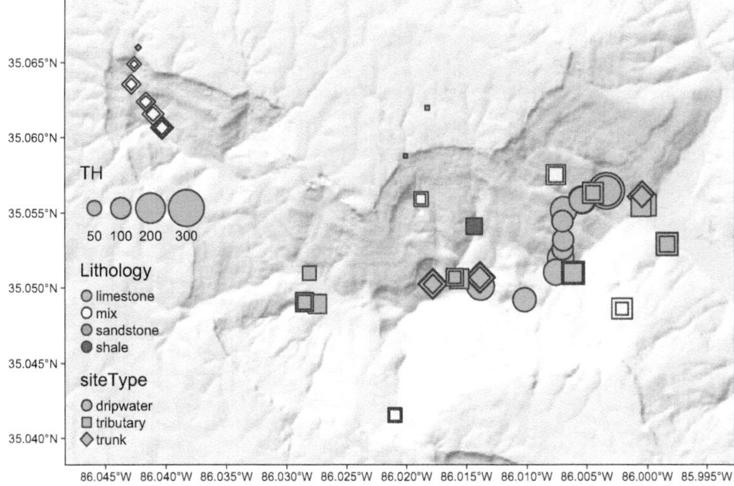

FIGURE 10.16 Total hardness from dissolved carbonates at water sampling sites in Upper Sinking Cove, TN

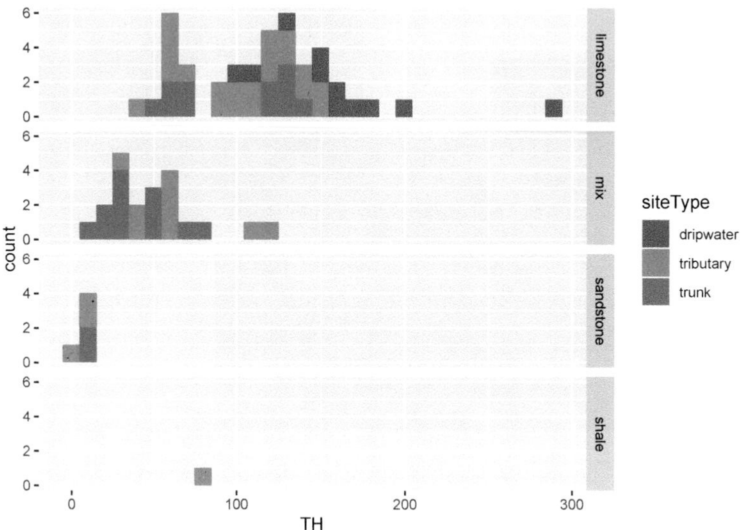

FIGURE 10.17 Sinking Cove dissolved carbonates as total hardness by lithology

```
  left_join(sites, by = c("Site" = "site")) %>%
  st_as_sf(coords = c("longitude", "latitude"), crs = 4326)
library(terra)
tmap_mode("plot")
DEM <- rast(ex("SinkingCove/DEM_SinkingCoveUTM.tif"))
slope <- terrain(DEM, v='slope')
aspect <- terrain(DEM, v='aspect')
hillsh <- shade(slope/180*pi, aspect/180*pi, angle=40, direction=330)
bounds <- st_bbox(wChem)
xrange <- bounds$xmax - bounds$xmin
yrange <- bounds$ymax - bounds$ymin
xMIN <- as.numeric(bounds$xmin - xrange/10)
xMAX <- as.numeric(bounds$xmax + xrange/10)
yMIN <- as.numeric(bounds$ymin - yrange/10)
yMAX <- as.numeric(bounds$ymax + yrange/10)
newbounds <- st_bbox(c(xmin=xMIN,xmax=xMAX,ymin=yMIN,ymax=yMAX),crs= st_crs(4326))
tm_shape(hillsh,bbox=newbounds) +
  tm_raster(palette="-Greys",legend.show=F,n=20, alpha=0.5) + tm_shape(wChem) +
  tm_symbols(size="TH", col="Lithology", scale=2, shape="siteType") +
  tm_legend() +
  tm_layout(legend.position = c("left", "bottom")) +
  tm_graticules(lines=F)
```

```
summary(aov(TH~siteType, data = wChemData))
```

```
##             Df Sum Sq Mean Sq F value   Pr(>F)
## siteType     2  79172   39586   20.59 1.07e-07 ***
## Residuals   67 128815    1923
```

```
## ---
## Signif. codes:  0 '***' 0.001 '**' 0.01 '*' 0.05 '.' 0.1 ' ' 1
```

```
summary(aov(TH~Lithology, data = wChemData))
```

```
##             Df Sum Sq Mean Sq F value   Pr(>F)
## Lithology    3  98107   32702   19.64 3.28e-09 ***
## Residuals   66 109881    1665
## ---
## Signif. codes:  0 '***' 0.001 '**' 0.01 '*' 0.05 '.' 0.1 ' ' 1
```

```
wChemData %>%
  ggplot(aes(x=TH, fill=siteType)) +
  geom_histogram() +
  facet_grid(Lithology ~ .)
```

Some observations and caveats from the above:

- There's pretty clearly a difference between surface waters (trunk and tributary) and cave dripwaters (from stalactites) in terms of solutes. Analysis of variance simply confirms the obvious.
- There's also pretty clearly a difference among lithologies on the basis of solutes, not surprising since limestone is much more soluble than sandstones. Similarly, analysis of variance confirms the obvious.
- The data may not be sufficiently normally distributed, and limestone hardness values are bimodal (largely due to the inaccessibility of waters in the trunk cave passages traveling 2 km through the Bangor limestone [1]), though analysis of variance may be less sensitive to this than a t test.
- While shale creates springs, shale strata are very thin, with most of them in the "mix" category, or form the boundary between the two major limestone formations. Tributary streams appear to cross the shale in caves that were inaccessible for sampling. We visually confirmed this in one cave, but this exploration required some challenging rappel work to access, so we were not able to sample.
- The geologic structure here is essentially flat, with sandstones on the plateau surface and the most massive limestones – the Bangor and Monteagle limestones – all below 400 m elevation (Figure 10.18).

- While the rapid increase in solutes happens when Cave Cove Creek starts draining through the much more soluble limestone until it reaches saturation, the distance traveled by the water (reflected by a drop in elevation) can be seen (Figure 10.19).

```
wChemData %>%
  ggplot(aes(x=Elevation, y=TH, col=Lithology)) +
  geom_point() +
  geom_smooth(method= "lm")
```

[1]We tried very hard to get into that cave that must extend from upper Cave Cove then under Farmer Cove to a spring in Wolf Cove – we have dye traces to prove it.

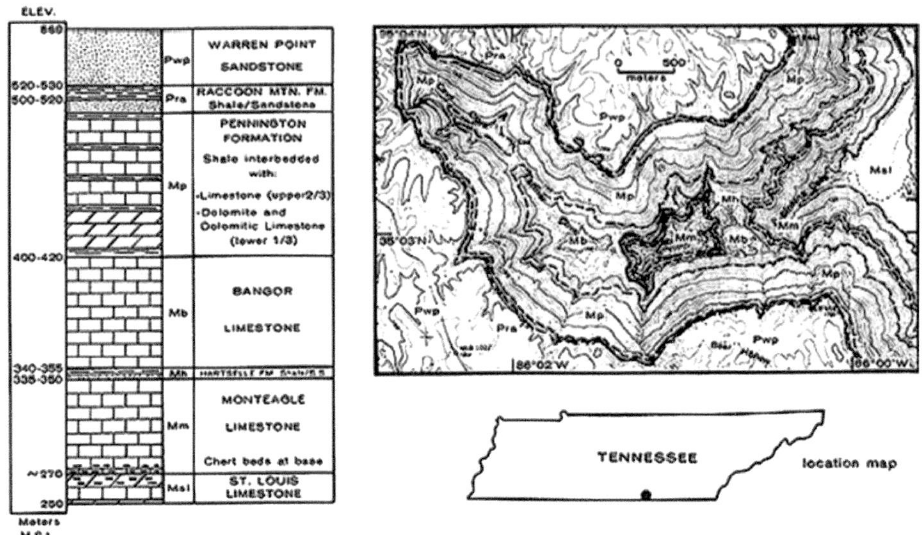

FIGURE 10.18 Upper Sinking Cove (Tennessee) stratigraphy

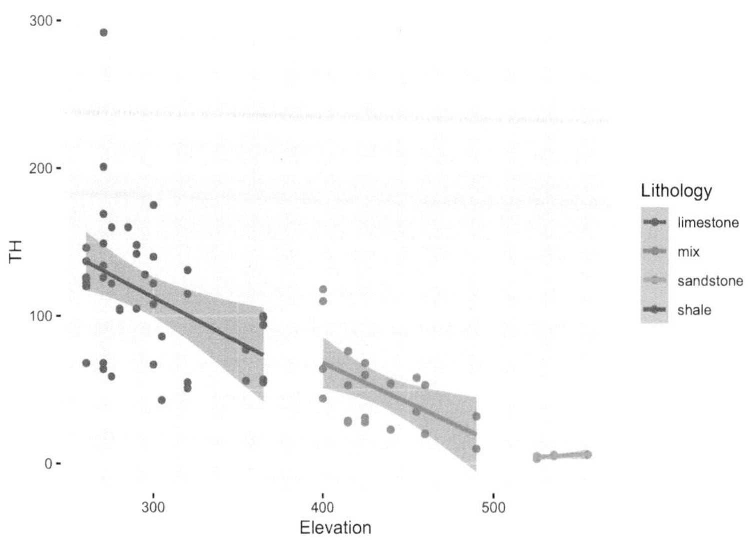

FIGURE 10.19 Sinking Cove dissolved carbonates as TH and elevation by lithology

10.4.3 Testing a correlation

We earlier looked at the correlation coefficient r. One test we can do is to see whether that correlation is significant:

```
cor.test(sierraFeb$TEMPERATURE, sierraFeb$ELEVATION)
```

```
##
##  Pearson's product-moment correlation
##
## data:  sierraFeb$TEMPERATURE and sierraFeb$ELEVATION
## t = -20.558, df = 60, p-value < 2.2e-16
## alternative hypothesis: true correlation is not equal to 0
## 95 percent confidence interval:
##  -0.9609478 -0.8952592
## sample estimates:
##        cor
## -0.9357801
```

So we can reject the null hypothesis that the correlation is not equal to zero: the probability of getting a correlation of -0.936 is less than 2.2×10^{-16} and thus the "true correlation is not equal to 0". So we can accept an alternative *working hypothesis* that they're negatively correlated, and that – no surprise – *it gets colder as we go to higher elevations*, at least in February in the Sierra, where our data come from.

In the next chapter, we'll use these data to develop a linear model and get a similar result comparing the slope of the model predicting temperature from elevation...

10.5 Exercises: Statistics

Exercise 10.1. Build a `soilvegJuly` data frame.

- Create a new RStudio project named Meadows.
- Create a soilveg tibble from "meadows/SoilVegSamples.csv" in the extdata.
- Have a look at this data frame and note that there is NA for SoilMoisture and NDVI for 8 of the records. These represent observations made in August, while the rest are all in July. While we have drone imagery and thus NDVI for these sites, as they are during the senescent period we don't want to compare these with the July samples.
- Filter the data frame to only include records that are not NA (`!is.na`) for SoilMoisture, and assign that to a new data frame `soilvegJuly`.

Exercise 10.2. Visualizations:

- Create a scatter plot of Soil Moisture vs NDVI, colored by veg (the three-character abbreviation of major vegetation types – `CAR` for Carex/sedge, `JUN` for Juncus/rush, `GRA` for mesic grasses and forbs, and `UPL` for more elevated (maybe by a meter) areas of sagebrush),

for `soilvegJuly`. What we can see is that this is a small sample. This study involved a lot of drone imagery, probably 100s of GB of imagery, which was the main focus of the study – to detect channels – but a low density of soil and vegetation ground samples.
- Create a histogram of soil moisture colored by veg, also for the July data. We see the same story, with not very many samples, though suggestive of a multimodal distribution overall.
- Create a density plot of the same, using alpha = 0.5 to see everything with transparency.

Exercise 10.3. Tests, July data

Using either `aov()` or `anova(lm())`, run an analysis of variance test of soil moisture ~ veg. Remember to specify the data, which should be just the July data., then do the same test for NDVI ~ veg.

Exercise 10.4. Meadow differences

- Now compare the meadows based on soil moisture in July, using a boxplot
- Then run an ANOVA test on this meadow grouping

Exercise 10.5. For the meadow data, create a pairs plot to find which variables are correlated along with their r values, and test the significance of that correlation and provide percentage of variation of one variable is explained by that of the other.

Exercise 10.6. Bulk density test, all data

We'll now look at bulk density for all samples (soilveg), including both July and August. Soil moisture and NDVI won't be a part of this analysis, only bulk density and vegetation. Look at the distribution of all bulk density values, using both a histogram with 20 bins and a density plot, then run an ANOVA test of bulk density predicted by (~) veg. What does the Pr(>F) value indicate?

Exercise 10.7. Carex or not?

Derive a bulkDensityCAR data frame by mutating a new variable CAR derived as a boolean result of `veg == "CAR"`. This will group the vegetation points as either Carex or not, then use that in another ANOVA test to predict bulk density. What does the Pr(>F) value indicate?

11

Modeling

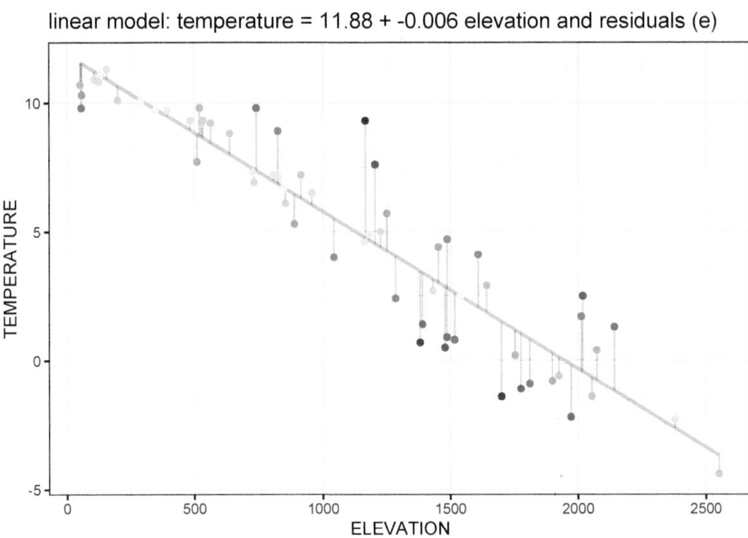

The purpose of a statistical model is to help understand what variables might best predict a phenomenon of interest, which ones have more or less influence, define a predictive equation with coefficients for each of the variables, and then apply that equation to predict values using the same input variables for other areas. This process requires samples with observations of the explanatory (or independent) and response (or dependent) variables in question.

11.1 Some Common Statistical Models

There are many types of statistical models. Variables may be nominal (categorical) or interval/ratio data. You may be interested in predicting a continuous interval/ratio variable from other continuous variables, or predicting the probability of an occurrence (e.g. of a species), or maybe the count of something (also maybe a species). You may be needing to classify your phenomena based on continuous variables. Here are some examples:

- `lm(y ~ x)` linear regression model with one explanatory variable
- `lm(y ~ x1 + x2 + x3)` multiple regression, a linear model with multiple explanatory variables

- `glm(y ~ x, family = poisson)` generalized linear model, poisson distribution; see ?family to see those supported, including binomial, gaussian, poisson, etc.
- `glm(y ~ x + y, family = binomial)` glm for logistic regression
- `aov(y ~ x)` analysis of variance (same as lm except in the summary)
- `gam(y ~ x)` generalized additive models
- `tree(y ~ x)` or `rpart(y ~ x)` regression/classification trees

11.2 Linear Model (`lm`)

If we look at the Sierra February climate example, the regression line in the graph above shows the response variable temperature predicted by the explanatory variable elevation. Here's the code that produced the model and graph:

```
library(igisci); library(tidyverse)
sierra <- sierraFeb %>%
  filter(!is.na(TEMPERATURE))
model1 = lm(TEMPERATURE ~ ELEVATION, data = sierra)
cc = model1$coefficients
sierra$resid = resid(model1)
sierra$predict = predict(model1)
eqn = paste("temperature =",
            (round(cc[1], digits=2)), "+",
            (round(cc[2], digits=3)), "elevation + e")
ggplot(sierra, aes(x=ELEVATION, y=TEMPERATURE)) +
  geom_smooth(method="lm", se=FALSE, color="lightgrey") +
  geom_segment(aes(xend=ELEVATION, yend=predict), alpha=.2) +
  geom_point(aes(color=resid)) +
  scale_color_gradient2(low="blue", mid="ivory2", high="red") +
  guides(color="none") +
  theme_bw() +
  ggtitle(paste("Residuals (e) from model: ",eqn))
```

The `summary()` function can be used to look at the results statistically:

```
summary(model1)
```

```
##
## Call:
## lm(formula = TEMPERATURE ~ ELEVATION, data = sierra)
##
## Residuals:
##     Min      1Q  Median      3Q     Max
## -2.9126 -1.0466 -0.0027  0.7940  4.5327
##
## Coefficients:
```

```
##                Estimate Std. Error t value Pr(>|t|)
## (Intercept) 11.8813804  0.3825302   31.06   <2e-16 ***
## ELEVATION   -0.0061018  0.0002968  -20.56   <2e-16 ***
## ---
## Signif. codes:  0 '***' 0.001 '**' 0.01 '*' 0.05 '.' 0.1 ' ' 1
##
## Residual standard error: 1.533 on 60 degrees of freedom
## Multiple R-squared:  0.8757, Adjusted R-squared:  0.8736
## F-statistic: 422.6 on 1 and 60 DF,  p-value: < 2.2e-16
```

Probably the most important statistic is the p value for the explanatory variable `ELEVATION`, which in this case is very small: $< 2.2 \times 10^{-16}$.

The graph shows not only the linear prediction line, but also the scatter plot of input points displayed with connecting offsets indicating the magnitude of the residual (error term) of the data point away from the prediction.

Making Predictions

We can use the linear equation and the coefficients derived for the intercept and the explanatory variable (and rounded a bit) to predict temperatures given the explanatory variable elevation, so:

$$Temperature_{prediction} = 11.88 - 0.006 elevation$$

So with these coefficients, we can apply the equation to a vector of elevations to predict temperature from those elevations:

```
a <- model1$coefficients[1]
b <- model1$coefficients[2]
elevations <- c(500, 1000, 1500, 2000)
elevations
```

```
## [1]  500 1000 1500 2000
```

```
tempEstimate <- a + b * elevations
tempEstimate
```

```
## [1]  8.8304692  5.7795580  2.7286468 -0.3222645
```

Next, we'll plot predictions and residuals on a map.

11.3 Spatial Influences on Statistical Analysis

Spatial statistical analysis brings in the spatial dimension to a statistical analysis, ranging from visual analysis of patterns to specialized spatial statistical methods. There are many applications for these methods in environmental research, since spatial patterns are generally highly relevant. We might ask:

- What patterns can we see?
- What is the effect of scale?
- Relationships among variables – do they vary spatially?

In addition to helping to identify variables not considered in the model, the effect of local conditions and tendency of nearby observations to be related is rich for further exploration. Readers are encouraged to learn more about spatial statistics, and some methods (such as spatial regression models) are explored at https://rspatial.org/terra/analysis/ (Hijmans (n.d.)).

But we'll use a fairly straightforward method that can help enlighten us about the effect of other variables and the influence of proximity – mapping residuals. Mapping residuals is often enlightening since they can illustrate the influence of other variables. There's even a recent blog on this: https://californiawaterblog.com/2021/10/17/sometimes-studying-the-variation-is-the-interesting-thing/?fbclid=IwAR1tfnj2SM_du6-jEd9h-AApRPO7w88TiPUbvTFz5dkagpQ1RMV04ZZYAr8.

11.3.1 Mapping residuals

If we start with a map of February temperatures in the Sierra and try to model these as predicted by elevation, we might see the following result. (First we'll build the model again, though this is probably still in memory.)

```
library(tidyverse)
library(igisci)
library(sf)
sierra <- st_as_sf(filter(sierraFeb, !is.na(TEMPERATURE)),
                   coords=c("LONGITUDE", "LATITUDE"), crs=4326)
modelElev <- lm(TEMPERATURE ~ ELEVATION, data = sierra)
summary(modelElev)
```

```
##
## Call:
## lm(formula = TEMPERATURE ~ ELEVATION, data = sierra)
##
## Residuals:
##     Min      1Q  Median      3Q     Max
## -2.9126 -1.0466 -0.0027  0.7940  4.5327
```

```
##
## Coefficients:
##               Estimate Std. Error t value Pr(>|t|)
## (Intercept) 11.8813804  0.3825302   31.06   <2e-16 ***
## ELEVATION   -0.0061018  0.0002968  -20.56   <2e-16 ***
## ---
## Signif. codes:  0 '***' 0.001 '**' 0.01 '*' 0.05 '.' 0.1 ' ' 1
##
## Residual standard error: 1.533 on 60 degrees of freedom
## Multiple R-squared:  0.8757, Adjusted R-squared:  0.8736
## F-statistic: 422.6 on 1 and 60 DF,  p-value: < 2.2e-16
```

We can see from the model summary that `ELEVATION` is significant as a predictor for `TEMPERATURE`, with a very low P value, and the R-squared value tells us that 87% of the variation in temperature might be "explained" by elevation. We'll store the residual (`resid`) and prediction (`predict`), and set up our mapping environment to create maps of the original data, prediction, and residuals.

```
cc <- modelElev$coefficients
sierra$resid <- resid(modelElev)
sierra$predict <- predict(modelElev)
eqn = paste("temperature =",
            (round(cc[1], digits=2)), "+",
            (round(cc[2], digits=3)), "elevation")
ct <- st_read(system.file("extdata","sierra/CA_places.shp",package="igisci"))
ct$AREANAME_pad <- paste0(str_replace_all(ct$AREANAME,'[A-Za-z]',' '),ct$AREANAME)
bounds <- st_bbox(sierra)
```

The original data and the prediction map

Figure 11.1 shows the original February temperature data, then Figure 11.2 displays the prediction of temperature by elevation at the same weather station locations using the linear model.

```
library(tmap); library(terra); library(igisci)
hillsh <- rast(ex("CA/ca_hillsh_WGS84.tif"))
tm_shape(hillsh, bbox=bounds) +
  tm_raster(palette="-Greys", style="cont", legend.show=F) +
  tm_shape(st_make_valid(CA_counties)) + tm_borders() +
  tm_shape(ct) + tm_dots() + tm_text("AREANAME", size=0.5, just=c("left","top")) +
  tm_shape(sierra) + tm_symbols(col="TEMPERATURE", palette="-RdBu", size=0.5) +
  tm_layout(legend.position=c("right","top")) +
  tm_graticules(lines=F) +
  tm_layout(main.title="February Normals",
            main.title.position = c("right"), main.title.size = 0.8)
```

```
library(tmap)
tm_shape(hillsh, bbox=bounds) +
  tm_raster(palette="-Greys", style="cont", legend.show=F) +
```

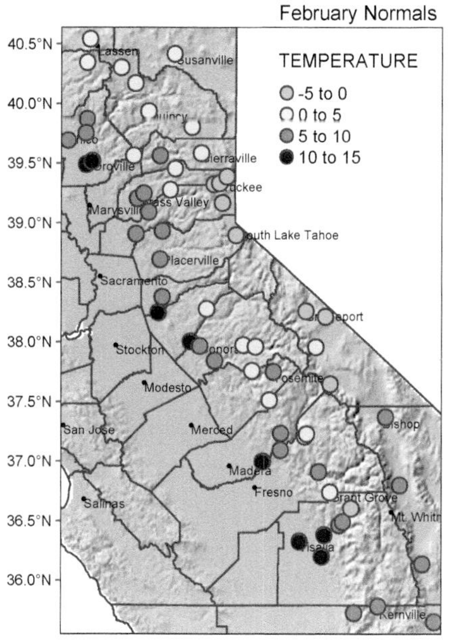

FIGURE 11.1 Original February temperature data

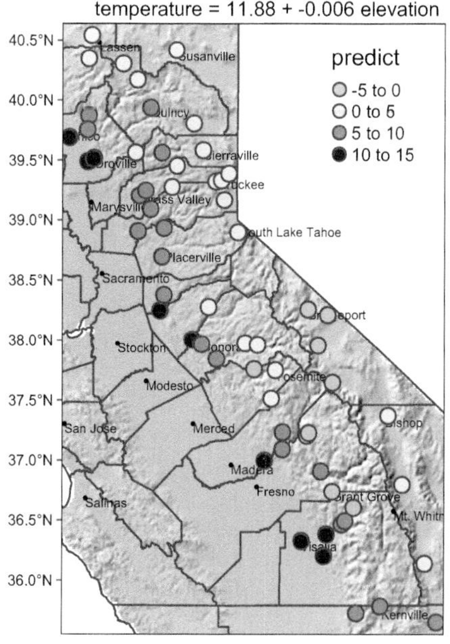

FIGURE 11.2 Temperature predicted by elevation model

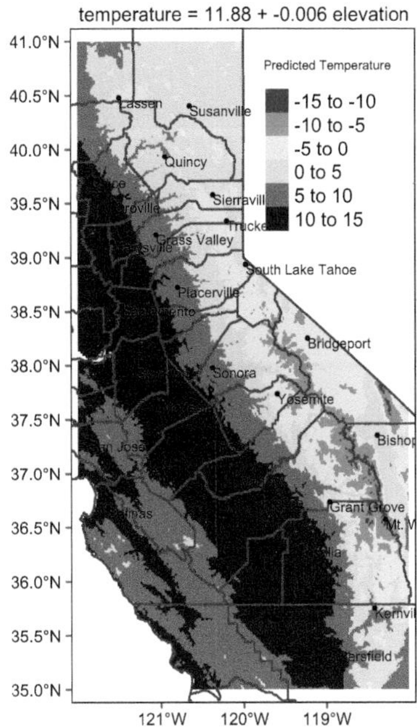

FIGURE 11.3 Temperature predicted by elevation raster

```
tm_shape(st_make_valid(CA_counties)) + tm_borders() +
tm_shape(ct) + tm_dots() + tm_text("AREANAME",size=0.5,just=c("left","top")) +
tm_shape(sierra) + tm_symbols(col="predict", palette="-RdBu", size=0.5) +
tm_layout(legend.position=c("right","top")) +
tm_graticules(lines=F) +
tm_layout(main.title=eqn,
          main.title.position = c("right"), main.title.size = 0.8)
```

We can also use the model to predict temperatures from a raster of elevations, which should match the elevations at the weather stations, allowing us to predict temperature for the rest of the study area. We should have more confidence in the area (the Sierra Nevada and nearby areas) actually sampled by the stations, so we'll crop with the extent covering that area (though this will include more coastal areas that are likely to differ in February temperatures from patterns observed in the area sampled).

We'll use GCS elevation raster data and map algebra to derive a prediction raster (Figure 11.3).

```
library(igisci)
library(tmap); library(terra)
elev <- rast(ex("CA/ca_elev_WGS84.tif"))
elevSierra <- crop(elev, ext(-122, -118, 35, 41))  # GCS
b0 <- coef(modelElev)[1]
b1 <- coef(modelElev)[2]
tempSierra <- b0 + b1 * elevSierra
```

```
names(tempSierra) = "Predicted Temperature"
#tempCA <- b0 + b1 * elev
tm_shape(tempSierra) + tm_raster(palette="-RdBu") + tm_graticules(lines=F) +
  tm_legend(position = c("right", "top")) +
  tm_shape(st_make_valid(CA_counties)) + tm_borders() +
  tm_shape(ct) + tm_dots() + tm_text("AREANAME", size=0.5, just=c("left","top")) +
  tm_layout(main.title=eqn,
            main.title.position = c("right"), main.title.size = 0.7)
```

The residuals

To consider patterns where the simple elevation model *doesn't* work well, we can map the model residuals at the stations (Figure 11.4).

```
library(tmap)
tm_shape(hillsh, bbox=bounds) +
  tm_raster(palette="-Greys", style="cont", legend.show=F) +
  tm_shape(st_make_valid(CA_counties)) + tm_borders() +
  tm_shape(ct) + tm_dots() + tm_text("AREANAME", size=0.5, just=c("left","top")) +
  tm_shape(sierra) + tm_symbols(col="resid", palette="-RdBu", size=0.5) +
  tm_layout(legend.position=c("right","top"),
            main.title = paste("e from",eqn,"+ e"),
            main.title.size = 0.7,
            main.title.position = "right") +
  tm_graticules(lines=F)
```

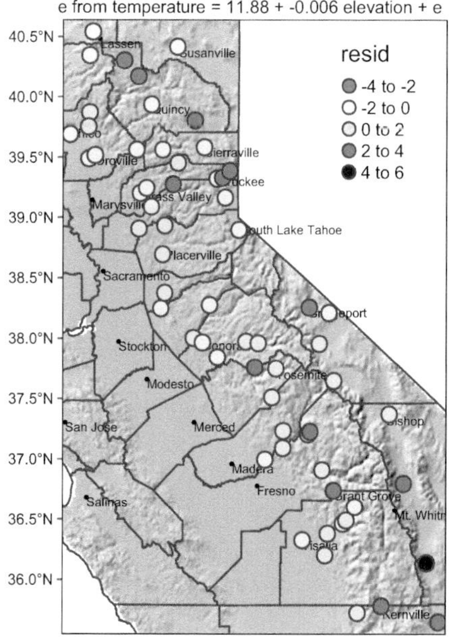

FIGURE 11.4 Residuals of temperature from model predictions by elevation

Statistically, if the *residuals* from regression are spatially autocorrelated, we should look for patterns in the residuals to find other explanatory variables. However, we can visually see, based upon the pattern of the residuals, that there's at least one important explanatory variable not included in the model: latitude. We'll leave that for you to look at in the exercises.

11.4 Analysis of Covariance

Analysis of covariance, or ANCOVA, has the same purpose as ANOVA that we looked at under tests earlier, but also takes into account the influence of other variables called covariates. In this way, analysis of covariance combines a linear model with an analysis of variance.

"Are water samples from streams draining sandstone, limestone, and shale different based on carbonate solutes, while taking into account elevation?"

The response variable is modeled from the factor (ANOVA) plus the covariate (regression):

- ANOVA: `TH ~ rocktype` where `TH` is total hardness, a measure of carbonate solutes
- Regression: `TH ~ elevation`
- ANCOVA: `TH ~ rocktype + elevation`
  - Yet shouldn't involve interaction between `rocktype` (factor) and `elevation` (covariate):

Example: stream types distinguished by discharge and slope

Three common river types are meandering (Figure 11.5), braided (Figure 11.6), and anastomosed (Figure 11.7).

For each, their slope varies by bankfull discharge in a relationship that looks something like Figure 11.8, which employs a log10 scale on both axes:

```
library(tidyverse)
csvPath <- system.file("extdata","streams.csv", package="igisci")
streams <- read_csv(csvPath)
streams$strtype <- factor(streams$type,
                          labels=c("Anastomosing","Braided","Meandering"))
library(scales) # needed for the trans_format function below
ggplot(streams, aes(Q, S, color=strtype)) +
  geom_point() + geom_smooth(method="lm", se = FALSE) +
  scale_x_continuous(trans=log10_trans(),
                     labels = trans_format("log10", math_format(10^.x))) +
```

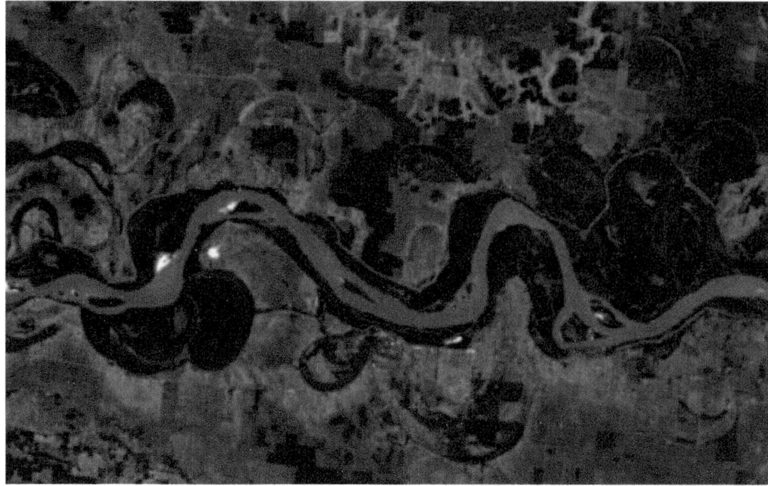

FIGURE 11.5 Meandering river

```
scale_y_continuous(trans=log10_trans(),
                   labels = trans_format("log10", math_format(10^.x)))
```

One requirement for the ANCOVA model is that there's no relationship between discharge (covariate) and channel type (factor). Another way of thinking about this is that the slope of the relationship between the covariate and response variable needs to be about the same for each group; only the intercept differs. The groups have reasonably parallel slopes in the graph, at least for log-transformed data, so that's looking good. Here are two model definitions in R and what they mean for the analysis:

`log10(S) ~ strtype * log10(Q)` ... interaction between covariate and factor

`log10(S) ~ strtype + log10(Q)` ... no interaction, parallel slopes

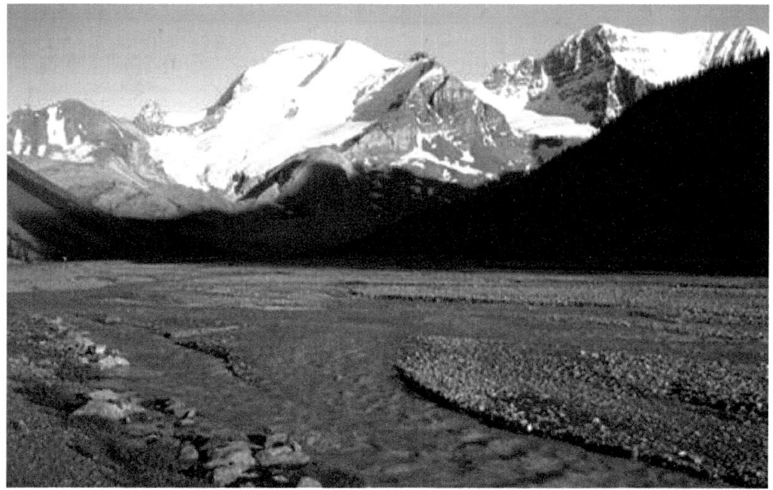

FIGURE 11.6 Braided river

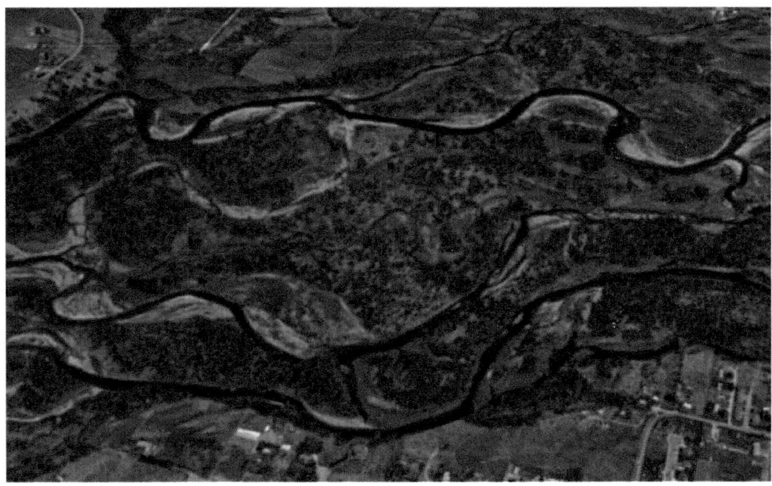

FIGURE 11.7 Anastomosed river

In the ANCOVA process, if these models are not significantly different, we can remove the interaction term due to parsimony, and thus satisfy this ANCOVA requirement. So we can go through the process:

```
ancova = lm(log10(S)~strtype*log10(Q), data=streams)
summary(ancova)
```

```
##
## Call:
## lm(formula = log10(S) ~ strtype * log10(Q), data = streams)
##
```

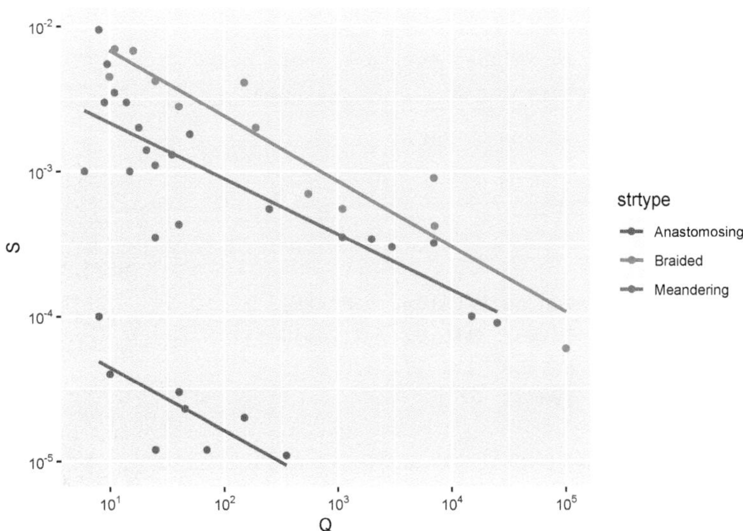

FIGURE 11.8 Q vs S with stream type

```
## Residuals:
##      Min       1Q   Median       3Q      Max
## -0.63636 -0.13903 -0.00032  0.12652  0.60750
##
## Coefficients:
##                            Estimate Std. Error t value Pr(>|t|)
## (Intercept)                -3.91819    0.31094 -12.601 1.45e-14 ***
## strtypeBraided              2.20085    0.35383   6.220 3.96e-07 ***
## strtypeMeandering           1.63479    0.33153   4.931 1.98e-05 ***
## log10(Q)                   -0.43537    0.18073  -2.409   0.0214 *
## strtypeBraided:log10(Q)    -0.01488    0.19102  -0.078   0.9384
## strtypeMeandering:log10(Q)  0.05183    0.18748   0.276   0.7838
## ---
## Signif. codes:  0 '***' 0.001 '**' 0.01 '*' 0.05 '.' 0.1 ' ' 1
##
## Residual standard error: 0.2656 on 35 degrees of freedom
## Multiple R-squared:  0.9154, Adjusted R-squared:  0.9033
## F-statistic: 75.73 on 5 and 35 DF,  p-value: < 2.2e-16
```

```
anova(ancova)
```

```
## Analysis of Variance Table
##
## Response: log10(S)
##                   Df  Sum Sq Mean Sq  F value    Pr(>F)
## strtype            2 18.3914  9.1957 130.3650 < 2.2e-16 ***
## log10(Q)           1  8.2658  8.2658 117.1821 1.023e-12 ***
## strtype:log10(Q)   2  0.0511  0.0255   0.3619    0.6989
## Residuals         35  2.4688  0.0705
## ---
## Signif. codes:  0 '***' 0.001 '**' 0.01 '*' 0.05 '.' 0.1 ' ' 1
```

```
# Now an additive model, which does not have that interaction
ancova2 = lm(log10(S)~strtype+log10(Q), data=streams)
anova(ancova2)
```

```
## Analysis of Variance Table
##
## Response: log10(S)
##           Df  Sum Sq Mean Sq F value    Pr(>F)
## strtype    2 18.3914  9.1957  135.02 < 2.2e-16 ***
## log10(Q)   1  8.2658  8.2658  121.37  3.07e-13 ***
## Residuals 37  2.5199  0.0681
## ---
## Signif. codes:  0 '***' 0.001 '**' 0.01 '*' 0.05 '.' 0.1 ' ' 1
```

```
anova(ancova,ancova2) # not significantly different, so simplification is justified
```

```
## Analysis of Variance Table
##
## Model 1: log10(S) ~ strtype * log10(Q)
## Model 2: log10(S) ~ strtype + log10(Q)
##   Res.Df    RSS Df Sum of Sq      F Pr(>F)
## 1     35 2.4688
## 2     37 2.5199 -2 -0.051051 0.3619 0.6989
```

```
# Now we remove the strtype term
ancova3 = update(ancova2, ~ . - strtype)
anova(ancova2,ancova3)  # Goes too far: removing strtype significantly different
```

```
## Analysis of Variance Table
##
## Model 1: log10(S) ~ strtype + log10(Q)
## Model 2: log10(S) ~ log10(Q)
##   Res.Df     RSS Df Sum of Sq      F    Pr(>F)
## 1     37  2.5199
## 2     39 25.5099 -2    -22.99 168.78 < 2.2e-16 ***
## ---
## Signif. codes:  0 '***' 0.001 '**' 0.01 '*' 0.05 '.' 0.1 ' ' 1
```

Alternatively, we can use R's step method for a more automated approach.

```
ancova = lm(log10(S)~strtype*log10(Q), data=streams)
step(ancova)
```

```
## Start:  AIC=-103.2
## log10(S) ~ strtype * log10(Q)
##
##                     Df Sum of Sq    RSS     AIC
## - strtype:log10(Q)  2  0.051051 2.5199 -106.36
## <none>                          2.4688 -103.20
##
## Step:  AIC=-106.36
## log10(S) ~ strtype + log10(Q)
##
##            Df Sum of Sq     RSS      AIC
## <none>                   2.5199 -106.364
## - log10(Q)  1    8.2658 10.7857  -48.750
## - strtype   2   22.9901 25.5099  -15.455

##
## Call:
## lm(formula = log10(S) ~ strtype + log10(Q), data = streams)
##
## Coefficients:
##       (Intercept)     strtypeBraided  strtypeMeandering           log10(Q)
##           -3.9583             2.1453             1.7294            -0.4109
```

ANOVA and ANCOVA are applications of a linear model. They use the `lm` model and the response variable is continuous and assumed normally distributed. The linear model for the stream types employing slope (S) and the covariate discharge ($\log_{10}$Q) was produced using:

```
mymodel <- lm(log10(S) ~ strtype + log10(Q))
```

Then as with standard ANOVA, we displayed the results using:

```
anova(mymodel)
```

11.5 Generalized linear model (GLM)

The glm in R allows you to work with various types of data using various distributions, described as families such as:

- `gaussian` : normal distribution – what is used with lm
- `binomial` : logit – used with probabilities
  - used for *logistic regression*
- `poisson` : for counts – commonly used for species counts

See `help(glm)` for other examples

11.5.1 Binomial family: logistic GLM with streams

When the glm family is binomial, we might do what's sometimes called *logistic regression* where we look at the probability of a categorical value (might be a species occurrence, for instance) given various independent variables.

A simple example of predicting a dog panting or not (a binomial, since the dog either panted or didn't) related to temperature can be found at: https://bscheng.com/2016/10/23/logistic-regression-in-r/

Let's try this with the streams data we just used for analysis of covariance, to predict the probability of a particular stream type from slope and discharge:

```
library(igisci)
library(tidyverse)
streams <- read_csv(ex("streams.csv"))
streams$strtype <- factor(streams$type,
                          labels=c("Anastomosing","Braided","Meandering"))
```

For logistic regression, we need probability data, so locations where we know the probability of something being true. If you think about it, that's pretty difficult to know for an observation, but we might know whether something is true or not, like we observed the phenomenon or not, and so we just assign a probability of 1 if something is observed, and 0 if it's not. Since in the case of the stream types, we really have three different probabilities: whether it's true that the stream is anastomosing, whether it's braided, and whether it's meandering, so we'll create three *dummy variables* for that with values of 1 or 0: if it's braided, the value for braided is 1, otherwise it's 0; the same applies to all three dummy variables:

```
streams <- streams %>%
  mutate(braided = ifelse(type == "B", 1, 0),
         meandering = ifelse(type == "M", 1, 0),
         anastomosed = ifelse(type == "A", 1, 0))
streams
```

```
## # A tibble: 41 x 7
##    type        Q       S strtype braided meandering anastomosed
##    <chr>   <dbl>   <dbl> <fct>     <dbl>      <dbl>       <dbl>
##  1 B         9.9 0.0045  Braided       1          0           0
##  2 B        11   0.007   Braided       1          0           0
##  3 B        16   0.0068  Braided       1          0           0
##  4 B        25   0.0042  Braided       1          0           0
##  5 B        40   0.0028  Braided       1          0           0
##  6 B       150   0.0041  Braided       1          0           0
##  7 B       190   0.002   Braided       1          0           0
##  8 B       550   0.0007  Braided       1          0           0
##  9 B      1100   0.00055 Braided       1          0           0
## 10 B      7100   0.00042 Braided       1          0           0
## # ... with 31 more rows
```

Then we can run the logistic regression on each one:

```
summary(glm(braided ~ log(Q) + log(S), family=binomial, data=streams))
```

```
##
## Call:
## glm(formula = braided ~ log(Q) + log(S), family = binomial, data = streams)
##
## Deviance Residuals:
##      Min        1Q    Median        3Q       Max
## -2.00386  -0.49007  -0.00961   0.15789   1.58088
```

```
##
## Coefficients:
##             Estimate Std. Error z value Pr(>|z|)
## (Intercept)  17.5853     6.7443   2.607  0.00912 **
## log(Q)        2.0940     0.8135   2.574  0.01005 *
## log(S)        4.3115     1.6173   2.666  0.00768 **
## ---
## Signif. codes:  0 '***' 0.001 '**' 0.01 '*' 0.05 '.' 0.1 ' ' 1
##
## (Dispersion parameter for binomial family taken to be 1)
##
##     Null deviance: 49.572  on 40  degrees of freedom
## Residual deviance: 23.484  on 38  degrees of freedom
## AIC: 29.484
##
## Number of Fisher Scoring iterations: 8
```

```
summary(glm(meandering ~ log(Q) + log(S), family=binomial, data=streams))
```

```
##
## Call:
## glm(formula = meandering ~ log(Q) + log(S), family = binomial,
##     data = streams)
##
## Deviance Residuals:
##     Min       1Q   Median       3Q      Max
## -1.5469  -1.2515   0.8197   1.1073   1.3072
##
## Coefficients:
##             Estimate Std. Error z value Pr(>|z|)
## (Intercept)  2.43255    1.38078   1.762   0.0781 .
## log(Q)       0.04984    0.13415   0.372   0.7103
## log(S)       0.34750    0.19471   1.785   0.0743 .
## ---
## Signif. codes:  0 '***' 0.001 '**' 0.01 '*' 0.05 '.' 0.1 ' ' 1
##
## (Dispersion parameter for binomial family taken to be 1)
##
##     Null deviance: 56.814  on 40  degrees of freedom
## Residual deviance: 53.074  on 38  degrees of freedom
## AIC: 59.074
##
## Number of Fisher Scoring iterations: 4
```

```
summary(glm(anastomosed ~ log(Q) + log(S), family=binomial, data=streams))
```

```
##
## Call:
## glm(formula = anastomosed ~ log(Q) + log(S), family = binomial,
```

```
##     data = streams)
##
## Deviance Residuals:
##        Min          1Q      Median          3Q         Max
## -2.815e-05  -2.110e-08  -2.110e-08  -2.110e-08   2.244e-05
##
## Coefficients:
##              Estimate Std. Error z value Pr(>|z|)
## (Intercept)   -176.09  786410.69       0        1
## log(Q)         -11.81   77224.35       0        1
## log(S)         -24.18   70417.40       0        1
##
## (Dispersion parameter for binomial family taken to be 1)
##
##     Null deviance: 4.0472e+01  on 40  degrees of freedom
## Residual deviance: 1.3027e-09  on 38  degrees of freedom
## AIC: 6
##
## Number of Fisher Scoring iterations: 25
```

So what we can see from this admittedly small sample is that:

- Predicting braided character from Q and S works pretty well, distinguishing these characteristics from the characteristics of all other streams (`braided==0`).
- Predicting anastomosed or meandering, however, isn't significant.

We really need a larger sample size, and North American geomorphology has recently discovered the "anabranching" stream type (of which anastomosed is a type) that has long been recognized in Australia, and its characteristics range across meandering at least, so muddying the waters...

11.5.2 Logistic landslide model

With its active tectonics, steep slopes and relatively common intense rain events during the winter season, coastal California has a propensity for landslides. Pacifica has been the focus of many landslide studies by the USGS and others, with events ranging from coastal erosion influenced by wave action to landslides (Figure 11.9) and debris flows on steep hillslopes driven by positive pore pressure (Ellen and Wieczorek (1988)).

Also see the relevant section in the San Pedro Creek Virtual Fieldtrip story map: Jerry Davis (n.d.).

Landslides detected in San Pedro Creek Watershed in Pacifica, CA, are in the SanPedro folder of extdata. Landslides were detected on aerial photography from 1941, 1955, 1975, 1983, and 1997 (Figure 11.10), and confirmed in the field during a study in the early 2000s (Sims (2004), Jerry Davis and Blesius (2015)) (Figure 11.11).

```
library(igisci); library(sf); library(tidyverse); library(terra)
slides <- st_read(ex("SanPedro/slideCentroids.shp")) %>%
```

FIGURE 11.9 Landslide in San Pedro Creek watershed

```
  mutate(IsSlide = 1)
slides$Visible <- factor(slides$Visible)
trails <- st_read(ex("SanPedro/trails.shp"))
streams <- st_read(ex("SanPedro/streams.shp"))
wshed <- st_read(ex("SanPedro/SPCWatershed.shp"))
```

```
ggplot(data=slides) + geom_sf(aes(col=Visible)) +
  scale_color_brewer(palette="Spectral") +
  geom_sf(data=streams, col="blue") +
  geom_sf(data=wshed, fill=NA)
```

We'll create a logistic regression to compare landslide probability to various environmental factors. We'll use elevation, slope, and other derivatives, distance to streams, distance to trails, and the results of a physical model predicting slope stability as a stability index (SI), using slope and soil transmissivity data.

At this point, we have data for *occurrences* of landslides, and so for each point we assign the dummy variable `IsSlide = 1`. We don't have any *absence* data, but we can create some using random points and assign them `IsSlide = 0`. It would probably be a good idea to avoid areas very close to the occurrences, but we'll just put random points everywhere and just remove the ones that don't have SI data.

For the random 10.2.3 locations (Figure 11.12), we'll set a seed so for the book the results will stay the same; and if we're good with that since it doesn't really matter what seed we use, we might as well use the answer to everything: 42.

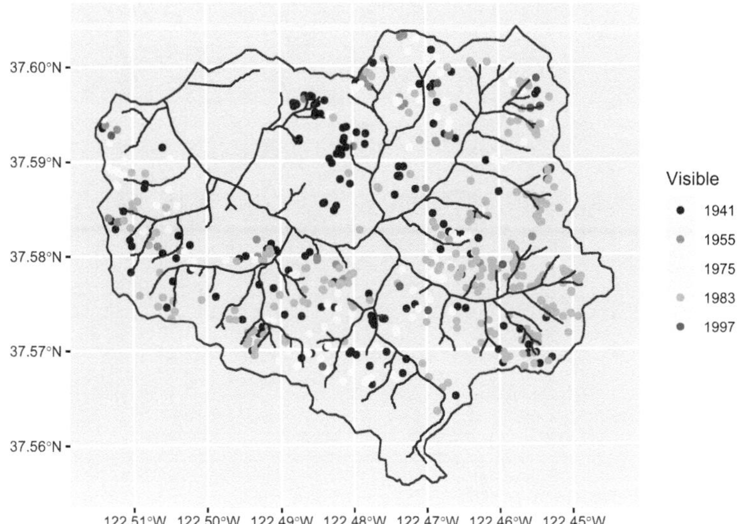

FIGURE 11.10 Landslides in San Pedro Creek watershed

FIGURE 11.11 Sediment source analysis

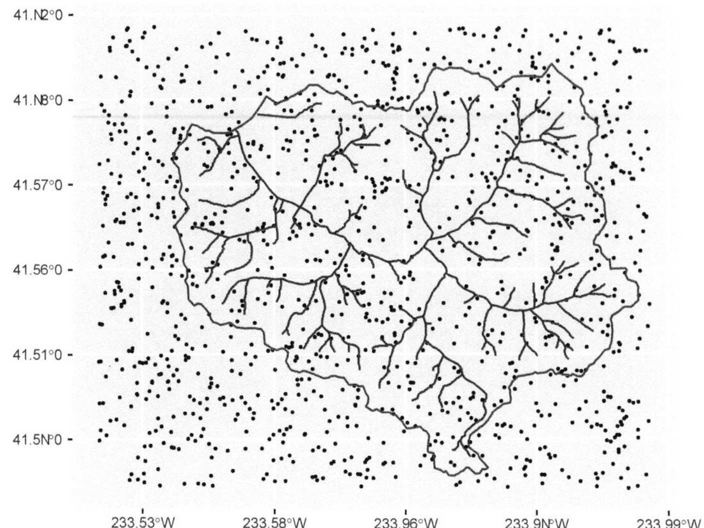

FIGURE 11.12 Raw random points

```
library(terra)
elev <- rast(ex("SanPedro/dem.tif"))
SI <- rast(ex("SanPedro/SI.tif"))
npts <- length(slides$Visible) * 2 # up it a little to get enough after clip
set.seed(42)
x <- runif(npts, min=xmin(elev), max=xmax(elev))
y <- runif(npts, min=ymin(elev), max=ymax(elev))
rsampsRaw <- st_as_sf(data.frame(x,y), coords = c("x","y"), crs=crs(elev))
ggplot(data=rsampsRaw) + geom_sf(size=0.5) +
  geom_sf(data=streams, col="blue") +
  geom_sf(data=wshed, fill=NA)
```

Now we have twice as many random points as we have landslides, within the rectangular extend of the watershed, as defined by the elevation raster. We doubled these because we need to exclude some points, yet still have enough to be roughly as many random points as slides. We'll exclude:

- all points outside the watershed
- points within 100 m of the landslide occurrences, the distance based on knowledge of the grain of landscape features (like spurs and swales) that might influence landslide occurrence

```
slideBuff <- st_buffer(slides, 100)
wshedNoSlide <- st_difference(wshed, st_union(slideBuff))
```

The 100 m buffer excludes all areas around all landslides in the data set (Figure 11.13). We're going to run the model on specific years, but the best selection of landslide absences will be to exclude areas around all landslides we know of (Figure 11.14).

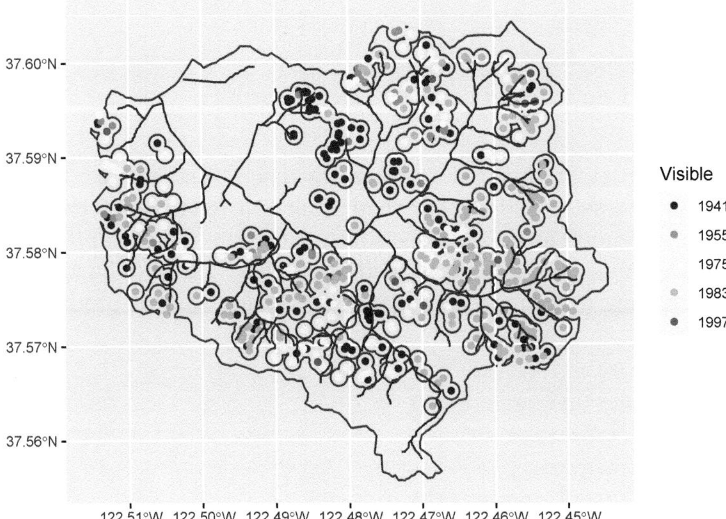

FIGURE 11.13 Landslides and buffers to exclude from random points

```
ggplot(data=slides) + geom_sf(aes(col=Visible)) +
  scale_color_brewer(palette="Spectral") +
  geom_sf(data=streams, col="blue") +
  geom_sf(data=wshedNoSlide, fill=NA)
```

```
rsamps <- st_intersection(rsampsRaw,wshedNoSlide) %>%
  mutate(IsSlide = 0)
allPts <- bind_rows(slides, rsamps)
```

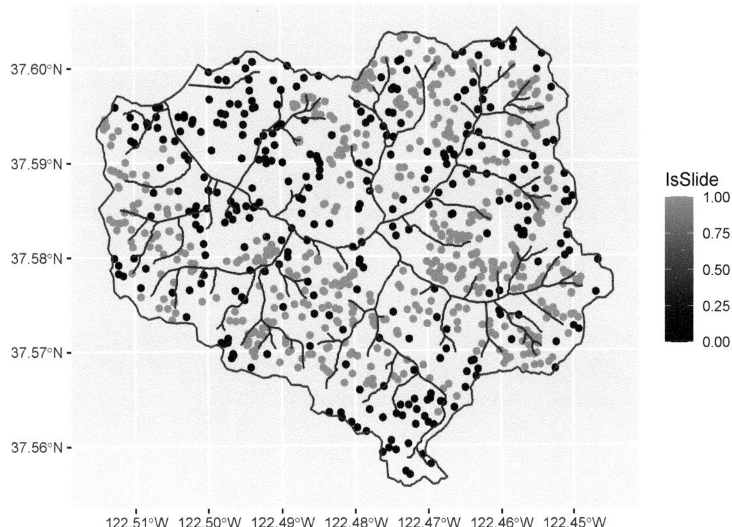

FIGURE 11.14 Landslides and random points (excluded from slide buffers)

```
ggplot(data=allPts) + geom_sf(aes(col=IsSlide)) +
  geom_sf(data=streams, col="blue") +
  geom_sf(data=wshed, fill=NA)
```

Now we'll derive some environmental variables as rasters and extract their values to the slide/noslide point data set. We'll use the shade function to create an exposure raster with the sun at the average noon angle from the south. Note the use of `dplyr::rename` to create meaningful variable names for rasters that had names like `"lyr1"` (who knows why?), which I determined by running it without the rename and seeing what it produced.

```
slidesV <- vect(allPts)
strms <- rasterize(vect(streams),elev)
tr <- rasterize(vect(trails),elev)
stD <- terra::distance(strms); names(stD) = "stD"
trD <- terra::distance(tr); names(trD) = "trD"
slope <- terrain(elev, v="slope")
curv <- terrain(slope, v="slope")
aspect <- terrain(elev, v="aspect")
exposure <- shade(slope/180*pi,aspect/180*pi,angle=53,direction=180)
elev_ex <- terra::extract(elev, slidesV) %>%
  rename(elev=dem) %>% dplyr::select(-ID)
SI_ex <- terra::extract(SI, slidesV) %>% dplyr::select(-ID)
slope_ex <- terra::extract(slope, slidesV) %>% dplyr::select(-ID)
curv_ex <- terra::extract(curv, slidesV) %>%
  rename(curv = slope) %>% dplyr::select(-ID)
exposure_ex <- terra::extract(exposure, slidesV) %>% rename(exposure=lyr1) %>%
  dplyr::select(-ID)
stD_ex <- terra::extract(stD, slidesV) %>% dplyr::select(-ID)
trD_ex <- terra::extract(trD, slidesV) %>% dplyr::select(-ID)
slidePts <- cbind(allPts,elev_ex,SI_ex,slope_ex,curv_ex,exposure_ex,stD_ex,trD_ex)
slidesData <- as.data.frame(slidePts) %>%
                              filter(!is.nan(SI)) %>% st_as_sf(crs=st_crs(slides))
```

We'll filter for just the slides in the 1983 imagery, and for the model also include all of the random points (detected with `NA` values for Visible):

```
slides83 <- slidesData %>% filter(Visible == 1983)
slides83plusr <- slidesData %>% filter(Visible == 1983 | is.na(Visible))
```

The logistic regression in the glm model uses the *binomial* family:

```
summary(glm(IsSlide ~ SI + curv + elev + slope + exposure + stD + trD,
           family=binomial, data=slides83plusr))
```

```
##
## Call:
## glm(formula = IsSlide ~ SI + curv + elev + slope + exposure +
##     stD + trD, family = binomial, data = slides83plusr)
##
```

```
## Deviance Residuals:
##     Min       1Q   Median       3Q      Max
## -2.3819  -0.5882  -0.0002   0.7378   3.9146
##
## Coefficients:
##               Estimate Std. Error z value Pr(>|z|)
## (Intercept)  0.1426071  1.0618610   0.134  0.89317
## SI          -1.8190286  0.3639202  -4.998 5.78e-07 ***
## curv         0.0059237  0.0113657   0.521  0.60223
## elev        -0.0025050  0.0014398  -1.740  0.08190 .
## slope        0.0623201  0.0211010   2.953  0.00314 **
## exposure     2.6971912  0.6603844   4.084 4.42e-05 ***
## stD         -0.0052639  0.0019765  -2.663  0.00774 **
## trD         -0.0020730  0.0006532  -3.174  0.00150 **
## ---
## Signif. codes:  0 '***' 0.001 '**' 0.01 '*' 0.05 '.' 0.1 ' ' 1
##
## (Dispersion parameter for binomial family taken to be 1)
##
##     Null deviance: 668.29  on 483  degrees of freedom
## Residual deviance: 402.90  on 476  degrees of freedom
## AIC: 418.9
##
## Number of Fisher Scoring iterations: 8
```

So while we might want to make sure that we've picked the right variables, some conclusions from this might be that stability index (`SI`) is not surprisingly a strong predictor of landslides, but also that other than elevation and curvature, the other variables are also significant.

11.5.2.1 Map the logistic result

We should be able to map the results; we just need to use the formula for the model, which in our case will have six explanatory variables SI (X_1), elev (X_2), slope (X_3), exposure (X_4), stD (X_5), and trD (X_6):

$$p = \frac{e^{b_0+b_1X_1+b_2X_2+b_3X_3+b_4X_4+b_5X_5}}{1+e^{b_0+b_1X_1+b_2X_2+b_3X_3+b_4X_4+b_5X_5}}$$

Then we retrieve the coefficients with `coef()` and use map algebra to create the prediction map, including dots for the landslides actually observed in 1983 (Figure 11.15).

```
library(tmap)
SI_stD_model <- glm(IsSlide ~ SI + slope + exposure + stD + trD, family=binomial,
                    data=slides83plusr)
b0 <- coef(SI_stD_model)[1]
b1 <- coef(SI_stD_model)[2]
b2 <- coef(SI_stD_model)[3]
b3 <- coef(SI_stD_model)[4]
```

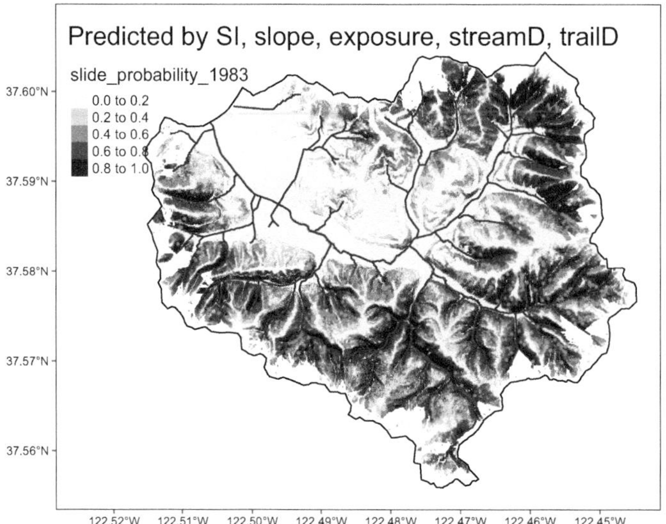

FIGURE 11.15 Logistic model prediction of 1983 landslide probability

```
b4 <- coef(SI_stD_model)[5]
b5 <- coef(SI_stD_model)[6]
predictionMap <- (exp(b0+b1*SI+b2*slope+b3*exposure+b4*stD+b5*trD))/
  (1 + exp(b0+b1*SI+b2*slope+b3*exposure+b4*stD+b5*trD))
names(predictionMap) <- "slide_probability_1983"
tm_shape(predictionMap) + tm_raster() +
  tm_shape(streams) + tm_lines(col="blue") +
  tm_shape(wshed) + tm_borders(col="black") +
  tm_shape(slides83) + tm_dots(col="green") +
  tm_layout(title="Predicted by SI, slope, exposure, streamD, trailD") +
  tm_graticules(lines=F)
```

11.5.3 Poisson regression

Poission regression (`family = poisson` in glm) is for analyzing counts. You might use this to determine which variables are significant predictors, or you might want to see where or when counts are particularly high, or higher than expected. Or it might be applied to compare groupings. For example, if instead of looking at differences among rock types in Sinking Cove based on chemical properties in water samples or measurements (where we applied ANOVA), we looked at differences in fish counts, that might be an application of poisson regression, as long as other data assumptions hold.

11.5.3.1 Seabird Poisson Model

This model is further developed in a separate case study where we'll map a prediction, but we'll look at the basics for applying a poisson family model in glm: *seabird observations*. The data is potentially suitable since its counts with known time intervals of observation effort and initial review of the data suggests that the mean of counts is reasonably similar to the variance. The data were collected through the ACCESS program (*Applied California Current Ecosystem Studies* (n.d.)) and a thorough analysis is provided by Studwell et al. (2017).

```
library(igisci); library(sf); library(tidyverse); library(tmap)
library(maptiles)
transects <- st_read(ex("SFmarine/transects.shp"))
```

One of the seabirds studies, the black-footed albatross, has the lowest counts (so a relatively rare bird), and studies of its distribution may be informative. We'll look at counts collected during transect cruises in July 2006 (Figure 11.16).

```
tmap_mode("plot")
oceanBase <- get_tiles(transects, provider="Esri.OceanBasemap")
transJul2006 <- transects %>%
  filter(month==7 & year==2006 & avg_tem>0 & avg_sal>0 & avg_fluo>0)
tm_shape(oceanBase) + tm_rgb() +
  tm_shape(transJul2006) + tm_symbols(col="bfal", size="bfal")
```

Prior studies have suggested that temperature, salinity, fluorescence, depth, and various distances might be good explanatory variables to use to look at spatial patterns, so we'll use these in the model.

```
summary(glm(bfal~avg_tem + avg_sal + avg_fluo + avg_dep +
            dist_land + dist_isla + dist_200m + dist_cord,
            data=transJul2006, family=poisson))
```

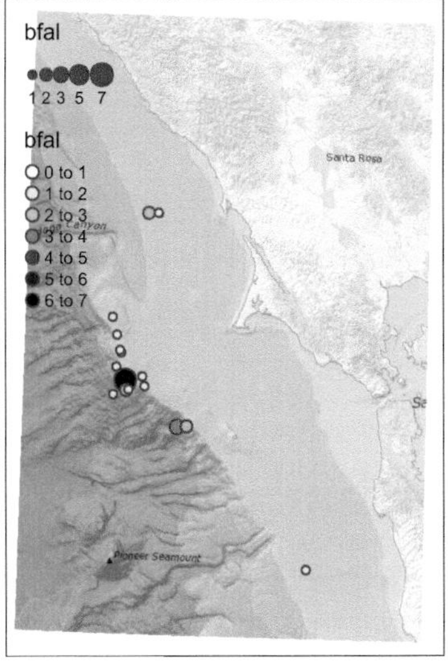

FIGURE 11.16 Black-footed albatross counts, July 2006

```
##
## Call:
## glm(formula = bfal ~ avg_tem + avg_sal + avg_fluo + avg_dep +
##     dist_land + dist_isla + dist_200m + dist_cord, family = poisson,
##     data = transJul2006)
##
## Deviance Residuals:
##     Min       1Q   Median       3Q      Max
## -1.2935  -0.4646  -0.2445  -0.0945   4.7760
##
## Coefficients:
##               Estimate Std. Error z value Pr(>|z|)
## (Intercept) -1.754e+02  7.032e+01  -2.495  0.01260 *
## avg_tem      7.341e-01  3.479e-01   2.110  0.03482 *
## avg_sal      5.202e+00  2.097e+00   2.480  0.01312 *
## avg_fluo    -6.978e-01  1.217e+00  -0.574  0.56626
## avg_dep     -3.561e-03  8.229e-04  -4.328 1.51e-05 ***
## dist_land   -2.197e-04  6.194e-05  -3.547  0.00039 ***
## dist_isla    1.372e-05  1.778e-05   0.771  0.44043
## dist_200m   -4.696e-04  1.037e-04  -4.528 5.95e-06 ***
## dist_cord   -5.213e-06  1.830e-05  -0.285  0.77570
## ---
## Signif. codes:  0 '***' 0.001 '**' 0.01 '*' 0.05 '.' 0.1 ' ' 1
##
## (Dispersion parameter for poisson family taken to be 1)
##
##     Null deviance: 152.14  on 206  degrees of freedom
## Residual deviance: 105.73  on 198  degrees of freedom
## AIC: 160.37
##
## Number of Fisher Scoring iterations: 7
```

We can see in the model coefficients table several predictive variables that appear to be significant, which we'll want to explore further, and also map the results as we did with Sierra temperature data.

- avg_tem
- avg_sal
- avg_dep
- dist_land
- dist_200m

Note R will allow you to run glm on any family, but it's up to you to make sure it's appropriate. For example, if you provide a response variable that is *not* count data or doesn't fit the requirements for a poisson model, R will create a result but it's bogus.

11.5.4 Models employing machine learning

Models using machine learning algorithms are commonly used in data science, fitting with its general exploratory and data-mining approach. There are many machine learning algorithms, and many resources for learning more about them, but they all share a basically black-box approach where a collection of variables are explored for patterns in input variables that help to explain a response variable. The latter is similar to more conventional statistical modeling describe above, with the difference being the machine learning approach – think of robots going through your data looking for connections.

We'll explore machine learning methods in the next chapter when we attempt to use them to classify satellite imagery, an important environmental application.

11.6 Exercises: Modeling

Exercise 11.1. Add `LATITUDE` as a second independent variable after `ELEVATION` in the model predicting February temperatures. Note that you'll need to change a setting in the `st_as_sf` to not remove `LATITUDE` and `LONGITUDE`, and then you'll need to change the `lm` regression formula to include both `ELEVATION + LATITUDE`. First derive and display the model summary results, and then answer the questions – is the explanation better based on the r^2?

Exercise 11.2. Now map the **predictions** (Figure 11.17) and **residuals**.

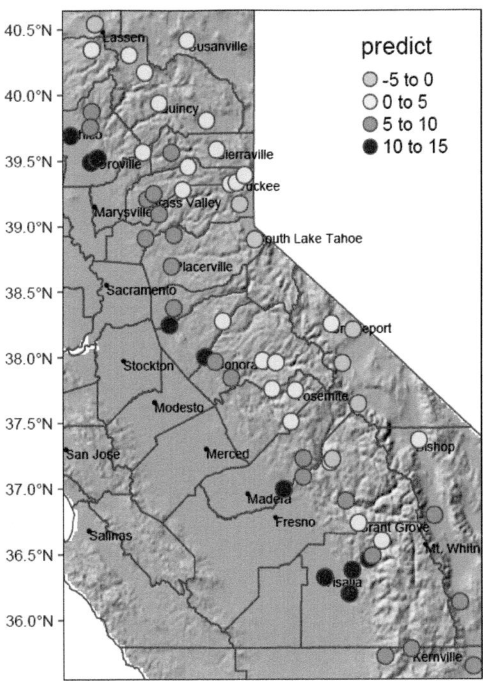

FIGURE 11.17 Prediction of temperature from elevation (one of two goals)

Exercise 11.3. Can you see any potential for additional variables? What evidence might there be for localized effects?

Exercise 11.4. Build a prediction raster from the above model. Start by reading in the `ca_elev_WGS84.tif` raster as was done earlier, with the same crop, and again assigning it to `elevSierra`, and get the three coefficients `b0`, `b1`, and `b2` (Figure 11.18).

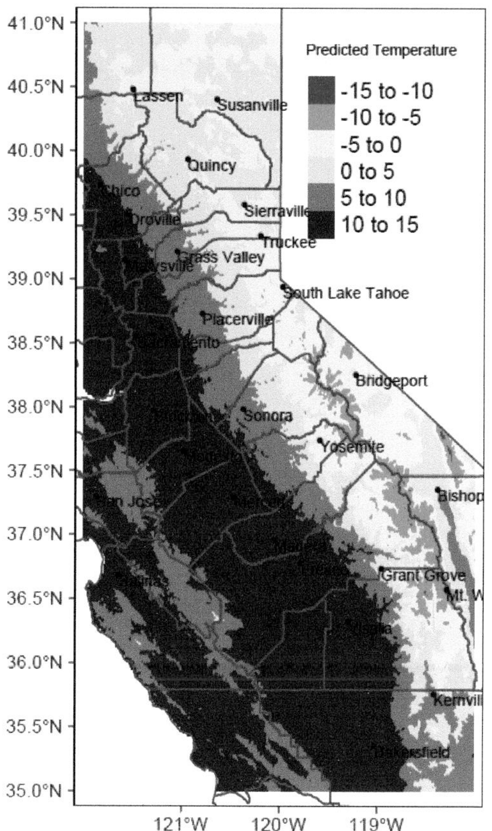

FIGURE 11.18 Prediction raster (goal)

Exercise 11.5. Now use the following code to create a raster of the same dimensions as `elevSierra`, but with latitude assigned as z values to the cells. Then build a raster **prediction of temperature** from the model and the two explanatory rasters.

Exercise 11.6. Modify the code for the San Pedro Creek Watershed logistical model to look at 1975 landslides. Are the conclusions different?

Exercise 11.7. Optionally, using your own data that should have location information as well as other attributes, create a statistical model of your choice, and provide your assessment of the results, a prediction map, and where appropriate mapped residuals. You'll probably want to create a new project for this with a `data` folder.

Exercise 11.8. Optionally, using your own data but with no location data needed, create a glm binomial model. If you don't have anything obvious, consider doing something like the study of a pet dog panting based on temperature. https://bscheng.com/2016/10/23/logistic-regression-in-r/

12

Imagery and Classification Models

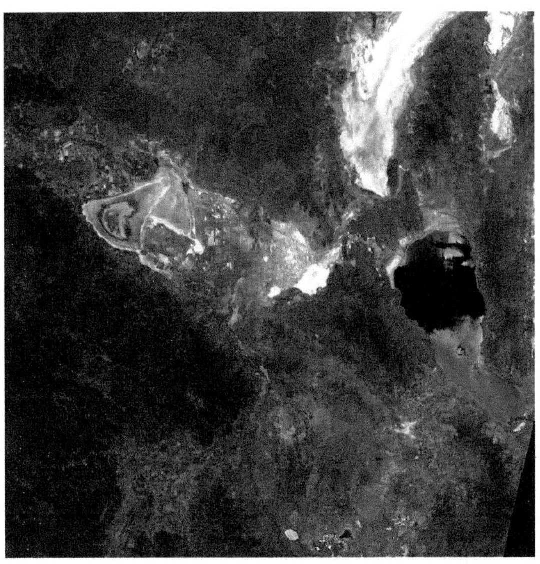

This chapter explores the use of satellite imagery, such as this Sentinel-2 scene (from 28 June 2021) of part of the northern Sierra Nevada, CA, to Pyramid Lake, NV, including display methods (such as the "false color" above) and imagery analysis methods. After we've learned how to read the data and cropped it to fit our study area in Red Clover Valley, we'll try some machine language classifiers to see if we can make sense of the patterns of hydric to xeric vegetation in a meadow under active restoration.

Image analysis methods are important for environmental research and are covered much more thoroughly in the discipline of *remote sensing*. Some key things to learn more about are the nature of the electromagnetic spectrum and especially bands of that spectrum that are informative for land cover and especially vegetation detection, and the process of imagery classification. We'll be using satellite imagery that includes a range of the electromagnetic spectrum from visible to short-wave infrared.

12.1 Reading and Displaying Sentinel-2 Imagery

We'll work with 10 and 20 m Sentinel-2 data downloaded from *Copernicus Open Access Hub* (n.d.) as an entire scene from 20210602 and 20210628, providing cloud-free imagery of Red

Clover Valley in the northern Sierra Nevada, where we are doing research on biogeochemical cycles and carbon sequestration from meadow restoration with The Sierra Fund (*Clover Valley Ranch Restoration, the Sierra Fund* (n.d.)).

Download the image zip files from...

https://sfsu.box.com/s/zx9rvbxdss03rammy7d72rb3uumfzjbf

...and extract into the folder `~/sentinel2/` – the `~` referring to your Home folder (typically `Documents`). (See the code below to see how one of the two images is read.) Make sure the folders stored there have names ending in `.SAFE`. The following are the two we'll be using:

- `S2B_MSIL2A_20210603T184919_N0300_R113_T10TGK_20210603T213609.SAFE`
- `S2A_MSIL2A_20210628T184921_N0300_R113_T10TGK_20210628T230915.SAFE`

```
library(stringr)
library(terra)
imgFolder <- paste0("S2A_MSIL2A_20210628T184921_N0300_R113_T10TGK_20210628T230915.",
                    "SAFE\\GRANULE\\L2A_T10TGK_A031427_20210628T185628")
img20mFolder <- paste0("~/sentinel2/",imgFolder,"\\IMG_DATA\\R20m")
imgDateTime <- str_sub(imgFolder,12,26)
imgDateTime
```

```
## [1] "20210628T184921"
```

```
imgDate <- str_sub(imgDateTime,1,8)
```

As documented at *Copernicus Open Access Hub* (n.d.), Sentinel-2 imagery is collected at three resolutions, with the most bands at the coarsest (60 m) resolution. The bands added at that coarsest resolution are not critical for our work as they relate to oceanographic and atmospheric research, and our focus will be on land cover and vegetation in a terrestrial area. So we'll work with four bands at 10 m and an additional six bands at 20 m resolution:

10 m bands

- B02 - Blue 0.490 μm
- B03 - Green 0.560 μm
- B04 - Red 0.665 μm
- B08 - NIR 0.842 μm

20 m bands

- B02 - Blue 0.490 μm
- B03 - Green 0.560 μm
- B04 - Red 0.665 μm
- B05 - Red Edge 0.705 μm
- B06 - Red Edge 0.740 μm
- B07 - Red Edge 0.783 μm
- B8A - NIR 0.865 μm
- B11 - SWIR 1.610 μm

- B12 - SWIR 2.190 μm

```
sentinelBands <- paste0("B",c("02","03","04","05","06","07","8A","11","12"))
sentFiles <- paste0(img20mFolder,"/T10TGK_",imgDateTime,"_",
                      sentinelBands,"_20m.jp2")
sentinel <- rast(sentFiles)
sentinelBands
```

```
## [1] "B02" "B03" "B04" "B05" "B06" "B07" "B8A" "B11" "B12"
```

```
names(sentinel) <- sentinelBands
```

12.1.1 Individual bands

The Sentinel-2 data are in a 12-bit format, ranging as integers from 0 to 4,095, but are preprocessed to represent reflectances at the base of the atmosphere.

We'll start by looking at four key bands – green, red, NIR, SWIR (B11) – and plot these in a gray scale (Figure 12.1).

```
par(mfrow=c(2,2))
plot(sentinel$B03, main = "Green", col=gray(0:4095 / 4095))
plot(sentinel$B04, main = "Red", col=gray(0:4095 / 4095))
plot(sentinel$B8A, main = "NIR", col=gray(0:4095 / 4095))
plot(sentinel$B11, main = "SWIR (B11)", col=gray(0:4095 / 4095))
```

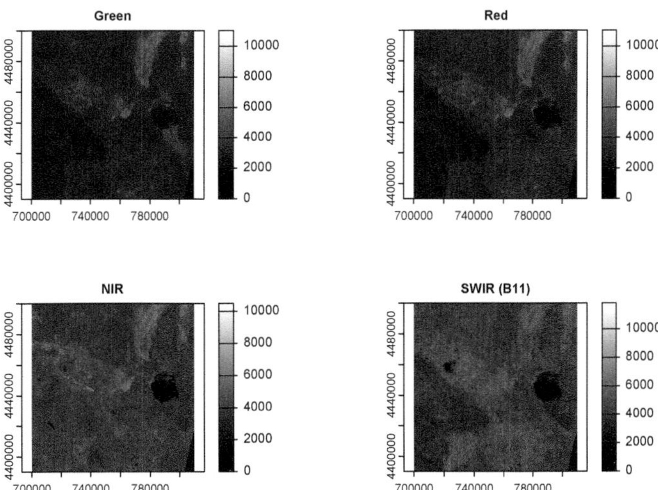

FIGURE 12.1 Four bands of a Sentinel-2 scene from 20210628.

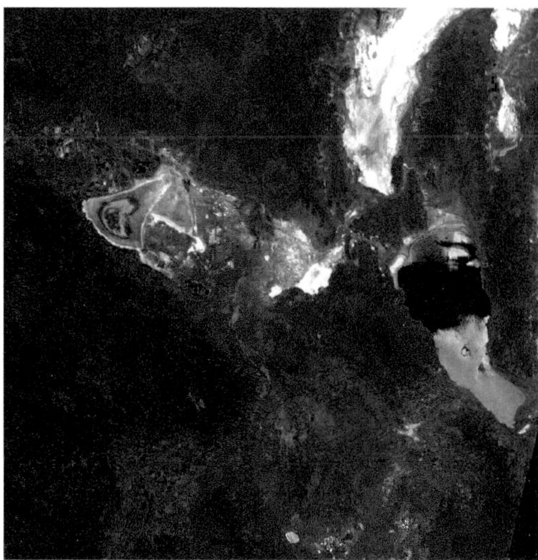

FIGURE 12.2 R-G-B image from Sentinel-2 scene 20210628.

12.1.2 Spectral subsets to create three-band R-G-B and NIR-R-G for visualization

Displaying the imagery three bands at a time (displayed as RGB on our computer screen) is always a good place to start (Figure 12.2), and two especially useful band sets are RGB itself – so looking like a conventional color aerial photography – and "false color" that includes a band normally invisible to our eyes, such as near infrared that reflects chlorophyll in healthy plants. In standard "false color", surface-reflected near-infrared is displayed as red, reflected red is displayed as green, and green is displayed as blue. Some advantages of this false color display include:

- Blue is not included at all, which helps reduce haze.
- Water bodies absorb NIR, so appear very dark in the image.
- Chlorophyll strongly reflects NIR, so healthy vegetation is bright red.

```
sentRGB <- subset(sentinel,3:1)
plotRGB(sentRGB,stretch="lin")
```

We can also spectrally subset to get NIR (B8A, the 7th band of the 20 m multispectral raster we've created). See the first figure in this chapter for the NIR-R-G image displayed with plotRGB.

```
sentFCC <- subset(sentinel,c(7,3,2))
```

12.1.3 Crop to study area extent

The Sentinel-2 scene covers a very large area, more than 100 km on each side, 12,000 km^2 in area. The area we are interested in – Red Clover Valley – is enclosed in an area of about 100 km^2, so we'll crop the scene to fit this area (Figures 12.3 and 12.4).

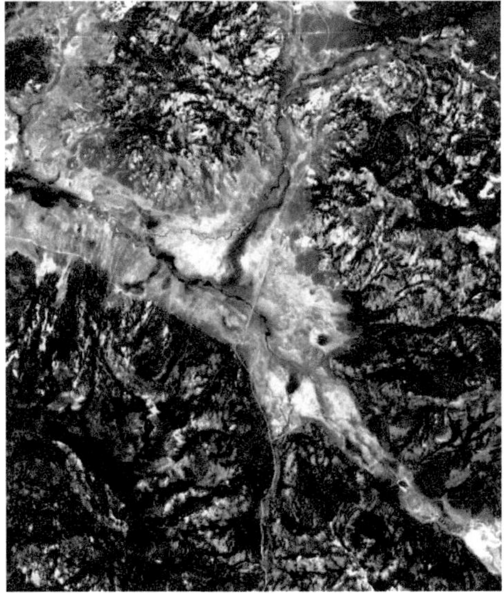

FIGURE 12.3 Color image from Sentinel-2 of Red Clover Valley, 20210628.

```
RCVext <- ext(715680,725040,4419120,4429980)
sentRCV <- crop(sentinel,RCVext)
sentRGB <- subset(sentRCV,3:1)
plotRGB(sentRGB,stretch="lin")
```

NIR-R-G:

```
plotRGB(subset(sentRCV,c(7,3,2)),stretch="lin")
```

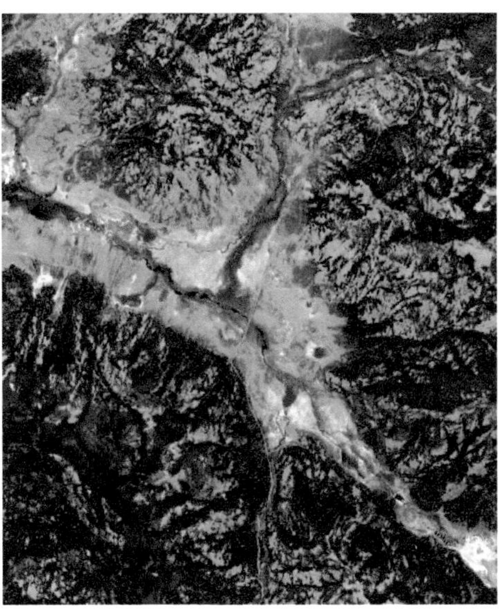

FIGURE 12.4 NIR-R-G image from Sentinel-2 of Red Clover Valley, 20210628.

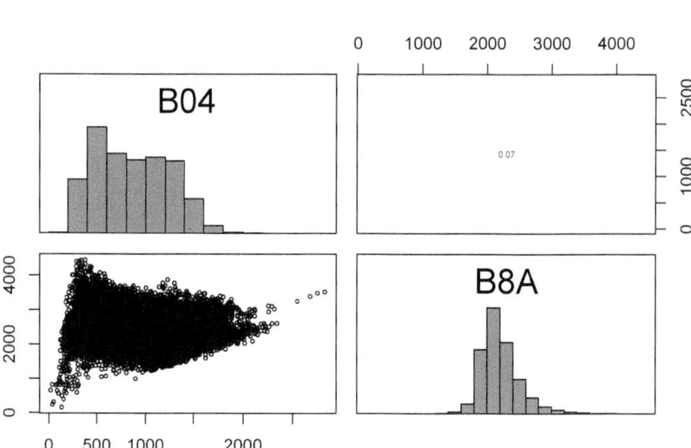

FIGURE 12.5 Relations between Red and NIR bands, Red Clover Valley Sentinel-2 image, 20210628

12.1.4 Saving results

Optionally, you may want to save this cropped image to call it up quicker at a later time. We'll build a simple name with the date and time as part of it.

```
writeRaster(sentRCV, filename=paste0("~/sentinel2/",imgDateTime,".tif"),overwrite=T)
```

12.1.5 Band scatter plots

Using scatter plots helps us to to understand how we might employ multiple bands in our analysis. Red vs NIR is particularly intriguing, commonly creating a triangular pattern reflecting the high absorption of NIR by water yet high reflection by healthy vegetation which is visually green, so low in red (Figure 12.5).

```
pairs(c(sentRCV$B04,sentRCV$B8A), main="Red vs NIR")
```

12.2 Spectral Profiles

A useful way to look at how our imagery responds to known vegetation/land-cover at *imagery training polygons* we'll use later for supervised classification, we can extract values

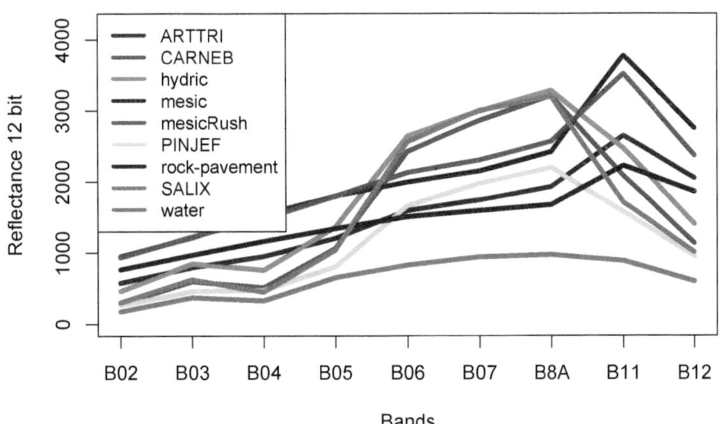

FIGURE 12.6 Spectral signature of nine-level training polygons, 20 m Sentinel-2 imagery from 20210628.

at points and create a *spectral profile* (Figure 12.6). We start by deriving points from the polygons we created based on field observations. One approach is to use polygon centroids (with `centroids()`), but to get more points from larger polygons, we'll instead sample the polygons to get a total of ~1,000 points.

```
library(igisci)
trainPolys9 <- vect(ex("RCVimagery/train_polys9.shp"))
ptsRCV <- spatSample(trainPolys9, 1000, method="random")
# for just centroids:  ptsRCV <- centroids(trainPolys9)
extRCV <- terra::extract(sentRCV, ptsRCV)
head(extRCV)
```

```
##   ID B02 B03 B04 B05  B06  B07  B8A  B11  B12
## 1  1 194 479 352 890 2467 2954 3312 1800  923
## 2  2 194 479 352 890 2467 2954 3312 1800  923
## 3  3 213 494 410 939 2404 2845 3231 1825  978
## 4  4 276 580 473 991 2436 2882 3247 1944 1060
## 5  5 213 494 410 939 2404 2845 3231 1825  978
## 6  6 228 500 377 909 2407 2862 3225 1780  970
```

```
specRCV <- aggregate(extRCV[,-1], list(ptsRCV$class), mean)
rownames(specRCV) <- specRCV[,1]
specRCV <- specRCV[,-1]
library(RColorBrewer)
LCcolors <- brewer.pal(length(unique(trainPolys9$class)),"Set1")
specRCV <- as.matrix(specRCV)
# Start with an emply plot
plot(0, ylim=c(0,4195), xlim=c(1,9),type='n',
```

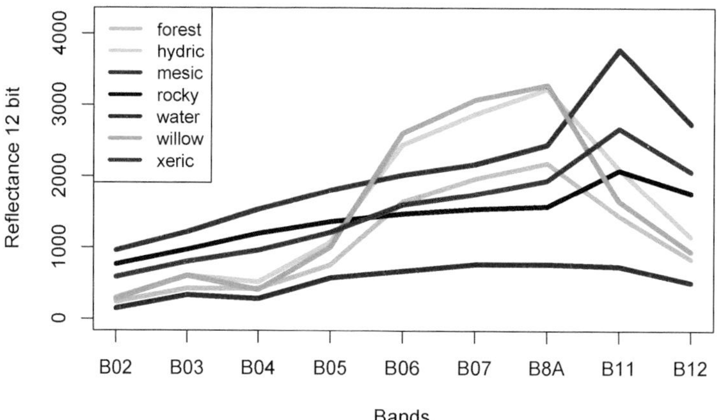

FIGURE 12.7 Spectral signature of seven-level training polygons, 20 m Sentinel-2 imagery from 20210628.

```
     xlab="Bands",ylab="Reflectance 12 bit", xaxt='n')
axis(side=1,at=1:9,labels=sentinelBands)
# Add the classes
for (i in 1:nrow(specRCV)){
  lines(specRCV[i,],type='l',lwd=4,lty=1, col=LCcolors[i])
}
title(main="Spectral signatures Sentinel-2 bands, Red Clover Valley")
legend('topleft',rownames(specRCV),cex=0.9,lwd=3,col=LCcolors) #bty='n'
```

One thing we can observe in the spectral signature is that some vegetation types are pretty similar; for instance `mesic` and `mesicRush` are very similar, as are `hydric` and `CARNEB`, so we might want to combine them, and have fewer categories. We also have a seven-level polygon classification, so we'll use that (later we'll combine values in code) (Figure 12.7).

```
trainPolys7 <- vect(ex("RCVimagery/train_polys7.shp"))
ptsRCV <- spatSample(trainPolys7, 1000, method="random")
extRCV <- terra::extract(sentRCV, ptsRCV)
specRCV <- aggregate(extRCV[,-1], list(ptsRCV$class), mean)
rownames(specRCV) <- specRCV[,1]
specRCV <- specRCV[,-1]
LCcolors <- c("cyan","gold","darkgreen","black","blue","lawngreen","red")
specRCV <- as.matrix(specRCV)
plot(0, ylim=c(0,4195), xlim=c(1,9),type='n',
     xlab="Bands",ylab="Reflectance 12 bit", xaxt='n')
axis(side=1,at=1:9,labels=sentinelBands)
for (i in 1:nrow(specRCV)){
  lines(specRCV[i,],type='l',lwd=4,lty=1, col=LCcolors[i])
```

```
}
title(main="Spectral signatures Sentinel-2 bands, seven land cover classes",
      font.main=2)
legend('topleft',rownames(specRCV),cex=0.9,lwd=3,col=LCcolors)
```

From this spectral signature, we can see that at least at this scale, there's really no difference between the spectral response `hydric` and `willow`. The structure of the willow copses can't be distinguished from the herbaceous hydric sedges with 20 m pixels. Fortunately, in terms of *hydrologic* response, willows are also hydric plants so we can just recode `willow` as `hydric` (Figure 12.8).

```
trainPolys6 <- trainPolys7
trainPolys6[trainPolys6$class=="willow"]$class <- "hydric"
ptsRCV <- spatSample(trainPolys6, 1000, method="random")
extRCV <- terra::extract(sentRCV, ptsRCV)
specRCV <- aggregate(extRCV[,-1], list(ptsRCV$class), mean)
rownames(specRCV) <- specRCV[,1]
specRCV <- specRCV[,-1]
specRCV <- as.matrix(specRCV)
plot(0, ylim=c(0,4195), xlim=c(1,9),type='n',
     xlab="Bands",ylab="Reflectance 12 bit", xaxt='n')
axis(side=1,at=1:9,labels=sentinelBands)
for (i in 1:nrow(specRCV)){
  lines(specRCV[i,],type='l',lwd=4,lty=1, col=LCcolors[i])
}
title(main="Spectral signatures, six land-cover classes, RCV", font.main=2)
legend('topleft',rownames(specRCV),cex=0.9,lwd=3,col=LCcolors)
```

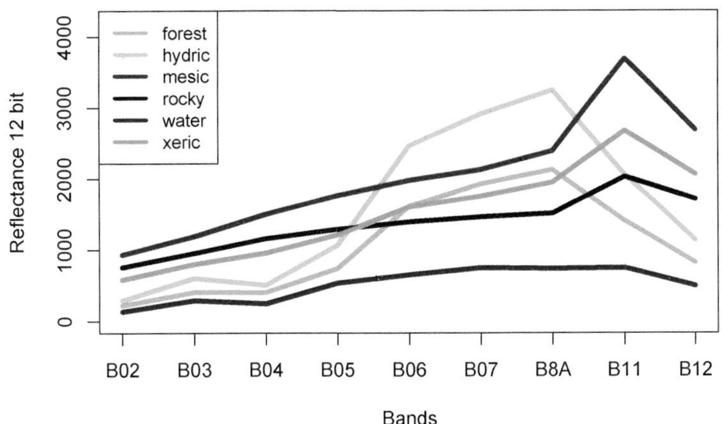

FIGURE 12.8 Spectral signature of six-level training polygons, 20 m Sentinel-2 imagery from 20210628

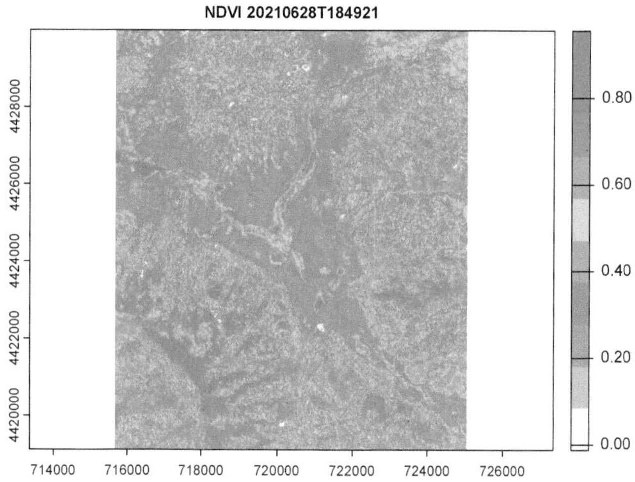

FIGURE 12.9 NDVI from Sentinel-2 image, 20210628

12.3 Map Algebra and Vegetation Indices

We can also do pixel-wise calculations, the *map algebra* method (Tomlin (1990)) used in raster GIS. We'll use it to create vegetation indices, which are widely used as a single quantitative measure to relate to phenological change, moisture, and nutrients.

12.3.1 Vegetation indices

The venerable *Normalized Difference Vegetation Index* (NDVI) uses NIR (Sentinel2 20m: B8A) and RED (Sentinel2 20m: B04) in a ratio that normalizes NIR with respect to visible (usually red) (Figure 12.9).

$$NDVI = \frac{NIR - RED}{NIR + RED}$$

```
ndviRCV <- (sentRCV$B8A - sentRCV$B04)/
           (sentRCV$B8A + sentRCV$B04)
plot(ndviRCV, col=rev(terrain.colors(10)), main=paste('NDVI',imgDateTime))
```

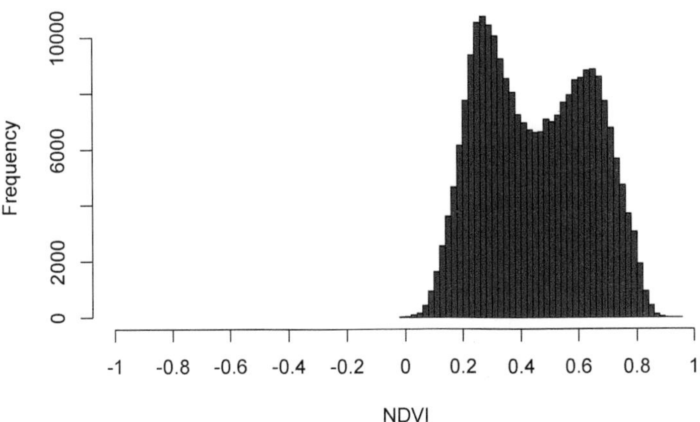

FIGURE 12.10 NDVI histogram, Sentinel-2 image, 20210628

12.3.2 Histogram

The histogram for an NDVI is ideally above zero as this one is (Figure 12.10), with the values approaching 1 representing pixels with a lot of reflection in the near infrared, so typically healthy vegetation. Standing water can cause values to go negative, but with 20 m pixels there's not enough water in the creeks to create a complete pixel of open water.

```
hist(ndviRCV, main='Normalized Difference Vegetation Index', xlab='NDVI',
     ylab='Frequency',col='#6B702B',xlim=c(-1,1),breaks=40,xaxt='n')
axis(side=1,at=seq(-1,1,0.2), labels=seq(-1,1,0.2))
```

12.3.3 Other vegetation indices

With the variety of spectral bands available on satellite imagery comes a lot of possibilities for other vegetation (and non-vegetation) indices.

One that has been used to detect moisture content in plants is the Normalized Difference Moisture Index $NDMI = \frac{NIR-SWIR}{NIR+SWIR}$, which for Sentinel-2 20 m imagery would be $\frac{B8A-B11}{B8A+B11}$ (Figure 12.11).

```
ndmiRCV <- (sentRCV$B8A - sentRCV$B11)/
           (sentRCV$B8A + sentRCV$B11)
plot(ndmiRCV, col=rev(terrain.colors(10)), main=paste('NDMI',imgDateTime))
```

In contrast to NDVI, the NDMI histogram extends to negative values (Figure 12.12).

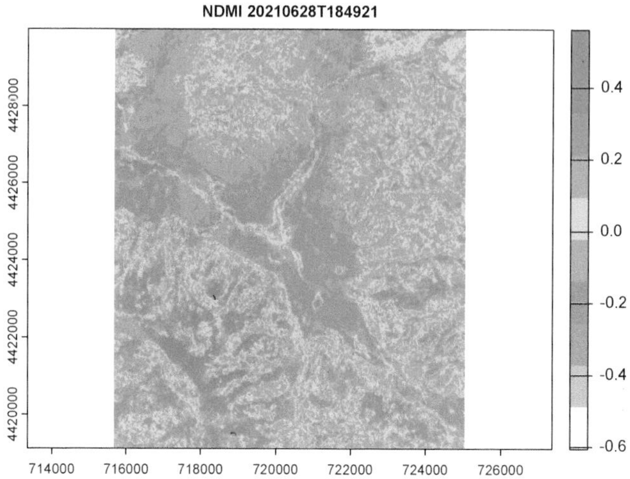

FIGURE 12.11 NDMI from Sentinel-2 image, 20210628

```
hist(ndmiRCV, main='Normalized Difference Moisture Index', xlab='NDMI',
     ylab='Frequency',col='#26547C',xlim=c(-1,1),breaks=40,xaxt='n')
axis(side=1,at=seq(-1,1,0.2), labels=seq(-1,1,0.2))
```

Another useful index – Normalized Difference Greenness Index (NDGI) uses three bands: NIR, Red, and Green (Yang et al. (2019)), and was first proposed for MODIS but can be varied slightly for Sentinel-2. It has the advantage of using the more widely available four-band imagery, such as is available in the 10 m Sentinel-2 product, 3 m PlanetScope,

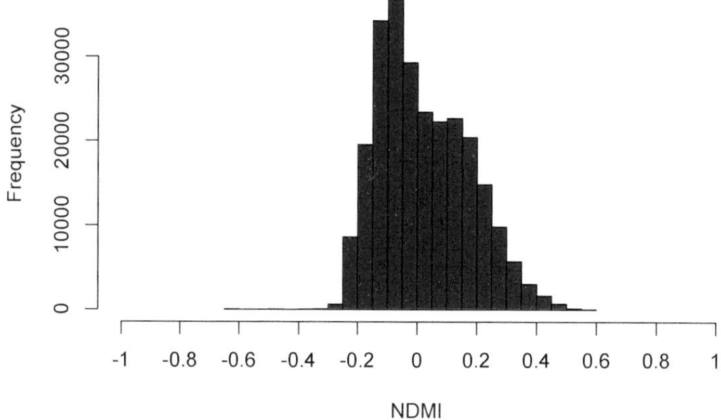

FIGURE 12.12 NDMI histogram, Sentinel-2 image, 20210628

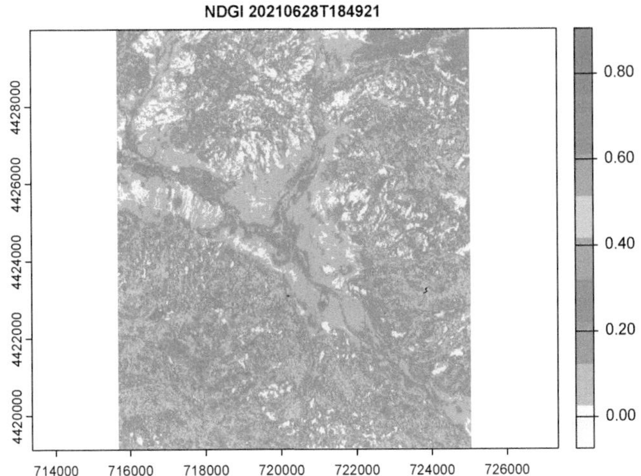

FIGURE 12.13 NDGI from Sentinel-2 image, 20210628

or multispectral drone imagery using a MicaSense, which for their Altum camera at 120 m above ground is 5.4 cm. We'll stick with the 20 m Sentinel-2 product for now and produce a plot (Figure 12.13) and a histogram (Figure 12.14).

$$NDGI = \frac{0.616GREEN + 0.384NIR - RED}{0.616GREEN + 0.384NIR + RED}$$

```
ndgiRCV <- (0.616 * sentRCV$B03 + 0.384 * sentRCV$B8A - sentRCV$B04)/
           (0.616 * sentRCV$B03 + 0.384 * sentRCV$B8A + sentRCV$B04)
```

```
plot(ndgiRCV, col=rev(terrain.colors(10)), main=paste('NDGI',imgDateTime))
```

```
hist(ndgiRCV, main='Normalized Difference Greenness Index',
     xlab='NDGI',ylab='Frequency',
     col='#38635A',xlim=c(-1,1),breaks=40,xaxt='n')
axis(side=1,at=seq(-1,1,0.2), labels=seq(-1,1,0.2))
```

12.4 Unsupervised Classification with k-means

We'll look at supervised classification next, where we'll provide the classifier known vegetation or land cover types, but with unsupervised we just provide the number of classes. We'll use the NDGI we just derived above and start by creating a vector from the values:

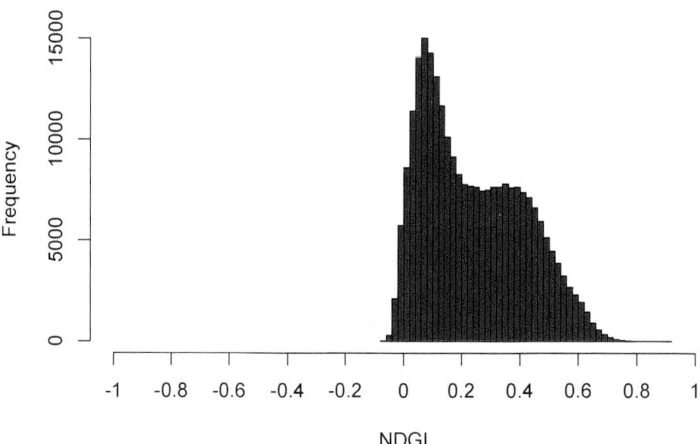

FIGURE 12.14 NDGI histogram, Sentinel-2 image, 20210628

```
vals <- values(ndgiRCV)
str(vals)
```

```
##  num [1:254124, 1] 0.356 0.429 0.4 0.359 0.266 ...
##  - attr(*, "dimnames")=List of 2
##   ..$ : NULL
##   ..$ : chr "B03"
```

We'll then run the kmeans function to derive clusters:

```
kmc <- kmeans(na.omit(vals), centers=5, iter.max=500, nstart=5, algorithm="Lloyd")
```

Now we convert the vector 2.6.1 back into a matrix 2.6.3 with the same dimensions as the original image...

```
kimg <- ndgiRCV
values(kimg) <- kmc$cluster
```

... and map the unsupervised classification (Figure 12.15). For the five classes generated, we'd need to follow up by overlaying with known land covers to see what it's picking up, but we'll leave that for the supervised classification next.

```
library(RColorBrewer)
LCcolors <- brewer.pal(length(unique(values(kimg))),"Set1")
plot(kimg, main = 'Unsupervised classification', col = LCcolors, type="classes")
```

FIGURE 12.15 Unsupervised k-means classification, Red Clover Valley, Sentinel-2, 20210628

12.5 Machine Learning Classification of Imagery

Using machine learning algorithms is one approach to classification, whether that classification is of observations in general or pixels of imagery. We're going to focus on a *supervised classification* method to identify land cover from imagery or other continuous raster variables (such as elevation or elevation-derived rasters such as slope, curvature, and roughness), employing samples of those variables called training samples.

It's useful to realize that the modeling methods we use for this type of imagery classification are really no different at the core from methods we'd use to work with continuous variables that might not even be spatial. For example, if we leave off the classification process for a moment, a machine learning algorithm might be used to predict a response result from a series of predictor variables, like predicting temperature from elevation and latitude, or acceleration might be predicted by some force acting on a body. So the first model might be used to predict the temperature of any location (within the boundary of the study area) given an elevation and latitude; or the second model might predict an acceleration given the magnitude of a force applied to the body (maybe of a given mass). A classification model varies on this by predicting a nominal variable like type of land cover; some other types of responses might be counts (using a Poisson model) or probabilities (using a logistic model).

The imagery classification approach adds to this model an input preparation process and an output prediction process:

- A training set of points and polygons are created that represent areas of known classification such as land cover like forest or wetland, used to identify values of the predictor variables (e.g. imagery bands).
- A predicted raster is created using the model applied to the original rasters.

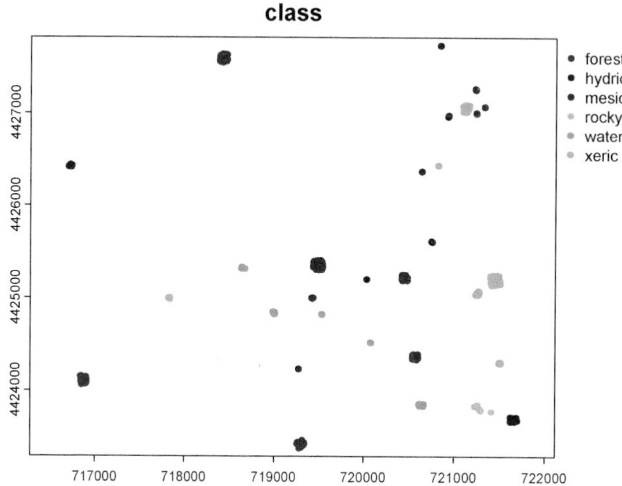

FIGURE 12.16 Training samples

We'll use a machine learning model called **Classification and Regression Trees** (CART) that is one of the simplest to understand in terms of the results it produces, which includes a display of the resulting classification decision "tree". We have collected field observations of vegetation/land cover types in our meadow, so can see if we can use these to train a classification model. You'll recognize our vegetation types (Figure 12.16) from the spectral signatures above. (We'll start by reading in the data again so this section doesn't depend on the above code.)

12.5.1 Read imagery and training data and extract sample values for training

```
library(terra); library(stringr)
imgFolder <- paste0("S2A_MSIL2A_20210628T184921_N0300_R113_T10TGK_20210628T230915.",
                    "SAFE\\GRANULE\\L2A_T10TGK_A031427_20210628T185628")
img20mFolder <- paste0("~/sentinel2/",imgFolder,"\\IMG_DATA\\R20m")
imgDateTime <- str_sub(imgFolder,12,26)
sentinelBands <- paste0("B",c("02","03","04","05","06","07","8A","11","12"))
sentFiles <- paste0(img20mFolder,"/T10TGK_",imgDateTime,"_",sentinelBands,
                    "_20m.jp2")
sentinel <- rast(sentFiles)
names(sentinel) <- sentinelBands
RCVext <- ext(715680,725040,4419120,4429980)
sentRCV <- crop(sentinel,RCVext)
trainPolys <- vect(ex("RCVimagery/train_polys7.shp"))
trainPolys[trainPolys$class=="willow"]$class <- "hydric"
```

```
ptsRCV <- spatSample(trainPolys, 1000, method="random")
plot(ptsRCV, "class")
```

```
LCclass <- c("forest","hydric","mesic","rocky","water","xeric")
classdf <- data.frame(value=1:length(LCclass),names=LCclass)
extRCV <- extract(sentRCV, ptsRCV)[,-1] # REMOVE ID
sampdataRCV <- data.frame(class = ptsRCV$class, extRCV)
```

12.5.2 Training the CART model

The *Recursive Partitioning and Regression Trees* package `rpart` fits the model.

```
library(rpart)
cartmodel <- rpart(as.factor(class)~., data = sampdataRCV,
                   method = 'class', minsplit = 5)
```

Now we can display the decision tree using the `rpart.plot` (Figure 12.17). This function clearly shows where each variable (in our case imagery spectral bands) contributes to the classification, and together with the spectral signature analysis above, helps us evaluate how the process works and possibly how to improve our training data where distinction among classes is unclear.

```
library(rpart.plot)
rpart.plot(cartmodel, fallen.leaves=F)
```

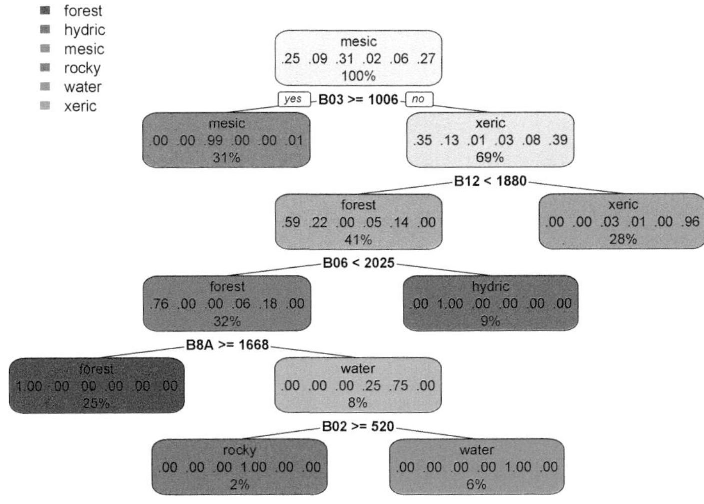

FIGURE 12.17 CART Decision Tree, Sentinel-2 20 m, date 20210628

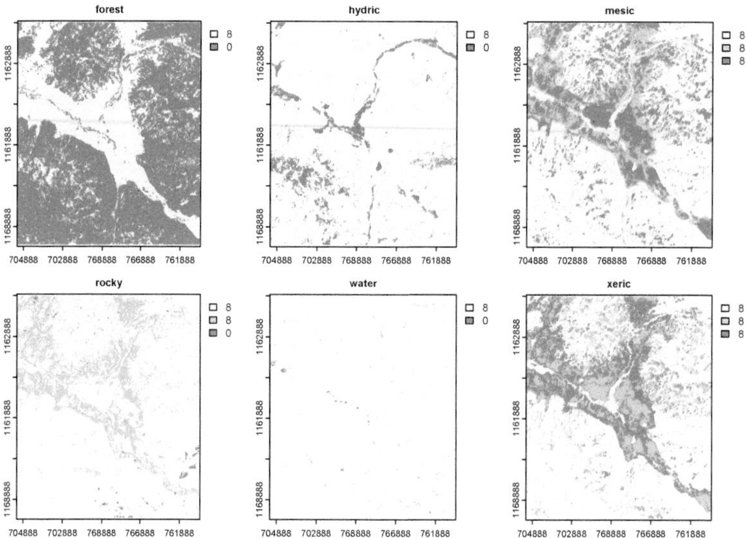

FIGURE 12.18 CART classification, probabilities of each class, Sentinel-2 20 m 20210628

12.5.3 Prediction using the CART model

Then we can use the model to predict the probabilities of each vegetation class for each pixel (Figure 12.18).

```
CARTpred <- predict(sentRCV, cartmodel, na.rm = TRUE)
CARTpred
```

```
## class       : SpatRaster
## dimensions  : 543, 468, 6  (nrow, ncol, nlyr)
## resolution  : 20, 20  (x, y)
## extent      : 715680, 725040, 4419120, 4429980  (xmin, xmax, ymin, ymax)
## coord. ref. : WGS 84 / UTM zone 10N (EPSG:32610)
## source      : memory
## names       : forest, hydric,    mesic, rocky, water, xeric
## min values  :      0,      0, 0.000000,     0,     0,  0.00
## max values  :      1,      1, 0.990991,     1,     1,  0.96
```

```
plot(CARTpred)
```

Note the ranges of values, each extending to values approaching 1.0. If we had used the nine-category training set described earlier, we would find some categories with very low scores, such that they will never dominate a pixel. We can be sure from the above that we'll get all categories used in the final output, and this will avoid potential problems in validation.

Now we'll make a single SpatRaster showing the vegetation/land cover with the *highest probability* (Figure 12.19).

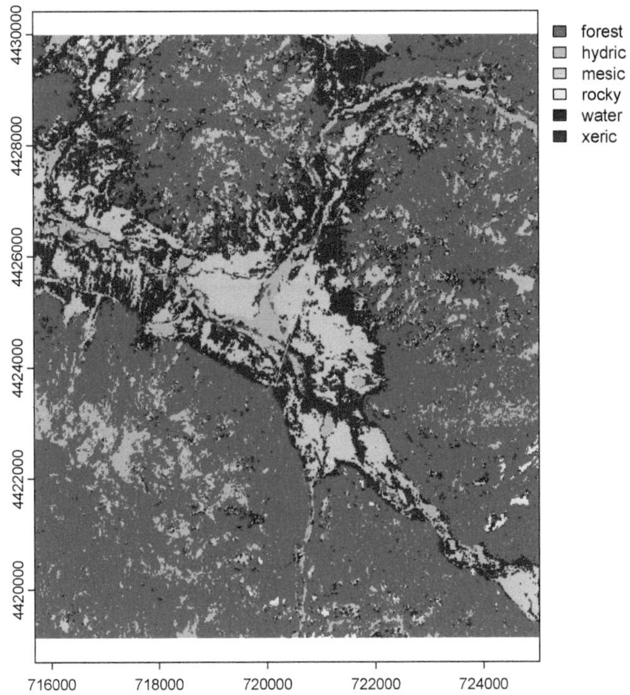

FIGURE 12.19 CART classification, highest probability class, Sentinel-2 20 m 20210628

```
LC20m <- which.max(CARTpred)
LC20m
```

```
## class       : SpatRaster
## dimensions  : 543, 468, 1  (nrow, ncol, nlyr)
## resolution  : 20, 20  (x, y)
## extent      : 715680, 725040, 4419120, 4429980  (xmin, xmax, ymin, ymax)
## coord. ref. : WGS 84 / UTM zone 10N (EPSG:32610)
## source      : memory
## name        : which.max
## min value   :         1
## max value   :         6
```

```
cls <- names(CARTpred)
df <- data.frame(id = 1:length(cls), class=cls)
levels(LC20m) <- df
LC20m
```

```
## class       : SpatRaster
## dimensions  : 543, 468, 1  (nrow, ncol, nlyr)
## resolution  : 20, 20  (x, y)
```

```
## extent      : 715680, 725040, 4419120, 4429980  (xmin, xmax, ymin, ymax)
## coord. ref. : WGS 84 / UTM zone 10N (EPSG:32610)
## source      : memory
## categories  : class
## name        :  class
## min value   : forest
## max value   :  xeric
```

```
LCcolors <- c("green4","cyan","gold","grey90","mediumblue","red")
plot(LC20m, col=LCcolors)
```

12.5.4 Validating the model

An important part of the imagery classification process is *validation*, where we look at how well the model works. The way this is done is pretty easy to understand, and requires having *testing* data in addition to the *training* data mentioned above. Testing data can be created in a variety of ways, commonly through field observations but also with finer resolution imagery like drone imagery. Since this is also how we *train* our data, often you're selecting some for training, and a separate set for testing.

12.5.4.1 The "overfit model" problem

It's important to realize that the accuracy we determine is *only* based on the training and testing data. The accuracy of the prediction of the classification elsewhere will likely be somewhat less than this, and if this is substantial our model is "overfit". We don't actually know how overfit a model truly is because that depends on how likely the conditions seen in our training and testing data also occur throughout the rest of the image; if those conditions are common, just not sampled, then the model might actually be pretty well fit.

In thinking about the concept of overfit models and selecting out training (and testing) sets, it's useful to consider the purpose of our classification and how important it is that our predictions are absolutely reliable. In choosing training sets, accuracy is also important, so we will want to make sure that they are good representatives of the land cover type (assuming that's our response variable). While some land covers are pretty clear (like streets or buildings), there's a lot of fuzziness in the world: you might be trying to identify wetland conditions based on the type of vegetation growing, but in a meadow you can commonly find wetland species mixed in with more mesic species – to pick a reliable wetland sample we might want to only pick areas with only wetland species (and this can get tricky since there are many challenges of "obligate wetland" or "facultative wetland" species.) The CART model applied a probability model for each response value, which we then used to derive a single prediction based on the maximum probability.

12.5.4.2 Cross validation : 20 m Sentinel-2 CART model

In *cross-validation*, we use one set of data and run the model multiple times, and for each validating with a part of the data not used for training the model. In *k-fold cross validation*, k represents the number of groups and number of models. The k value can be up to the number of observations, but you do need to consider processing time, and you may not get much more reliable assessments with using all of the observations. We'll use five folds.

The method we'll use to cross-validate the model and derive various accuracy metrics is based on code provided in the "Remote Sensing with *terra*" section of https://rspatial.org by Aniruddha Ghosh and Robert J. Hijmans (Hijmans (n.d.)).

```
set.seed(42)
k <- 5 # number of folds
j <- sample(rep(1:k, each = round(nrow(sampdataRCV))/k))
table(j)
```

```
## j
##  1  2  3  4  5
## 71 71 71 71 71
```

```
x <- list()
for (k in 1:5) {
    train <- sampdataRCV[j!= k, ]
    test <- sampdataRCV[j == k, ]
    cart <- rpart(as.factor(class)~., data=train, method = 'class',
                  minsplit = 5)
    pclass <- predict(cart, test, na.rm = TRUE)
    # assign class to maximum probability
    pclass <- apply(pclass, 1, which.max)
    # create a data.frame using the reference and prediction
    x[[k]] <- cbind(test$class, as.integer(pclass))
}
```

12.5.4.3 Accuracy and error metrics

Now that we have our test data with observed and predicted values, we can derive accuracy metrics. There have been many metrics developed to assess model classification accuracy and error. We can look at accuracy either in terms of how well a model works or what kinds of errors it has; these are complements. We'll look at *Producer's* and *User's* accuracies (and their complements *omission* and *commission* errors) below, and we'll start by building a *confusion matrix*, which will allow us to derived the various metrics.

Confusion Matrix One way to look at (and, in the process, *derive*) these metrics is to build a *confusion matrix*, which puts the **observed** *reference data* and **predicted** *classified data* in rows and columns. These can be set up in either way, but in both cases the diagonal shows those correctly classified, with the row and column names from the same list of land cover classes that we've provided to train the model.

```
y <- do.call(rbind, x)
y <- data.frame(y)
```

```
colnames(y) <- c('observed', 'predicted')
conmat <- table(y) # confusion matrix, a contingency table
# change the name of the classes
colnames(conmat) <- LCclass # sort(unique(sampdataRCV$class))
rownames(conmat) <- LCclass
print(conmat)
```

```
##          predicted
## observed forest hydric mesic rocky water xeric
##    forest     87      0     0     0     1     0
##    hydric      0     32     0     0     0     0
##    mesic       0      0   107     0     0     6
##    rocky       0      0     0     7     0     1
##    water       1      0     0     0    20     0
##    xeric       0      0     2     0     0    95
```

Overall Accuracy The overall accuracy is simply the overall ratio of the number correctly classified divided by the total, so the sum of the diagonal divided by n.

```
n <- sum(conmat) # number of total cases
diag <- diag(conmat) # number of correctly classified cases per class
OA <- sum(diag) / n # overall accuracy
OA
```

```
## [1] 0.9693593
```

kappa Another overall measure of accuracy is *Cohen's kappa* statistic (Cohen (1960)), which measures *inter-rater reliability* for categorical data. It's somewhat similar to the goal of chi square in that it compares *expected* vs. *observed* conditions, only chi square is limited to two cases or Boolean data.

```
rowsums <- apply(conmat, 1, sum)
p <- rowsums / n # observed (true) cases per class
colsums <- apply(conmat, 2, sum)
q <- colsums / n # predicted cases per class
expAccuracy <- sum(p*q)
kappa <- (OA - expAccuracy) / (1 - expAccuracy)
kappa
```

```
## [1] 0.9594579
```

Table of Producer's and User's accuracies

We can look at accuracy either in terms of how well a model works or what kinds of errors it has; these are complements. Both of these allow us to look at accuracy of classification *for each class.*

- The so-called "Producer's Accuracy" refers to how often real conditions on the ground are displayed in the classification, from the perspective of the map producer, and is the complement of errors of *omission.*

- The so-called "User's Accuracy" refers to how often the class will actually occur on the ground, so it is from the perspective of the user (though this is a bit confusing) and is the complement of errors of *commission*.

```
PA <- diag / colsums # Producer accuracy
UA <- diag / rowsums # User accuracy
outAcc <- data.frame(producerAccuracy = PA, userAccuracy = UA)
outAcc
```

```
##        producerAccuracy userAccuracy
## forest        0.9886364    0.9886364
## hydric        1.0000000    1.0000000
## mesic         0.9816514    0.9469027
## rocky         1.0000000    0.8750000
## water         0.9523810    0.9523810
## xeric         0.9313725    0.9793814
```

12.6 Classifying with 10 m Sentinel-2 Imagery

The 10 m product of Sentinel-2 has much fewer bands (4), but the finer resolution improves its visual precision. The four bands are the most common baseline set used in multispectral imagery: blue (B02, 490 nm), green (B03, 560 nm), red (B04, 665 nm), and NIR (B08, 842 nm). From this we can create color and NIR-R-G images, and the vegetation indices NDVI and NDGI, among others.

For the 10 m product, we won't repeat many of the exploration steps we looked at above, such as the histograms, but go right to training and classification using CART.

```
imgFolder <- paste0("S2A_MSIL2A_20210628T184921_N0300_R113_T10TGK_20210628T230915.",
                    "SAFE\\GRANULE\\L2A_T10TGK_A031427_20210628T185628")
img10mFolder <- paste0("~/sentinel2/",imgFolder,"\\IMG_DATA\\R10m")
imgDateTime <- str_sub(imgFolder,12,26)
imgDateTime
```

```
## [1] "20210628T184921"
```

```
imgDate <- str_sub(imgDateTime,1,8)
```

12.6.1 Subset bands (10 m)

The 10-m Sentinel-2 data includes only four bands blue (B02), green (B03), red (B04), and NIR (B08).

```
sentinelBands <- paste0("B",c("02","03","04","08"))
sentFiles <- paste0(img10mFolder,"/T10TGK_",imgDateTime,"_",sentinelBands,
                    "_10m.jp2")
sentinel <- rast(sentFiles)
names(sentinel) <- sentinelBands
```

12.6.2 Crop to RCV extent and extract pixel values

This is similar to what we did for the 20 m data.

```
RCVext <- ext(715680,725040,4419120,4429980)
sentRCV <- crop(sentinel,RCVext)
extRCV <- extract(sentRCV, ptsRCV)[,-1] # REMOVE ID
sampdataRCV <- data.frame(class = ptsRCV$class, extRCV)
```

12.6.3 Training the CART model (10 m) and plot the tree

Again, we can plot the regression tree for the model using 10 m data (Figure 12.20).

```
library(rpart); library(rpart.plot)
cartmodel <- rpart(as.factor(class)~., data = sampdataRCV,
                   method = 'class', minsplit = 5)
rpart.plot(cartmodel, fallen.leaves=F)
```

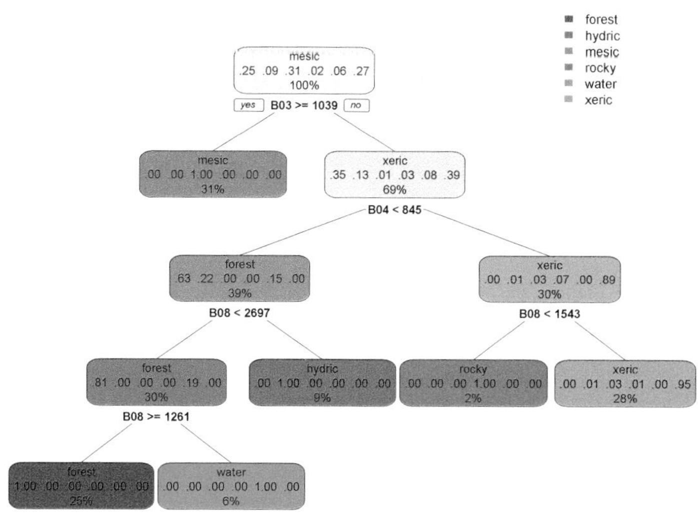

FIGURE 12.20 10 m CART regression tree

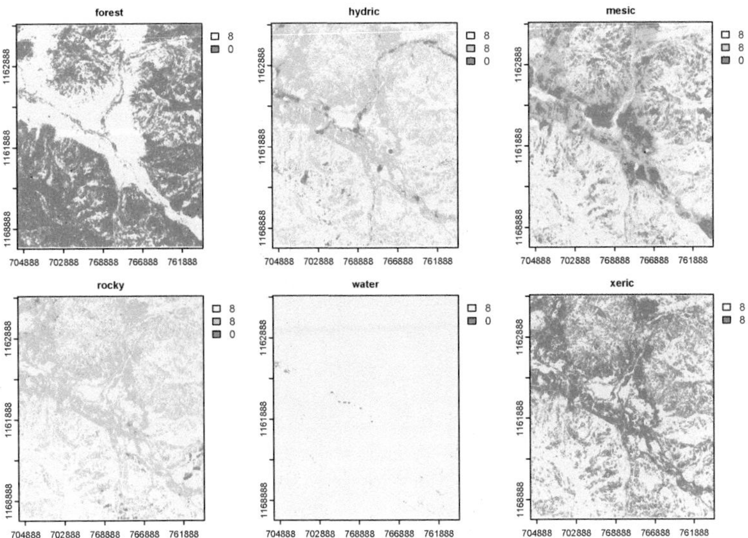

FIGURE 12.21 CART classification, probabilities of each class, Sentinel-2 10 m 20210628

12.6.4 Prediction using the CART model (10 m)

As with the 20 m prediction, we'll start by producing a composite of all class probabilities by pixel (Figure 12.21)...

```
CARTpred <- predict(sentRCV, cartmodel, na.rm = TRUE)
plot(CARTpred)
```

```
LC10m <- which.max(CARTpred)
cls <- names(CARTpred)
df <- data.frame(id = 1:length(cls), class=cls)
levels(LC10m) <- df
```

... and then map the class with the highest probability for each pixel (Figure 12.22).

```
plot(LC10m, col=LCcolors)
```

12.6.4.1 Cross-validation : 10 m Sentinel-2 CART model

```
set.seed(42)
k <- 5 # number of folds
j <- sample(rep(1:k, each = round(nrow(sampdataRCV))/k))
x <- list()
for (k in 1:5) {
```

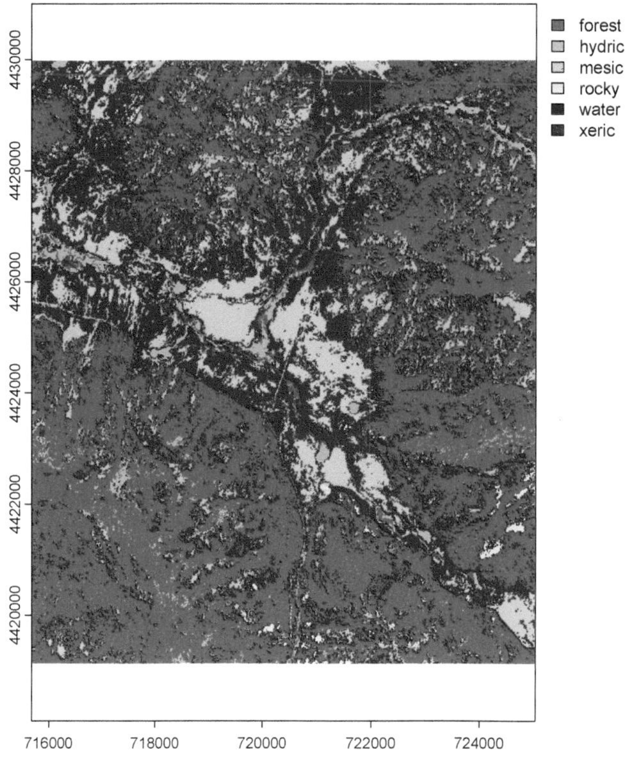

FIGURE 12.22 CART classification, highest probability class, Sentinel-2 10 m 20210628

```
    train <- sampdataRCV[j!= k, ]
    test <- sampdataRCV[j == k, ]
    cart <- rpart(as.factor(class)~., data=train, method = 'class',
                  minsplit = 5)
    pclass <- predict(cart, test, na.rm = TRUE)
    # assign class to maximum probablity
    pclass <- apply(pclass, 1, which.max)
    # create a data.frame using the reference and prediction
    x[[k]] <- cbind(test$class, as.integer(pclass))
}
```

12.6.4.2 Accuracy metrics : 10 m Sentinel-2 CART model

As with the 20 m model, we'll look at the confusion matrix and the various accuracy metrics.

Confusion Matrix (10 m)

```
y <- do.call(rbind, x)
y <- data.frame(y)
colnames(y) <- c('observed', 'predicted')
conmat <- table(y) # confusion matrix
# change the name of the classes
colnames(conmat) <- LCclass # sort(unique(sampdataRCV$class))
rownames(conmat) <- LCclass # sort(unique(sampdataRCV$class))
print(conmat)
```

```
##         predicted
## observed forest hydric mesic rocky water xeric
##   forest     87      0     0     0     0     1
##   hydric      0     30     0     0     0     2
##   mesic       0      0   109     0     0     4
##   rocky       0      0     1     5     0     2
##   water       1      0     0     0    20     0
##   xeric       0      0     0     0     0    97
```

Overall Accuracy (10 m)

```
n <- sum(conmat) # number of total cases
diag <- diag(conmat) # number of correctly classified cases per class
OA <- sum(diag) / n # overall accuracy
OA
```

```
## [1] 0.9693593
```

kappa (10 m)

```
rowsums <- apply(conmat, 1, sum)
p <- rowsums / n # observed (true) cases per class
colsums <- apply(conmat, 2, sum)
q <- colsums / n # predicted cases per class
expAccuracy <- sum(p*q)
kappa <- (OA - expAccuracy) / (1 - expAccuracy)
kappa
```

```
## [1] 0.9592908
```

User and Producer accuracy (10 m)

```
PA <- diag / colsums # Producer accuracy
UA <- diag / rowsums # User accuracy
outAcc <- data.frame(producerAccuracy = PA, userAccuracy = UA)
outAcc
```

```
##        producerAccuracy userAccuracy
## forest        0.9886364    0.9886364
```

```
## hydric        1.0000000   0.9375000
## mesic         0.9909091   0.9646018
## rocky         1.0000000   0.6250000
## water         1.0000000   0.9523810
## xeric         0.9150943   1.0000000
```

12.7 Classification Using Multiple Images Capturing Phenology

Most plants change their spectral response seasonally, a process referred to as *phenology*. In these montane meadows, phenological changes start with snow melt, followed by "green-up", then senescence. 2019 was a particularly dry year, so the period from green-up to senescence was short. We'll use a 3 June image as green-up, then a 28 June image as senescence.

Spring

```
library(terra); library(stringr)
imgFolderSpring <- paste0("S2B_MSIL2A_20210603T184919_N0300_R113_T10TGK_20210603",
                    "T213609.SAFE\\GRANULE\\L2A_T10TGK_A022161_20210603T185928")
img10mFolderSpring <- paste0("~/sentinel2/",imgFolderSpring,"\\IMG_DATA\\R10m")
imgDateTimeSpring <- str_sub(imgFolderSpring,12,26)
```

Summer

```
imgFolderSummer <- paste0("S2A_MSIL2A_20210628T184921_N0300_R113_T10TGK_20210628",
                    "T230915.SAFE\\GRANULE\\L2A_T10TGK_A031427_20210628T185628")
img10mFolderSummer <- paste0("~/sentinel2/",imgFolder,"\\IMG_DATA\\R10m")
imgDateTimeSummer <- str_sub(imgFolderSummer,12,26)
```

```
sentinelBands <- paste0("B",c("02","03","04","08"))
sentFilesSpring <- paste0(img10mFolderSpring,"/T10TGK_",
                          imgDateTimeSpring,"_",sentinelBands,"_10m.jp2")
sentinelSpring <- rast(sentFilesSpring)
names(sentinelSpring) <- paste0("spring",sentinelBands)

sentFilesSummer <- paste0(img10mFolderSummer,"/T10TGK_",
                          imgDateTimeSummer,"_",sentinelBands,"_10m.jp2")
sentinelSummer <- rast(sentFilesSummer)
names(sentinelSummer) <- paste0("summer",sentinelBands)
```

```
RCVext <- ext(715680,725040,4419120,4429980)
sentRCVspring <- crop(sentinelSpring,RCVext)
sentRCVsummer <- crop(sentinelSummer,RCVext)
```

```
blueSp <- sentRCVspring$springB02
greenSp <- sentRCVspring$springB03
redSp <- sentRCVspring$springB04
nirSp <- sentRCVspring$springB08
ndgiSp <- (0.616 * greenSp + 0.384 * nirSp - redSp)/
          (0.616 * greenSp + 0.384 * nirSp + redSp)
blueSu <- sentRCVsummer$summerB02
greenSu <- sentRCVsummer$summerB03
redSu <- sentRCVsummer$summerB04
nirSu <- sentRCVsummer$summerB08
ndgiSu <- (0.616 * greenSu + 0.384 * nirSu - redSu)/
          (0.616 * greenSu + 0.384 * nirSu + redSu)
```

12.7.1 Create a 10-band stack from both images

Since we have two imagery dates and also a vegetation index (NDGI) for each, we have a total of 10 variables to work with. So we'll put these together as a multi-band stack simply using `c()` with the various bands.

```
sentRCV <- c(blueSp,greenSp,redSp,nirSp,ndgiSp,blueSu,greenSu,redSu,nirSu,ndgiSu)
names(sentRCV) <- c("blueSp","greenSp","redSp","nirSp","ndgiSp",
                    "blueSu","greenSu","redSu","nirSu","ndgiSu")
```

12.7.2 Extract the training data (10 m spring + summer)

We'll repeat some of the steps where we read in the training data and set up the land cover classes, reducing the 7 to 6 classes, then extract from our new 10-band image stack the training data.

```
trainPolys <- vect(ex("RCVimagery/train_polys7.shp"))
ptsRCV <- spatSample(trainPolys, 1000, method="random")
LCclass <- c("forest","hydric","mesic","rocky","water","willow","xeric")
classdf <- data.frame(value=1:length(LCclass),names=LCclass)
trainPolys <- vect(ex("RCVimagery/train_polys7.shp"))
trainPolys[trainPolys$class=="willow"]$class <- "hydric"
ptsRCV <- spatSample(trainPolys, 1000, method="random")
LCclass <- c("forest","hydric","mesic","rocky","water","xeric")
classdf <- data.frame(value=1:length(LCclass),names=LCclass)

extRCV <- terra::extract(sentRCV, ptsRCV)[,-1] # REMOVE ID
sampdataRCV <- data.frame(class = ptsRCV$class, extRCV)
```

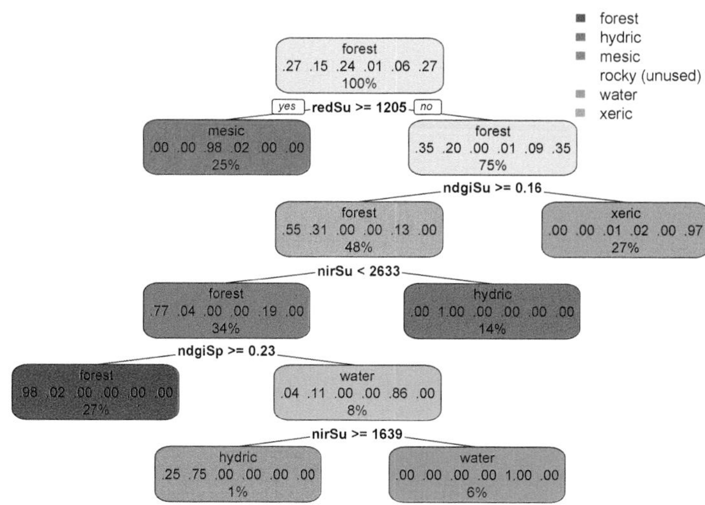

FIGURE 12.23 CART decision tree, Sentinel 10-m, spring and summer 2021 images

12.7.3 CART model and prediction (10 m spring + summer)

```
library(rpart)
# Train the model
cartmodel <- rpart(as.factor(class)~.,
                   data = sampdataRCV, method = 'class', minsplit = 5)
```

The CART model tree for the two-season 10 m example (Figure 12.23) interestingly does not end up with anything in the "rocky" pixels...

```
library(rpart.plot)
rpart.plot(cartmodel, fallen.leaves=F)
```

... and we can see in the probability map set that all "rocky" pixel probabilities are very low, below 0.03 (Figure 12.24) ...

```
CARTpred <- predict(sentRCV, cartmodel, na.rm = TRUE)
plot(CARTpred)
```

```
LCpheno <- which.max(CARTpred)
cls <- names(CARTpred)
df <- data.frame(id = 1:length(cls), class=cls)
levels(LCpheno) <- df
```

... and thus no "rocky" classes are assigned (Figure 12.25).

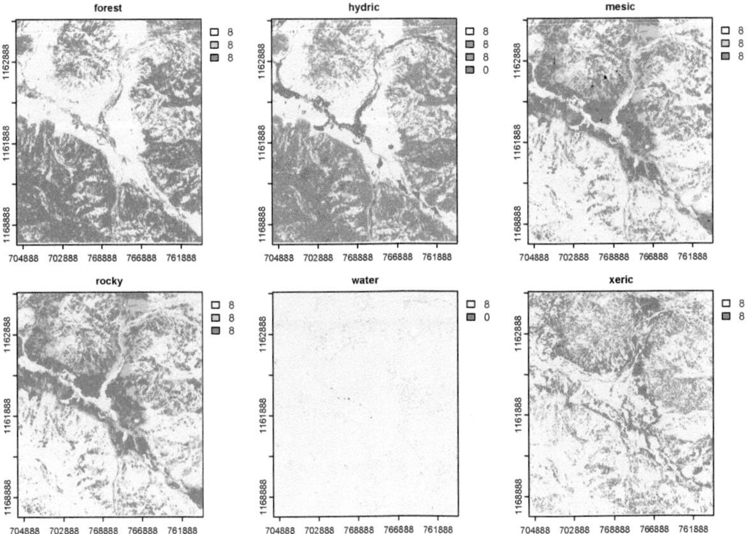

FIGURE 12.24 CART classification, probabilities of each class, Sentinel-2 10 m, 2021 spring and summer phenology

```
LCcolors <- c("green4","cyan","gold","grey90","mediumblue","red")
plot(LCpheno, col=LCcolors)
```

We might compare this plot with the two previous – the 20 m (Figure 12.26) and 10 m (Figure 12.27) classifications from 28 June.

```
plot(LC20m, col=LCcolors)
```

```
plot(LC10m, col=LCcolors)
```

12.7.3.1 Cross validation : CART model for 10 m spring + summer

```
set.seed(42)
k <- 5 # number of folds
j <- sample(rep(1:k, each = round(nrow(sampdataRCV))/k))
# table(j)
```

```
x <- list()
for (k in 1:5) {
    train <- sampdataRCV[j!= k, ]
    test <- sampdataRCV[j == k, ]
    cart <- rpart(as.factor(class)~., data=train, method = 'class',
```

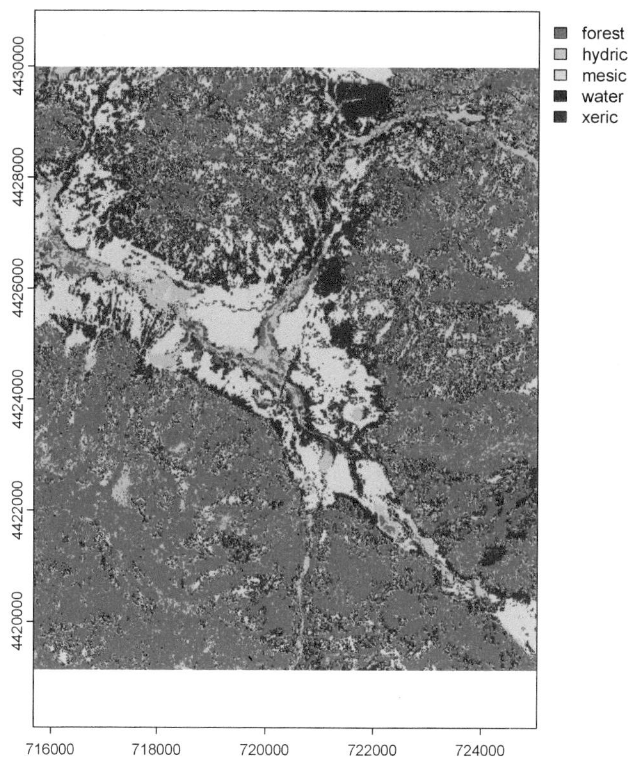

FIGURE 12.25 CART classification, highest probability class, Sentinel-2 10 m, 2021 spring and summer phenology

```
                 minsplit = 5)
    pclass <- predict(cart, test, na.rm = TRUE)
    # assign class to maximum probablity
    pclass <- apply(pclass, 1, which.max)
    # create a data.frame using the reference and prediction
    x[[k]] <- cbind(test$class, as.integer(pclass))
}
```

12.7.3.2 Accuracy metrics : CART model for 10 m spring + summer

Finally, as we did above, we'll look at the various accuracy metrics for the classification using the two-image (peak-growth spring and senescent summer) source.

Confusion Matrix

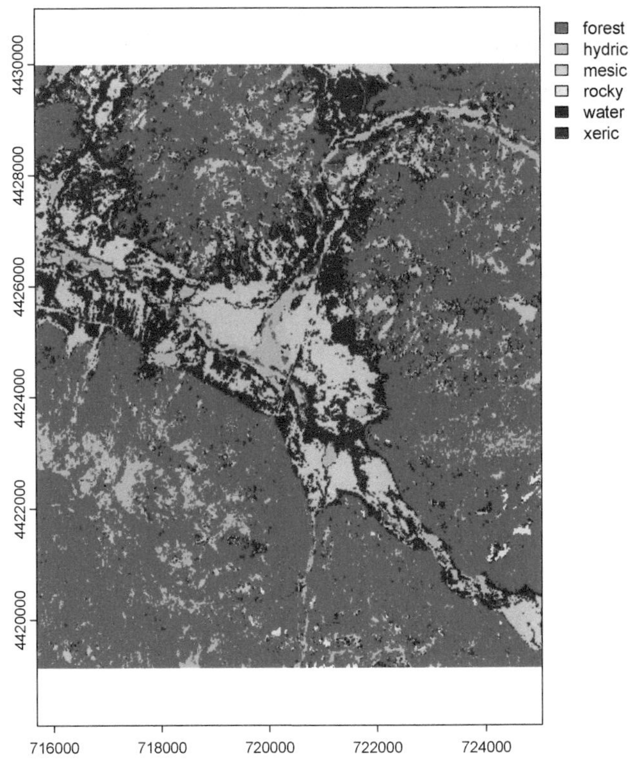

FIGURE 12.26 Classification of Sentinel-2 20 m image

```
y <- do.call(rbind, x)
y <- data.frame(y)
colnames(y) <- c('observed', 'predicted')
conmat <- table(y) # confusion matrix
# change the name of the classes
colnames(conmat) <- LCclass # sort(unique(sampdataRCV$class))
rownames(conmat) <- LCclass # sort(unique(sampdataRCV$class))
print(conmat)
```

```
##          predicted
## observed forest hydric mesic rocky water xeric
##   forest     96      2     0     0     1     0
##   hydric      2     53     0     0     1     0
##   mesic       0      0    89     1     0     1
##   rocky       0      0     2     0     0     2
##   water       1      1     0     0    22     0
##   xeric       0      0     0     0     0    99
```

Overall Accuracy

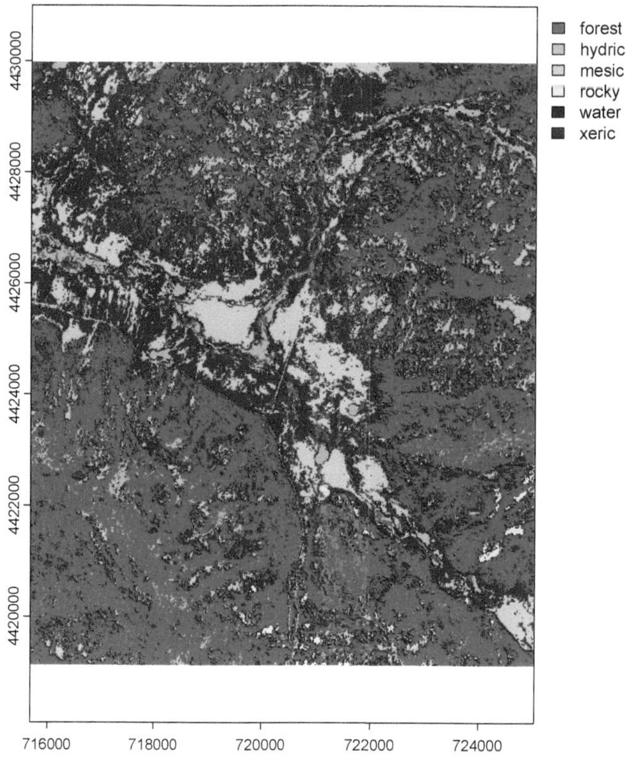

FIGURE 12.27 Classification of Sentinel-2 10 m spring and summer images

```
n <- sum(conmat) # number of total cases
diag <- diag(conmat) # number of correctly classified cases per class
OA <- sum(diag) / n # overall accuracy
OA
```

```
## [1] 0.9624665
```

kappa

```
rowsums <- apply(conmat, 1, sum)
p <- rowsums / n # observed (true) cases per class
colsums <- apply(conmat, 2, sum)
q <- colsums / n # predicted cases per class
expAccuracy <- sum(p*q)
kappa <- (OA - expAccuracy) / (1 - expAccuracy)
kappa
```

```
## [1] 0.9513023
```

User and Producer accuracy

```
PA <- diag / colsums # Producer accuracy
UA <- diag / rowsums # User accuracy
outAcc <- data.frame(producerAccuracy = PA, userAccuracy = UA)
outAcc
```

```
##          producerAccuracy userAccuracy
## forest          0.9696970    0.9696970
## hydric          0.9464286    0.9464286
## mesic           0.9780220    0.9780220
## rocky           0.0000000    0.0000000
## water           0.9166667    0.9166667
## xeric           0.9705882    1.0000000
```

12.8 Conclusions and Next Steps for Imagery Classification

In this chapter, we've seen that we can effectively classify imagery (at least produce convincing maps!), but you can probably sense from the varying results that classification isn't always as consistent as we might expect. We saw something similar with interpolation models; you can get very different results with different methods, and so you need to consider what method and settings might be most appropriate, based on the nature of your data – its scale and density, for instance – and the goals of your research.

The next steps might be to compare some of these classifications with our field data (basically go back to the training set), but I would be tempted to do that in ArcGIS, ERDAS, or maybe QGIS after writing out our classification rasters to geotiffs. We could then also take advantage of the better cartographic capabilities of GIS and remote sensing software when working with multi-band imagery.

We've really only scratched the surface on *remote sensing* methods for imagery classification, and as noted above, readers are encouraged to enroll in remote sensing classes at your local university to learn more of its theory and relevant technology. For instance, why did we look at Sentinel-2, and not Landsat, PlanetScope, ASTER, Spot, ... (the list goes on)? There are lots of satellites out there, more all the time, and understanding the the spectral-sensing capabilities of their sensors, and the spatial and temporal properties of their missions, is essential for making the right choice for your research.

There's also the consideration of relative cost, both for the imagery itself and for your time in processing it. We chose Sentinel-2 because it's free and served the purpose well, since the 10 and 20 m resolution worked at the scale of the study area in question. It also actually classifies more readily than finer resolution imagery. But there's a lot more to that decision, and we also employ 1 and 5 cm pixel imagery from drone surveys of the same study area, but for a different purpose (erosion studies) and at much lower image capture frequency, given the significant field time required.

Finally, we focused on pixel classifiers. One advantage of these is that they're reasonably similar to general classification models, whether using machine learning or not. But there are other imagery classification methods developed in the remote-sensing as well as medical-imaging communities, such as object-based classifiers which use vector geometry to aid in the classification of features. As with so much else in remote sensing methods, we'll leave those and other methods for you to explore.

12.9 Exercises: Imagery Analysis and Classification Models

The exercises for this chapter are to apply some of the same methods above to your own downloaded Sentinel2 scene. We won't use training sets for spectral signatures or supervised classification, but that would be the next step once you've collected training data, either in the field, from high-resolution imagery, or from a combination of sources.

Exercise 12.1. Download another Sentinel-2 image from Copernicus (you'll need to create a free account to get SciHub credentials), and use it to create RGB and NIR-R-G image displays of the entire imagery scene.

For downloading (and other processing), you might also try the `sen2r` package. You'll also need SciHub credentials, which you can get from https://scihub.copernicus.eu/. And while `sen2r` will download the entire images without it, Google Cloud SDK must be installed (see https://cloud.google.com/sdk/docs/install), which in turn will require a Python installation to do other processing.

Caution: These files can be very large, with each scene about a gigabyte, so be careful to limit how many images you're downloading. It's a good idea to be selective about which scenes to acquire, and the Copernicus Open Access Hub is a good interface for using visual inspection, which helps you decide if an image shows your area of interest well. For instance, while automated selection of percentage cloud cover is useful as a first step, your area of interest may be an anomaly: even a 50% cloud coverage for the overall scene may not cover your area of interest, or a 5% cloud coverage may sit right over it.

Exercise 12.2. From your imagery, find coordinates to crop with for an area of interest, then as we've done for Red Clover Valley, create clipped RGB and NIR-R-G displays.

Exercise 12.3. Create and display NDVI and NDGI indices from your cropped image.

Exercise 12.4. From the same cropped image, create a k-means unsupervised classification, with your choice of number of classes.

Part IV

Time Series

13

Time Series Visualization and Analysis

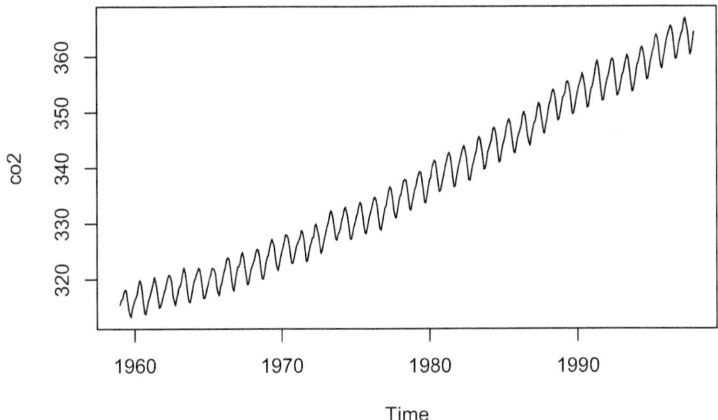

We've been spending a lot of time in the spatial domain in this book, but environmental data clearly also exist in the time domain, such as what's been collected about CO_2 on Mauna Loa, Hawai'i, since the late 1950s (shown above). Environmental conditions change over time; and we can sample and measure air, water, soil, geology, and biotic systems to investigate explanatory and response variables that change over time. And just as we've looked at multiple geographic scales, we can look at multiple temporal scales, such as the long-term Mauna Loa CO_2 data above, or much finer-scale data from data loggers, such as eddy covariance flux towers (Figures 13.1 and 13.2).

FIGURE 13.1 Red Clover Valley eddy covariance flux tower installation

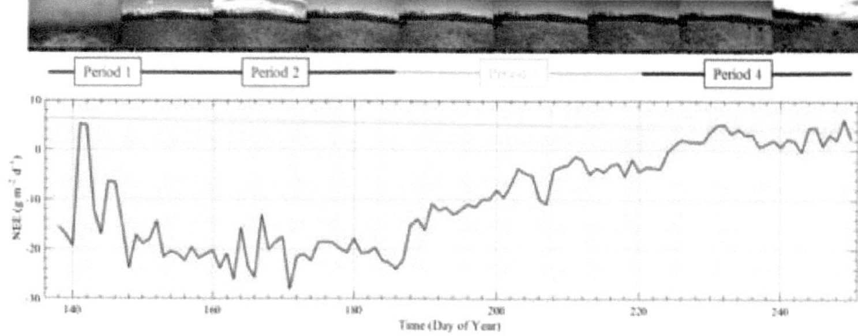

FIGURE 13.2 Loney Meadow net ecosystem exchange (NEE) results (Blackburn et al. 2021)

In this chapter, we'll look at a variety of methods for working with data over time. Many of these methods have been developed for economic modeling, but have many applications in environmental data science. The *time series* data type has been used in R for some time, and in fact a lot of the built-in data sets are time series, like the Mauna Loa CO_2 data or the flow of the Nile (Figure 13.3).

```
plot(Nile)
```

So what's the benefit of a time series data object? Why not just use a data frame with time stamps as a variable?

You could just work with these data as data frames, and often this is all you need. In this chapter, for instance, we won't work exclusively with time series data objects; our data may just include variation over time. However there are advantages in R understanding how to deal with actual time series in a temporal dimension, including knowing what time

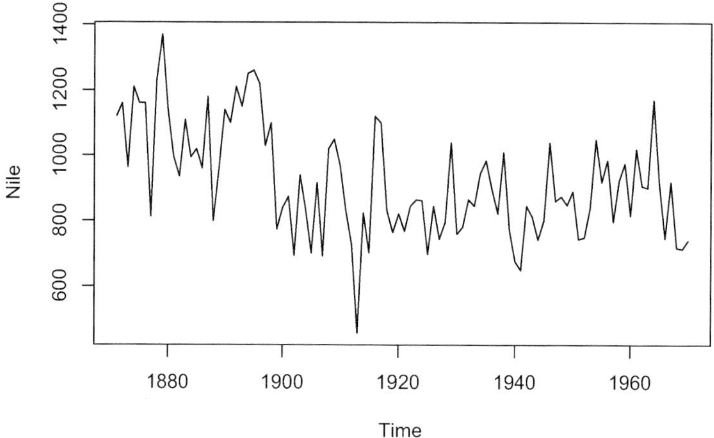

FIGURE 13.3 Time series of Nile River flows

period we're dealing with, especially if there are regular fluctuations over that period. This is similar to spatial data where it's useful to maintain the geometry of our data and the associated coordinate system. So we should learn how time series objects work and how they're built, just as we did with spatial data.

13.1 Structure, Seasonality, and Decomposition of Time Series

A time series is comprised of a series of samples of a phenomenon at a specific time **interval**, such as 10 seconds, 4 hours, 1 day, or 20 years (the unit of time is important), where that interval might be the time spacing of observations such as temperature, precipitation, solar radiation, or water quality, to name a few of the many environmental parameters commonly monitored by government agencies. These are organized into longer regular **periods** of time, such as days or years, by specifying the **frequency** of samples in each period. This is especially useful for data that have periodicity, such as diurnal or seasonal climatic variables.

We'll thus create the time series with a period in mind by defining its sampling frequency, and considering the time unit. For example, if your data are *monthly* samples and you specify a frequency of 12, your period is one year; similarly a frequency of 365 with *daily* observations will also create a yearly period. But we don't have to limit our intervals to single units of time: a frequency of 6 two-month samples will also create a yearly period, just as a frequency of 12 two-hour interval samples will create a daily period.

With periodic data, we have the concept of *seasonality* where we want to consider values changing as a cycle over a regular time period, whether it's seasons of the year or "seasons" of the day or week or other time period. There may also be a **trend** over time. *Decomposing* the time series will display all of these things.

A good data set to observe seasonality and trends is the Mauna Loa CO_2 time series, with regular annual cycles, yet with a regularly increasing trend over the years, as shown above in the graphic at the start of the chapter. One time-series graphic, a **decomposition**, shows the original observations, followed by a trend line that removes the seasonal and local (short-term) random irregularities, a detrended *seasonal* picture which removes that trend to just show the seasonal cycles, followed by the random irregularities (Figure 13.4).

```
plot(decompose(co2))
```

In reading this graph, note the vertical scale: the units are all the same – parts per million – so the actual amplitude of the seasonal cycle should be the same as the annual amplitude of the observations. It's just scaled to the chart height, which tends to exaggerate the seasonal cycles and random irregularities.

So we'll look at a more advanced decomposition called *Seasonal Decomposition of Time Series with Loess* in the `stl` package that also provides a scale bar, allowing us to compare their amplitudes; we might, for instance, see where random variation is greater than the seasonal amplitude. If `s.window = "periodic"`, the mean is used for smoothing: the seasonal

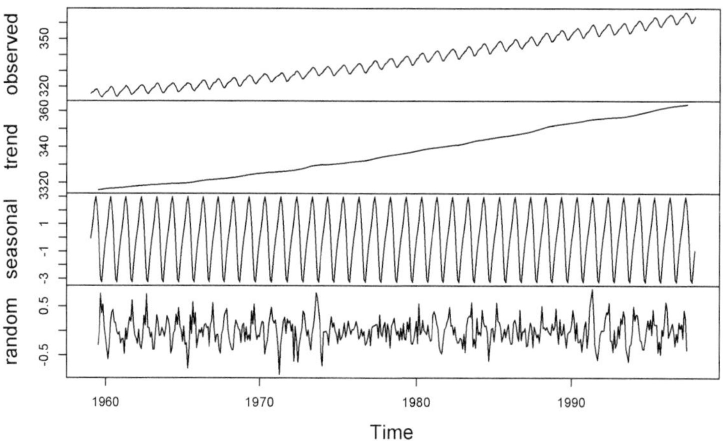

FIGURE 13.4 Decomposition of Mauna Loa CO2 data

values are removed and the remainder smoothed to find the trend, as we'll see in the Mauna Loa CO_2 data (Figure 13.5).

```
plot(stl(co2, s.window="periodic"))
```

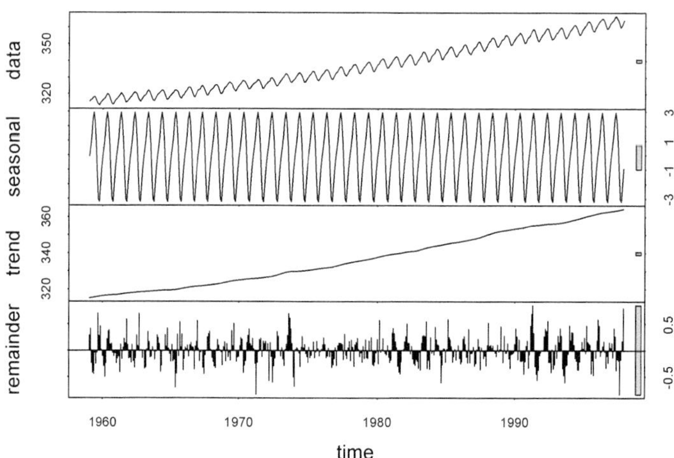

FIGURE 13.5 Seasonal deomposition of time series using loess (stl) applied to CO2

13.2 Creation of Time Series (ts) Data

So far, we've been using built-in time series data, and there are many such available data sets, but we will need to understand how to create them. A time series (ts) can be created with the `ts()` function:

- the period of time we're organizing our data into (e.g. daily, yearly) can be anything
- observations must be a regularly spaced series
- time values are not normally included as a variable in the time series

We'll build a simple time series with monthly high and low temperatures in San Francisco I just pulled off a Google search. We'll start with just building a data frame, first converting the temperatures from Fahrenheit to Celsius (Figure 13.6).

```
library(tidyverse)
SFhighF <- c(58,61,62,63,64,67,67,68,71,70,64,58)
SFlowF  <- c(47,48,49,50,51,53,54,55,56,55,51,47)
SFhighC <- (SFhighF-32)*5/9
SFlowC  <- (SFlowF-32)*5/9
SFtempC <- bind_cols(high=SFhighC,low=SFlowC)
plot(ts(SFtempC), main="SF temperature, monthly time unit")
```

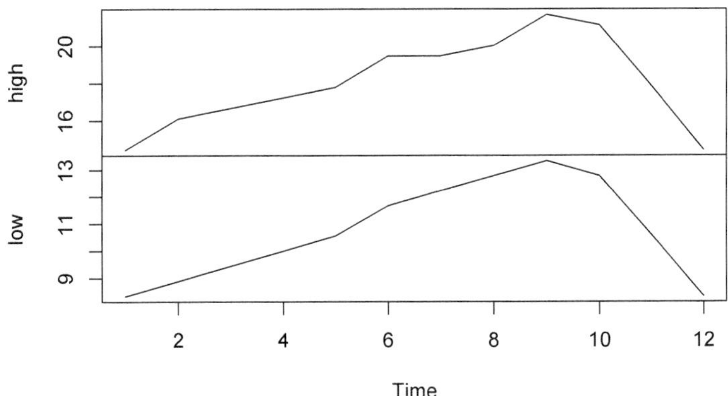

FIGURE 13.6 San Francisco monthly highs and lows as time series

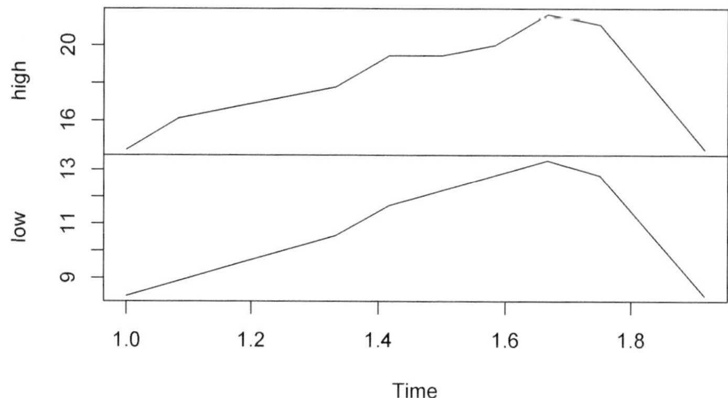

FIGURE 13.7 SF data with yearly period

13.2.1 Frequency, start, and end parameters for `ts()`

To convert this data frame into a time series, we'll need to provide at least a `frequency` setting (the default `frequency=1` was applied above), which sets how many observations per period. The `ts()` function doesn't seem to care what the period is in reality, however some functions figure it out, at least for an annual period, e.g. that 1-12 means months when there's a frequency of 12 (Figure 13.7).

```
plot(ts(SFtempC, frequency=12), main="yearly period")
```

frequency < 1

You can have a frequency less than 1. For instance, for greenhouse gases (CO_2, CH_4, N_2O) captured from ice cores in Antarctica (Law Dome Ice Core, just south of Cape Poinsett, Antarctica (Irizarry (2019))), there are values for every 20 years, so if we want year to be the period, the frequency would need to be to be 1/20 years = 0.05 (Figure 13.8). Note the use of a pivot to transform the data.

```
library(dslabs)
data("greenhouse_gases")
GHGwide <- pivot_wider(greenhouse_gases, names_from = gas,
                       values_from = concentration)
GHG <- ts(GHGwide, frequency=0.05, start=20)
plot(GHG)
```

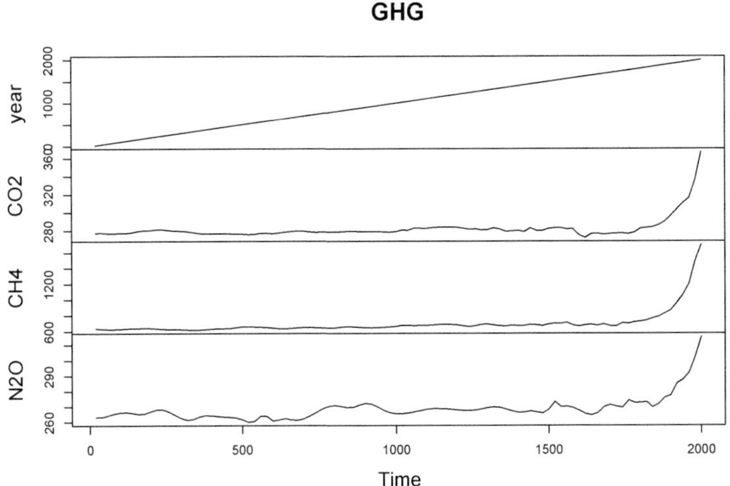

FIGURE 13.8 Greenhouse gases with 20 year observations, so 0.05 annual frequency

13.2.2 Associating times with time series

In the greenhouse gas data above, we not only specified `frequency` to determine the period, we also provided a `start` value of 20 to provide an actual year to start the data with. Time series don't normally include time stamps, so we need to specify **`start`** and **`end`** parameters. This can be either a single number (if it's the first reading during the period) or a vector of two numbers (the second of which is an integer), which specify a natural period and a number of samples into that period (e.g. which month during the year, or which hour, minute or second during that day, etc.)

Example with year as the period and monthly data, starting July 2019 and ending June 2020:

```
frequency=12, start=c(2019,7), end=c(2020,6)
```

13.2.3 Subsetting time series by times

Time series can be queried for their `frequency`, `start`, and `end` properties, and we'll look at these for a built-in time series data set, `sunspot.month`, which has monthly values of sunspot activity from 1749 to 2013 (Figure 13.9).

```
plot(sunspot.month)
```

```
frequency(sunspot.month)
start(sunspot.month)
end(sunspot.month)
```

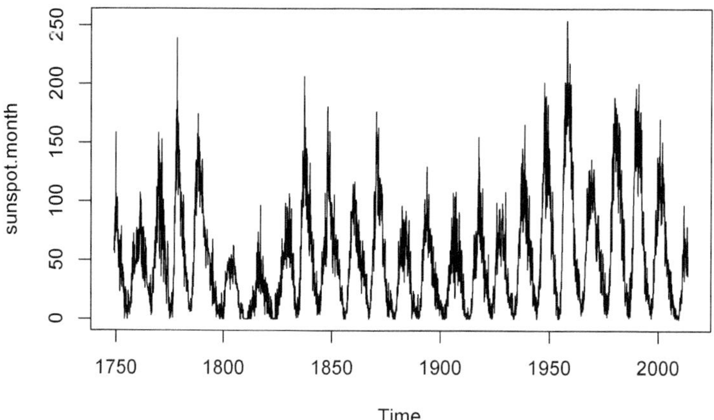

FIGURE 13.9 Monthly sunspot activity from 1749 to 2013

```
## [1] 12
## [1] 1749    1
## [1] 2013    9
```

To subset a section of a time series, we can use the `window` function (Figure 13.10).

```
plot(window(sunspot.month, start = 1940, end = 1970))
```

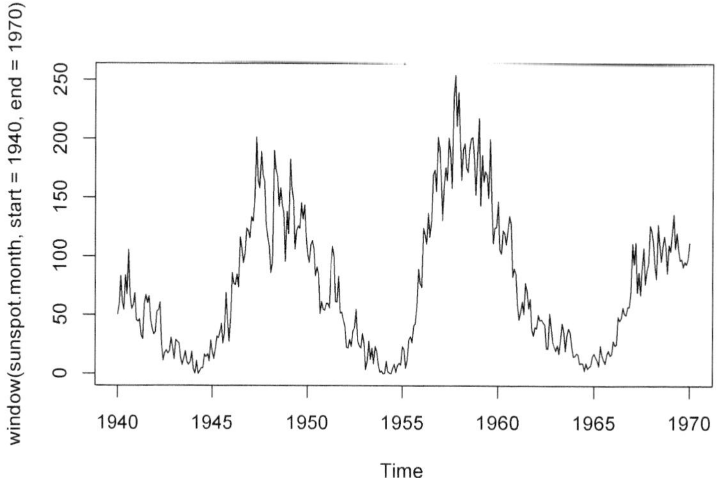

FIGURE 13.10 Monthly sunspot activity from 1940 to 1970

13.2.4 Changing the frequency to use a different period

We commonly use periods of days or years, but other options are possible. During the COVID-19 pandemic, weekly cycles of cases were commonly reported due to the weekly work cycle and weekend activities. Lunar tidal cycles are another.

Sunspot activity is known to fluctuate over about a 11-year cycle, and other than the well-known effects on radio transmission is also part of short-term climate fluctuations (so must be considered when looking at climate change). We can see the 11-year cycle in the above graphs, and we can perhaps see it more clearly in a decomposition where we use an 11-year cycle as the period.

To do this, we need to change the frequency. This is pretty easy to do, but to understand the process, first realize that we can take the original time series and create an identical copy to see how we can set the same parameters that already exist:

```
sunspot <- ts(sunspot.month, frequency=12, start=c(1749,1))
```

To create one with an 11-year sunspot cycle would then require just setting an appropriate frequency:

```
sunspotCycles <- ts(sunspot.month, frequency=11*12, start=c(1749,1))
```

But we might want to figure out when to start, so let's window it for the first 20 years so we can visualize it (Figure 13.11).

```
plot(window(sunspot.month, end=c(1769,12)))
```

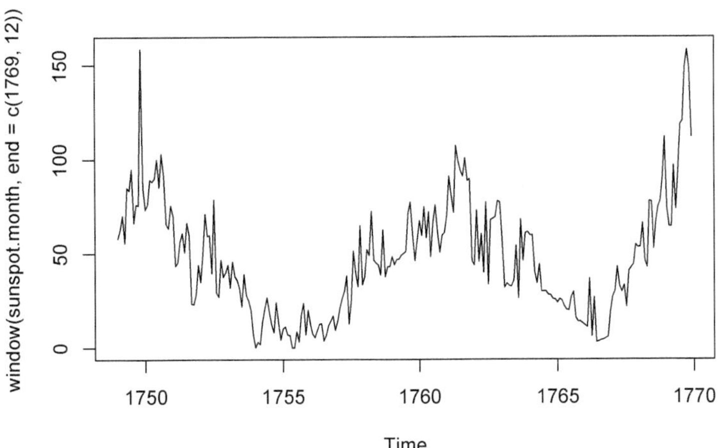

FIGURE 13.11 Sunspots of the first 20 years of data

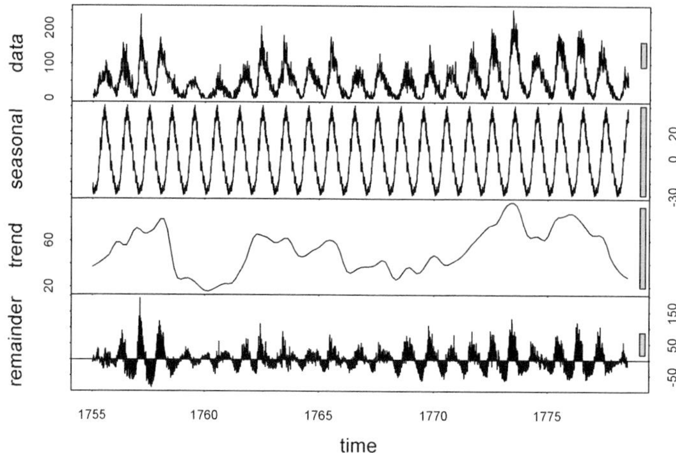

FIGURE 13.12 11-year sunspot cycle decomposition

Looks like a low point about 1755, so let's start there:

```
sunspotCycles <- ts(window(sunspot.month, start=1755), frequency=11*12, start=1755)
```

Now this is assuming that the sunspot cycle is *exactly* 11 years, which is probably not right, but a look at the decomposition might be useful (Figure 13.12).

```
plot(stl(sunspotCycles,s.window="periodic"))
```

The seasonal picture makes things look simple, but it's deceptive. The sunspot cycle varies significantly; and since we can see (using the scale bars) that the random variation is much greater than the seasonal, this suggests that we might not be able to use this periodicity reliably. But it was worth a try, and perhaps a good demonstration of an unconventional type of period.

13.2.5 Time stamps and extensible time series

What we've been looking at so far are data in regularly spaced series where we just provide the date and time as when the series starts and ends. The data don't include what we'd call *time stamps* as a variable, and these aren't really needed if our data set is regularly spaced. However, there are situations where there may be missing data or imperfectly spaced data. Missing data for regularly spaced data can simply be entered as `NA`, but imperfectly spaced data may need a time stamp. One way is with *extensible time series*, which we'll look at below, but first let's look at a couple of `lubridate` functions for working with days of the year and Julian dates, which can help with time stamps.

13.2.5.1 Lubridate and Julian dates

As we saw earlier, the `lubridate` package makes dealing with dates a lot easier, with many routines for reading and manipulating dates in variables. *If you didn't really get a good understanding of using dates and times with lubridate, you may want to review that section earlier in the book. Dates and times can be difficult to work with, so you'll need to make sure you know the best tools, and lubridate has the best tools.*

One variation on dates it handles is the number of days since some start date, using either the year day **`yday()`** (for any given year, starting with `1` on the first of January) or the **`julian()`** date for any starting date. The default origin for `julian()` is `1970-01-01`, which surprisingly yields a Julian date of **`0`**, but the same date would be a year day of **`1`**.

For climatological work where the year is very important, we usually want the year day, ranging from `1` on the first day of January to `365` or `366` on 31 December, of any given year. It's the same thing as setting the origin for `julian()` as 12-31 of the previous year, and is useful in time series when the year is the period and observations are by days of the year.

```
library(lubridate)
Jan1_1970 <- ymd("1970-01-01")
Jan1_1970
paste("Julian date:",julian(Jan1_1970))
paste("Year day   :",yday(Jan1_1970))
```

```
## [1] "1970-01-01"
## [1] "Julian date: 0"
## [1] "Year day   : 1"
```

```
Jan1_2020 <- ymd("2020-01-01")
Jan1_2020
paste("Year day           :",yday(Jan1_2020))
paste("Julian date        :",julian(Jan1_2020))
paste("  with origin set:",julian(Jan1_2020, origin=ymd("2019-12-31")))
```

```
## [1] "2020-01-01"
## [1] "Year day           : 1"
## [1] "Julian date        : 18262"
## [1] "  with origin set: 1"
```

To create a decimal `yday` including fractional days, you could use a function like the following, which simply uses the number of hours in a day, minutes in a day, and seconds in a day.

```
ydayDec <- function(d) {
           yday(d)-1 + hour(d)/24 + minute(d)/1440 + second(d)/86400}
datetime <- now()
print(datetime)
```

```
## [1] "2022-12-01 17:32:07 PST"
```

```
print(ydayDec(datetime), digits=12)
```

```
## [1] 334.730644588
```

13.2.5.2 Extensible time series (xts)

We'll look at an example where we need to provide time stamps. Weekly water quality data (*E. coli*, total coliform, and *Enterococcus* bacterial counts) were downloaded from San Mateo County Health Department for San Pedro Creek, collected weekly since 2012, usually on Monday but sometimes on Tuesday, and occasionally with long gaps in the data, but with the date recorded in the variable `SAMPLE_DATE`.

First, we'll just look at this with the data frame, selecting date, total coliform, and *E. coli*:

```
library(igisci)
library(tidyverse)
SanPedroCounts <- read_csv(ex("SanPedro/SanPedroCreekBacterialCounts.csv")) %>%
  filter(!is.na(`Total Coliform`)) %>%
  rename(Total_Coliform = `Total Coliform`, Ecoli = `E. Coli`) %>%
  dplyr::select(SAMPLE_DATE,Total_Coliform,Ecoli)
```

To create a time series from the San Pedro Creek *E. coli* data, we need to use the package **xts** (Extensible Time Series), using `SAMPLE_DATE` to order the data (Figure 13.13).

```
library(xts); library(forecast)
Ecoli <- xts(SanPedroCounts$Ecoli, order.by = SanPedroCounts$SAMPLE_DATE)
attr(Ecoli, 'frequency') <- 52
plot(as.ts(Ecoli))
```

Then a decomposition can be produced of the E. coli data (Figure 13.14). Note the use of `as.ts` to produce a normal time series.

```
plot(stl(as.ts(Ecoli),s.window="periodic"))
```

What we can see from the above decomposition is no real signal from the annual cycle (52 weeks), since while there is some tendency for peak levels ("first-flush" events) happening late in the calendar year, the timing of those first heavy rains varies significantly from year to year.

As we saw above with the decomposition, *E. coli* values don't exhibit a clear annual cycle; the same is apparent with the `stl` result here. The trend may, however, be telling us something about the general trend of bacterial counts over time.

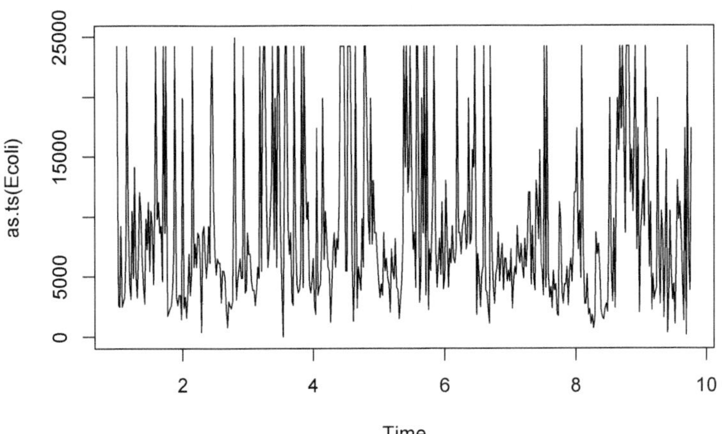

FIGURE 13.13 San Pedro Creek E. coli time series

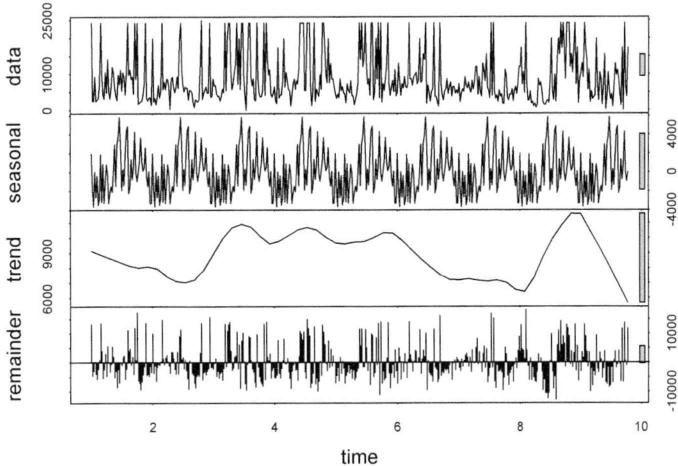

FIGURE 13.14 Decomposition of weekly E. coli data, annual period (frequency 52)

13.2.5.3 The tsibble

The tsibble package provides an alternative way to build time series as "a data- and model-oriented object" (Wang, Cook, and Hyndman (2020)), following the principles of the tidyverse. It uses an **index** to order records, one or more **key** variables that define observational units, with each observation uniquely identified by an index-key combination. In contrast to regular time series, tsibbles preserve the time index as a data variable. You can convert either regular time series or data frames with dates to tsibbles with the `as_tsibble` function, as long as things are set up right. We'll stick with regular time series (and the use of `xts` to handle time stamps), but the reader might consider exploring the tsibble package, as described at https://tsibble.tidyverts.org/, to learn more.

13.3 Data smoothing: moving average (ma)

The great amount of fluctuation in bacterial counts we see above makes it difficult to interpret the significant patterns. A moving average is a simple way to generalize data, using an order parameter to define the size of the moving window, which should be an odd number.

To see how it works, let's apply a `ma()` to the SF data we created a few steps back. This isn't a data set we need to smooth, but we can see how the function works by comparing the original and averaged data.

```
library(forecast)
ts(SFtempC, frequency=12)
```

```
##             high       low
## Jan 1 14.44444  8.333333
## Feb 1 16.11111  8.888889
## Mar 1 16.66667  9.444444
## Apr 1 17.22222 10.000000
## May 1 17.77778 10.555556
## Jun 1 19.44444 11.666667
## Jul 1 19.44444 12.222222
## Aug 1 20.00000 12.777778
## Sep 1 21.66667 13.333333
## Oct 1 21.11111 12.777778
## Nov 1 17.77778 10.555556
## Dec 1 14.44444  8.333333
```

```
ma(ts(SFtempC,frequency=12),order=3)
```

```
##             [,1]      [,2]
## Jan 1         NA        NA
```

```
## Feb 1 15.74074  8.888889
## Mar 1 16.66667  9.444444
## Apr 1 17.22222 10.000000
## May 1 18.14815 10.740741
## Jun 1 18.88889 11.481481
## Jul 1 19.62963 12.222222
## Aug 1 20.37037 12.777778
## Sep 1 20.92593 12.962963
## Oct 1 20.18519 12.222222
## Nov 1 17.77778 10.555556
## Dec 1       NA        NA
```

A moving average can also be applied to a vector or a variable in a data frame – doesn't have to be a time series. For the water quality data, we'll use a large order (15) to best clarify the significant pattern of *E. coli* counts (Figure 13.15).

```
ggplot(data=SanPedroCounts) + geom_line(aes(x=SAMPLE_DATE, y=ma(Ecoli, order=15)))
```

Moving average of CO_2 data

We'll look at another application of a moving average looking at CO_2 data from Antarctic ice cores (Irizarry (2019)), first the time series itself (Figure 13.16)...

```
library(dslabs)
data("greenhouse_gases")
GHGwide <- pivot_wider(greenhouse_gases,
           names_from = gas,
           values_from = concentration)
CO2 <- ts(GHGwide$CO2,
```

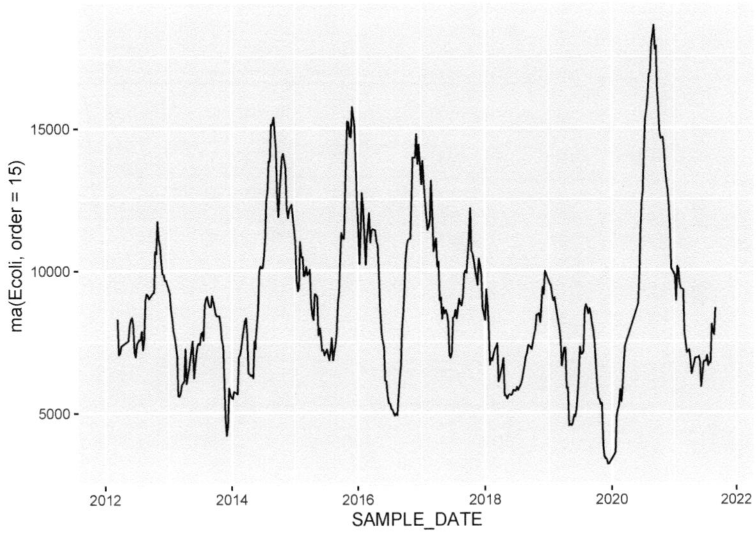

FIGURE 13.15 Moving average (order=15) of E. coli data

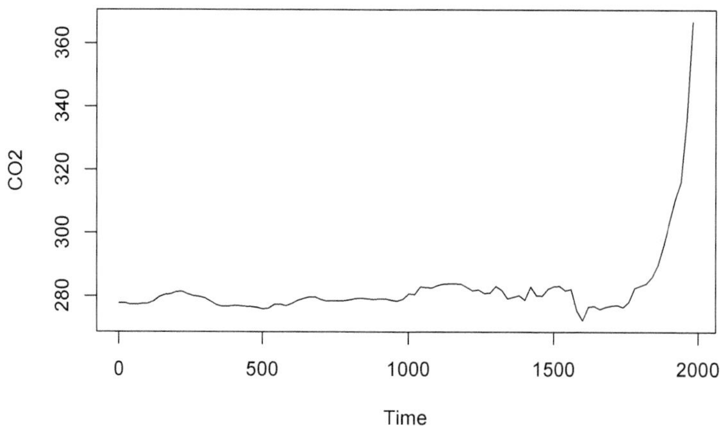

FIGURE 13.16 GHG CO2 time series

```
          frequency = 0.05)
plot(CO2)
```

... and then the moving average with an order of 7 (Figure 13.17).

```
library(forecast)
CO2ma <- ma(CO2, order=7)
plot(CO2ma)
```

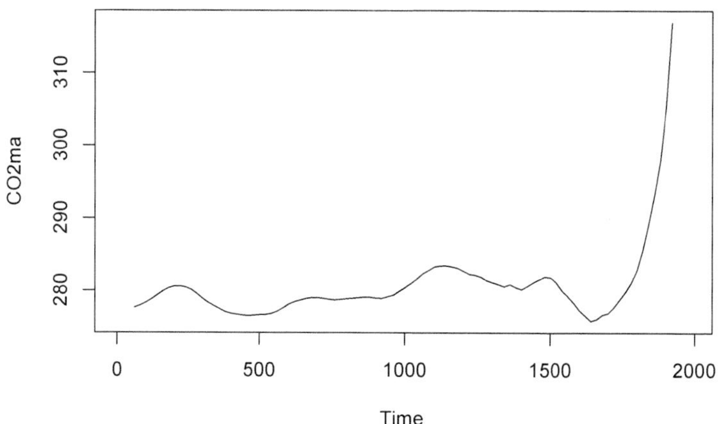

FIGURE 13.17 Moving average (order=7) of CO2 time series

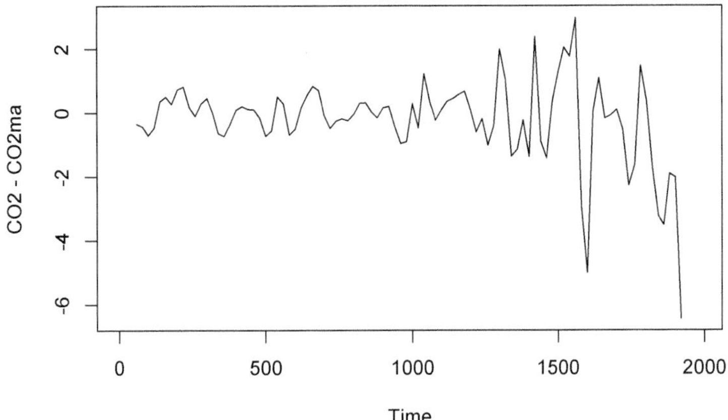

FIGURE 13.18 Random variation seen by subtracting moving average

For periodic data we can decompose time series into seasonal and trend signals, with the remainder left as random variation. But even for non-periodic data, we can consider random variation by simply subtracting the major signal of the moving average from the original data (Figure 13.18).

```
plot(CO2-CO2ma)
```

For the *E. coli*, maybe we can use a moving average to try to see a clearer pattern? We'll need to use `tseries::na.remove` to remove the `NA` at the beginning and end of the data, and I'd probably recommend making sure we're not creating any time errors doing this (Figure 13.19).

```
ecoli15 <- tseries::na.remove(as.ts(ma(Ecoli, order=15)))
plot(stl(ecoli15, s.window="periodic"))
```

13.4 Decomposition of data logger data: Marble Mountains

Data loggers can create a rich time series data set, with readings repeated at very short time frames (depending on the capabilities of the sensors and data logging system) and extending as far as the instrumentation allows. For a study of chemical water quality and hydrology of a karst system in the Marble Mountains of California, a spring *resurgence* was instrumented to measure water level, temperature, and specific conductance (a surrogate for total dissolved solids) over a four-year time frame (JD Davis and Davis 2001) (Figures 13.20 and 13.21). The frequency was quite coarse – one sample every two hours – but still

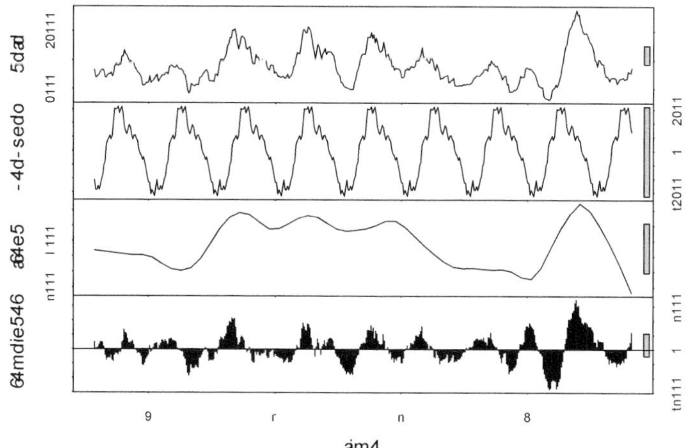

FIGURE 13.19 Decomposition using stl of a 15th-order moving average of E. coli data

provides a useful data set to consider some of the time series methods we've been exploring (Figure 13.20).

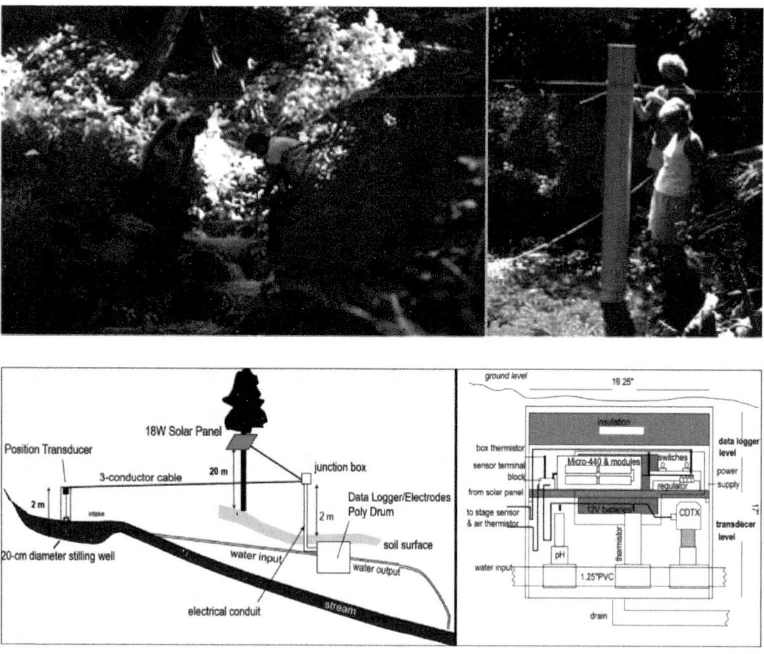

FIGURE 13.20 Marble Mountains resurgence data logger design

FIGURE 13.21 Marble Mountains resurgence data logger equipment

As always, you might want to start by just exploring the data...

```
read_csv(ex("marbles/resurgenceData.csv"))
```

```
## # A tibble: 17,537 x 9
##    date      time   wlevelm WTemp ATemp    EC InputV BTemp precip_mm
##    <chr>     <time>   <dbl> <dbl> <dbl> <dbl>  <dbl> <dbl>     <dbl>
##  1 9/23/1994 14:00    0.325  5.69  24.7  204.   12.5  24.8        NA
##  2 9/23/1994 16:00    0.325  5.69  20.4  207.   12.5  18.7        NA
##  3 9/23/1994 18:00    0.322  5.64  16.6  207.   12.5  16.3        NA
##  4 9/23/1994 20:00    0.324  5.62  16.1  207.   12.5  14.9        NA
##  5 9/23/1994 22:00    0.324  5.65  15.6  207.   12.5  14.0        NA
##  6 9/24/1994 00:00    0.323  5.66  14.4  207.   12.5  13.3        NA
##  7 9/24/1994 02:00    0.323  5.63  13.9  207.   12.5  12.8        NA
##  8 9/24/1994 04:00    0.322  5.65  13.1  207.   12.5  12.3        NA
##  9 9/24/1994 06:00    0.319  5.61  11.8  207.   12.5  11.9        NA
## 10 9/24/1994 08:00    0.321  5.66  13.1  207.   12.4  11.6        NA
## # ... with 17,527 more rows
```

What we can see is that the date_time stamps show that measurements are every two hours. This will help us set the frequency for an annual period. The start time is `1994-09-23 14:00:00`, so we need to provide that information as well. Note the computations used to do this (I first worked out that the `yday` for 9/23 was 266) (Figure 13.22).

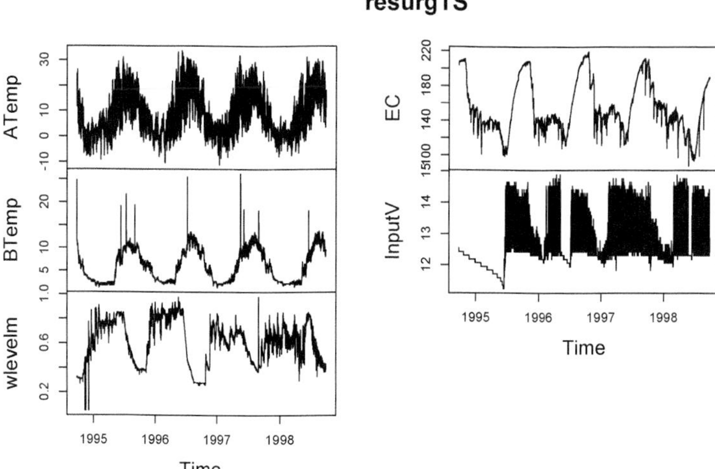

FIGURE 13.22 Data logger data from the Marbles resurgence

```
resurg <- read_csv(ex("marbles/resurgenceData.csv"))
resurgVars <- resurg %>%
  dplyr::select(ATemp, BTemp, wlevelm, EC, InputV)#%>% # removed date & time
resurgTS <- ts(resurgVars, frequency = 12*365, start = c(1994, 266*12+7))
plot(resurgTS)
```

In this study, water level water level has an annual seasonality that might be best understood by decomposition (Figure 13.23).

```
library(tidyverse); library(lubridate)
resurg <- read_csv(ex("marbles/resurgenceData.csv"))
wlevelTS <- ts(resurg$wlevelm, frequency = 12*365, start = c(1994, 266*12+7))
fit <- stl(wlevelTS, s.window="periodic")
plot(fit)
```

13.5 Facet Graphs for Comparing Variables over Time

We've just been looking at a data set with multiple variables, and often what we're just needing to do is to create a combination of graphs of these variables over a common timeframe on the x axis. We'll look at some flux data from Loney Meadow (Figure 13.24).

The flux tower data were collected at a high frequency for eddy covariance processing where 3D wind speed data are used to model the movement of atmospheric gases, including CO_2

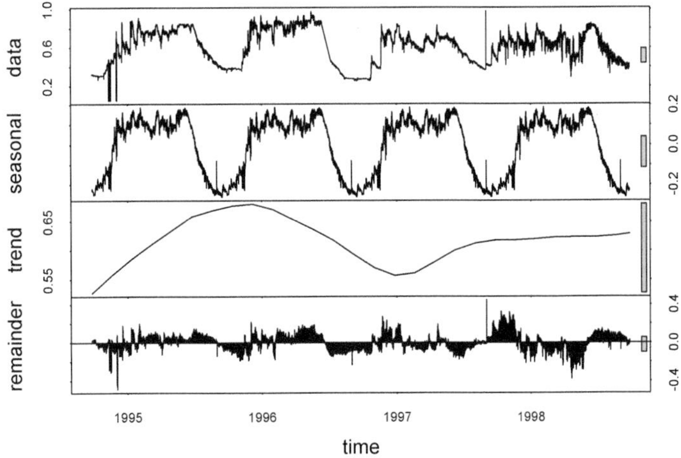

FIGURE 13.23 stl decomposition of Marbles water level time series

flux driven by photosynthesis and respiration processes. Note that the sign convention of CO_2 flux is that positive flux is released to the atmosphere, which might happen when less photosynthesis is happening but respiration and other CO_2 releases continue, while a negative flux might happen when more photosynthesis is capturing more CO_2.

A spreadsheet of 30-minute summaries from 17 May to 6 September can be found in the `igisci` extdata folder as `"meadows/LoneyMeadow_30minCO2fluxes_Geog604.xls"`, and includes data on photosynthetically active radiation (PAR), net radiation (Qnet), air temperature, relative humidity, soil temperature at 2 and 10 cm depth, wind direction, wind speed, rainfall, and soil volumetric water content (VWC). There's clearly a lot more we can do

FIGURE 13.24 Flux tower installed at Loney Meadow, 2016. Photo credit: Darren Blackburn

with these data (see Blackburn, Oliphant, and Davis (2021)), but for now we'll just create a facet graph from the multiple variables collected.

```
library(igisci)
library(readxl); library(tidyverse); library(lubridate)
vnames <- read_xls(ex("meadows/LoneyMeadow_30minCO2fluxes_Geog604.xls"),
                   n_max=0) %>% names()
vunits <- read_xls(ex("meadows/LoneyMeadow_30minCO2fluxes_Geog604.xls"),
                   n_max=0, skip=1) %>% names()
Loney <- read_xls(ex("meadows/LoneyMeadow_30minCO2fluxes_Geog604.xls"),
                  skip=2, col_names=vnames) %>%
  rename(DecimalDay = `Decimal time`,
         YDay = `Day of Year`,
         Hour = `Hour of Day`,
         CO2flux = `CO2 Flux`,
         Tsoil2cm = `Tsoil 2cm`,
         Tsoil10cm = `Tsoil 10cm`) %>%
  mutate(datetime = as_date(DecimalDay, origin="2015-12-31"))
```

We're going to want to analyze changes over a daily period, but we can also look at the data changes over the entire collection duration. A `group_by-summarize` process *by days* will give us a generalized picture of changes over that duration, with the diurnal fluctuations removed, reflecting phenological changes from first exposure after snowmelt through the maximum growth period and through the major senescence period of late summer. We'll look at a faceted graph (with `free_y` scales setting) from a `pivot_longer` table (Figure 13.25).

```
LoneyDaily <- Loney %>%
  group_by(YDay) %>%
  summarize(CO2flux = mean(CO2flux),
            PAR = mean(PAR),
            Qnet = mean(Qnet),
            Tair = mean(Tair),
            RH = mean(RH),
            Tsoil2cm = mean(Tsoil2cm),
            Tsoil10cm = mean(Tsoil10cm),
            wdir = mean(wdir),
            wspd = mean(wspd),
            Rain = mean(Rain),
            VWC = mean(VWC))
LoneyDailyLong <- LoneyDaily %>%
  pivot_longer(cols = CO2flux:VWC,
               names_to="parameter",
               values_to="value") %>%
  filter(parameter %in% c("CO2flux", "Qnet", "Tair", "RH", "Tsoil10cm", "VWC"))
p <- ggplot(data = LoneyDailyLong, aes(x=YDay, y=value)) +
  geom_line()
p + facet_grid(parameter ~ ., scales = "free_y")
```

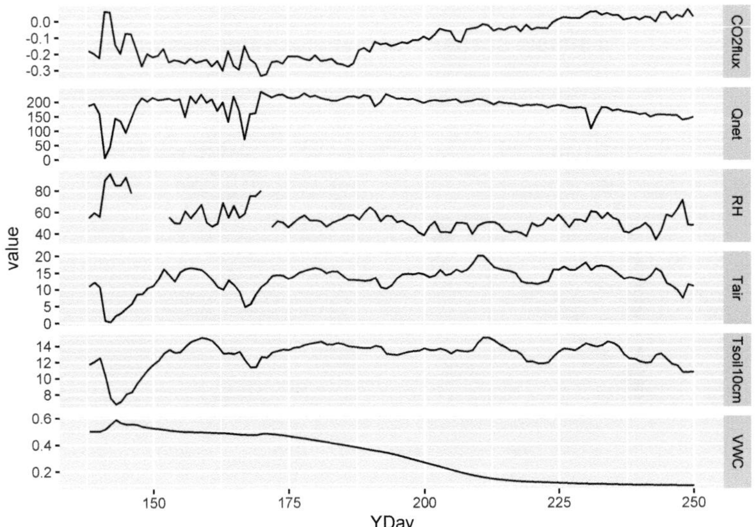

FIGURE 13.25 Facet plot with free y scale of Loney flux tower parameters

13.6 Lag Regression

Lag regression is like a normal linear model, but there's a timing difference, or lag, between the explanatory and response variables. We have used this method to compare changes in snow melt in a Sierra Nevada watershed and changes in reservoir inflows downstream (Powell et al. (2011)).

Another good application of lag regression is comparing solar radiation and temperature. We'll explore solar radiation and temperature data over an 8-day period for Bugac, Hungary, and Manaus, Brazil, maintained by the European Fluxes Database (*European Fluxes Database*, n.d.). These data are every half hour, but let's have a look. You could navigate to the spreadsheet and open it in Excel, and I would recommend it – you should know where the extdata are stored on your computer – since this is the easiest way to see that there are two worksheets. With that worksheet naming knowledge, we can also have a quick look in R:

```
library(readxl)
read_xls(ex("ts/SolarRad_Temp.xls"), sheet="BugacHungary")
```

```
## # A tibble: 17,521 x 5
##    Year  `Day of Yr` Hr      SolarRad Tair
##    <chr> <chr>       <chr>   <chr>    <chr>
##  1 YYYY  DOY         0-23.5  W m^-2   deg C
##  2 2006  1           0.5     0        0.55000001192092896
##  3 2006  1           1       0        0.56999999284744263
##  4 2006  1           1.5     0        0.82999998331069946
```

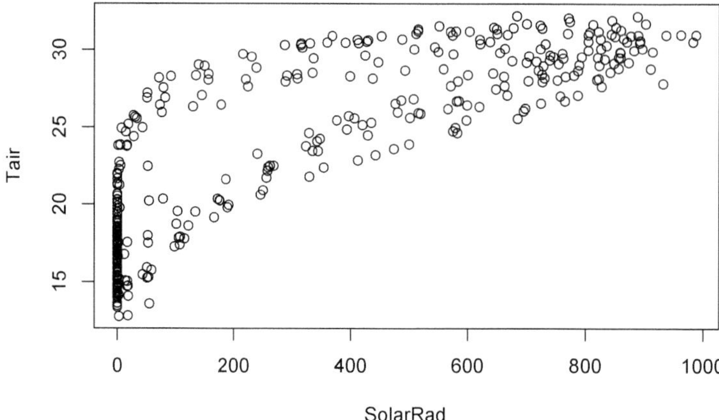

FIGURE 13.26 Scatter plot of Bugac solar radiation and air temperature

```
## 5 2006  1          2     0         1.0399999618530273
## 6 2006  1          2.5   0         1.1699999570846558
## 7 2006  1          3     0         1.2000000476837158
## 8 2006  1          3.5   0         1.2300000190734863
## 9 2006  1          4     0         1.2599999904632568
## 10 2006 1          4.5   0         1.2599999904632568
## # ... with 17,511 more rows
```

Indeed the readings are every half hour, so for a daily period, we should use a frequency of **48**. We need to remove the second line of the input, which has the units of the measurements, by using the accessor `[-1,]`. We'll filter for a year day range and select the solar radiation and air temperature variables (Figure 13.26).

```
library(readxl); library(tidyverse)
BugacSolstice <- read_xls(ex("ts/SolarRad_Temp.xls"),
                  sheet="BugacHungary", col_types = "numeric")[-1,] %>%
  filter(`Day of Yr` < 177 & `Day of Yr` > 168) %>%
  dplyr::select(SolarRad, Tair)
BugacSolsticeTS <- ts(BugacSolstice, frequency = 48)
plot(BugacSolstice, main="a simple scatter plot that illustrates hysteresis")
```

Hysteresis is "the dependence of a state of a system on its history" ("Hysteresis" (n.d.)), and one place this can be seen is in periodic data, when an explanatory variable's effect on a response variable differs on an increasing limb from what it produces on a decreasing limb. Two classic situations where this occur are in (a) water quality, where the rising limb of the hydrograph will carry more sediment load than a falling limb; and in (b) the daily or seasonal pattern of temperature in response to solar radiation, which we can see in the graph above with two apparent groups of scatterpoints representing the varying impact of solar radiation in spring vs fall, where temperatures lag behind the increase in solar radiation (Figure 13.27).

```
plot(BugacSolsticeTS)
```

13.6.1 The lag regression, using a lag function in a linear model

The lag function uses a lag number of sample intervals to determine which value in the response variable corresponds to a given value in the explanatory variable. So we compare air temperature `Tair` to a lagged solar radiation `SolarRad` as `lm(Tair~lag(SolarRad,i)`, where `i` is the number of units to lag. The following code loops through six values (0:5) for i, starting with assigning the `RMSE` (root mean squared error) for no lag (0), then appending to `RMSE` for `i` of `1:5` to hopefully hit the minimum value.

```
getRMSE <- function(i) {sqrt(sum(lm(Tair~lag(SolarRad,i),
                                   data=BugacSolsticeTS)$resid**2))}
RMSE <- getRMSE(0)
for(i in 1:5){RMSE <- append(RMSE,getRMSE(i))}
lagTable <- bind_cols(lags = 0:5,RMSE = RMSE)
knitr::kable(lagTable)
```

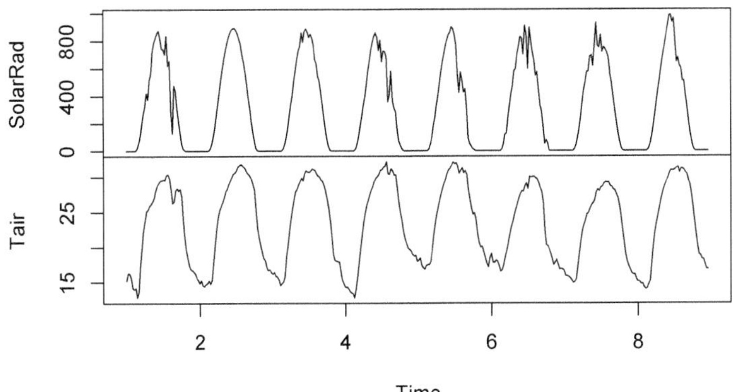

FIGURE 13.27 Solstice 8-day time series of solar radiation and temperature

lags	RMSE
0	64.01352
1	55.89803
2	50.49347
3	48.13214
4	49.64892
5	54.36344

Minimum RMSE is for a lag of 3, so 1.5 hours. Let's go ahead and use that minimum RMSE lag (using `which.min` with the RMSE to automate that selection), then look at the summarized model.

```
n <- lagTable$lags[which.min(lagTable$RMSE)]
paste("Lag: ",n)
```

```
## [1] "Lag:  3"
```

```
summary(lm(Tair~lag(SolarRad,n), data=BugacSolsticeTS))
```

```
##
## Call:
## lm(formula = Tair ~ lag(SolarRad, n), data = BugacSolsticeTS)
##
## Residuals:
##     Min      1Q  Median      3Q     Max
## -5.5820 -1.7083 -0.2266  1.8972  7.9336
##
## Coefficients:
##                   Estimate Std. Error t value Pr(>|t|)
## (Intercept)      18.352006   0.174998  104.87   <2e-16 ***
## lag(SolarRad, n)  0.016398   0.000386   42.48   <2e-16 ***
## ---
## Signif. codes:  0 '***' 0.001 '**' 0.01 '*' 0.05 '.' 0.1 ' ' 1
##
## Residual standard error: 2.472 on 379 degrees of freedom
##   (3 observations deleted due to missingness)
## Multiple R-squared:  0.8265, Adjusted R-squared:  0.826
## F-statistic:  1805 on 1 and 379 DF,  p-value: < 2.2e-16
```

So, not surprisingly, air temperature has a significant relationship to solar radiation. Confirmatory statistics should never be surprising, but just give us confidence in our conclusions. What we didn't know before was just what amount of lag produces the strongest relationship, with the least error (root mean squared error, or RMSE), which turns out to be 3 observations or 1.5 hours. Interestingly, the *peak* solar radiation and temperature are separated by 5 units, so 2.5 hours, which we can see by looking at the generalized seasonals from decomposition (Figure 13.28). There's a similar lag for Manaus, though the peak solar radiation and temperatures coincide.

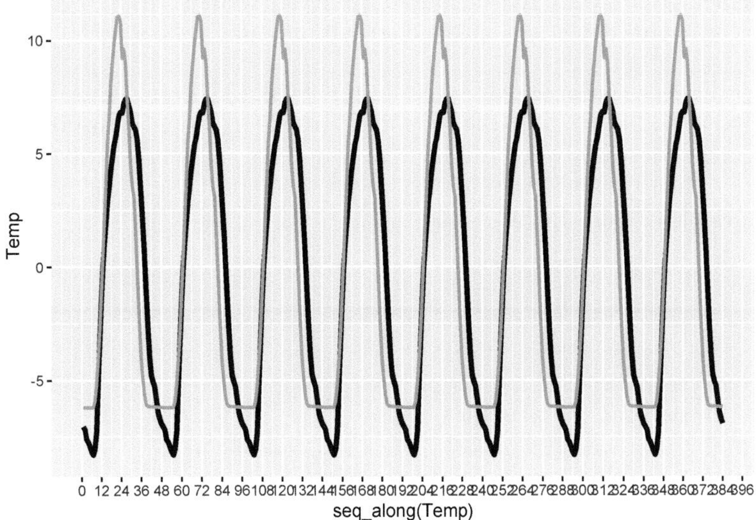

FIGURE 13.28 Bugac solar radiation and temperature

```
SolarRad_comp <- decompose(ts(BugacSolstice$SolarRad, frequency = 48))
Tair_comp <- decompose(ts(BugacSolstice$Tair, frequency = 48))
seasonals <- bind_cols(Rad = SolarRad_comp$seasonal, Temp = Tair_comp$seasonal)
```

```
ggplot(seasonals) +
  geom_line(aes(x=seq_along(Temp), y=Temp), col="black",size=1.5) +
  geom_line(aes(x=seq_along(Rad), y=Rad/50), col="gray", size=1) +
  scale_x_continuous(breaks = seq(0,480,12))
```

```
paste(which.max(seasonals$Rad), which.max(seasonals$Temp))
```

```
## [1] "23 28"
```

13.7 Ensemble Summary Statistics

For any time series data with meaningful periods like days or years (or even weeks when the observations vary by days of the week – this has been the case with COVID-19 data), ensemble averages (along with other summary statistics) are a good way to visualize changes in a variable over that period.

As you could see from the decomposition and sequential graphics we created for the Bugac and Manaus solar radiation and temperature data, looking at what happens over one day as an "ensemble" of all of the days – where the mean value is displayed along with error bars based on standard deviations of the distribution – might be a useful figure.

We've used `group_by` - `summarize` elsewhere in this book (3.4.4) and this is yet another use of that handy method. Note that the grouping variable `Hr` is not categorical, not a factor, but simply a numeric variable of the hour as a decimal number – so 0.0, 0.5, 1.0, etc. – representing half-hourly sampling times, so since these repeat each day they can be used as a grouping variable (Figure 13.29).

```
library(cowplot)
Manaus <- read_xls(system.file("extdata","ts/SolarRad_Temp.xls",package="igisci"),
                   sheet="ManausBrazil", col_types = "numeric")[-1,] %>%
  dplyr::select(Year:Tair)
ManausSum <- Manaus %>% group_by(Hr) %>%
  summarize(meanRad = mean(SolarRad), meanTemp = mean(Tair),
            sdRad = sd(SolarRad), sdTemp = sd(Tair))
px <- ManausSum %>% ggplot(aes(x=Hr)) + scale_x_continuous(breaks=seq(0,24,3))
p1 <- px + geom_line(aes(y=meanRad), col="blue") +
  geom_errorbar(aes(ymax = meanRad + sdRad, ymin = meanRad - sdRad)) +
  ggtitle("Manaus 2005 ensemble solar radiation")
p2 <- px + geom_line(aes(y=meanTemp),col="red") +
  geom_errorbar(aes(ymax=meanTemp+sdTemp, ymin=meanTemp-sdTemp)) +
  ggtitle("Manaus 2005 ensemble air temperature")
plot_grid(plotlist=list(p1,p2), ncol=1, align='v') # from cowplot
```

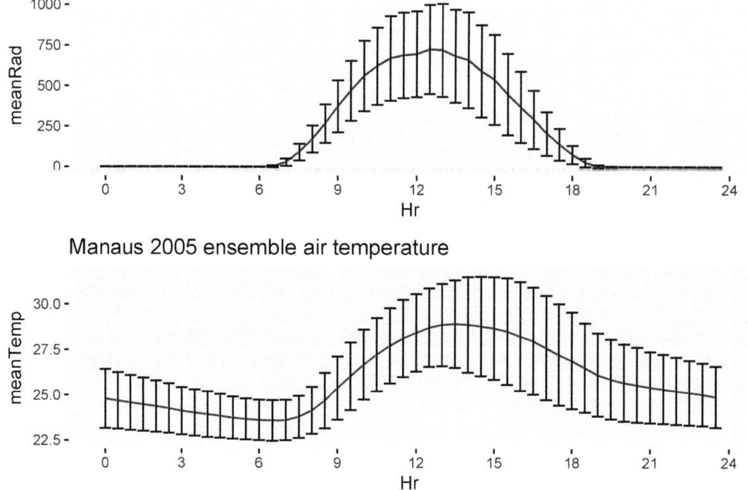

FIGURE 13.29 Manaus ensemble averages with error bars

13.8 Learning more about Time Series in R

There has been a lot of work on time series in R, especially in the financial sector where time series are important for forecasting models. Some of these methods should have applications in environmental research. Here are some resources for learning more:

- At CRAN: https://cran.r-project.org/web/views/TimeSeries.html
- The time series data library: https://pkg.yangzhuoranyang.com/tsdl/
- https://a-little-book-of-r-for-time-series.readthedocs.io/
- The tsibble: https://tsibble.tidyverts.org/

13.9 Exercises: Time Series

Exercise 13.1. Create an half-hourly ensemble plot of Bugac, Hungary, similar to the one we created for Manaus, Brazil.

Exercise 13.2. Create a ManausJan tibble for just the January data, then decompose a ts of `ManausJan$SolarRad`, and plot it. Consider carefully what frequency to use to get a daily period. Also, remember that "seasonal" in this case means "diurnal". Code and plot.

Exercise 13.3. Then do the same for air temperature (`Tair`) for Manaus in January, providing code and plot. What observations can you make from this on conditions in Manaus in January.

Exercise 13.4. Now do both of the above for Bugac, Hungary, and reflect on what it tells you, and how these places are distinct (other than the obvious).

Exercise 13.5. Now do the same for June. You'll probably need to use the `yday` and `ymd` (or similar) functions from `lubridate` to find the date range for 2005 or 2006 (neither are leap years, so they should be the same).

Exercise 13.6. Do a lag regression over an 8-day period over the March equinox (about 3/21), when the sun's rays should be close to vertical since it's near the Equator, but it's very likely to be cloudy at the Intertropical Convergence Zone. As we did with the Bugac data, work out the best lag value using the minimum-RMSE model process we used for Bugac, and produce a summary of that model.

Exercise 13.7. Create an ensemble plot of monthly air and box temperature from the Marble Mountains Resurgence data. The air temperature was measured from a thermistor mounted near the ground but outside the box, and the box temperature was measured within the buried box (see the pictures to visualize this.) You'll want to start with reading the dates and times – which we didn't actually use when we built a time series from the data – by building `date_time` and `month` variables:

```
library(tidyverse); library(lubridate); library(igisci)
resurg <- read_csv(ex("marbles/resurgenceData.csv")) %>%
  mutate(date = mdy(date),
         date_time = date+time,
         month = month(date_time)) %>%
  dplyr::select(date_time, month, date, time, ATemp, BTemp, wlevelm, EC, InputV)
```

Exercise 13.8. How about looking at daily cycles? The Marbles resurgence data were collected every two hours, so group by hour (created using `hour(time)`, which will return integers 0,2,...,22). Interpret what you get, and what the air vs box temperature patterns might show, considering where the sensors are installed.

Exercise 13.9. Create a **facet graph** of *daily* (summarize by Date, after reading it in with `mdy(date)`, air temperature (ATemp), water level (wlevelm), conductance (EC), and input voltage from the solar panel (InputV) from the Marble Mountains resurgence data (Figure 13.30). Use the `free_y` scales setting.

Exercise 13.10. Tree-ring data: Using `forecast::ma` and the built-in `treering` time series, find a useful order value for a curve that communicates a pattern of tree rings over the last 8,000 years, with high values around -3,500 and 0 years. Note: the order may be pretty big, and you might want to loop through a sequence of values to see what works. I found that creating a sequence in a for loop with seq(from=100,to=1000,by100) useful, and any further created some odd effects.

Exercise 13.11. Sunspot activity is known to fluctuate over about a 11-year cycle, and other than the well-known effects on radio transmission is also part of short-term climate fluctuations (so must be considered when looking at climate change). The built-in data set `sunspot.month` has monthly values of sunspot activity from 1749 to 2014. Use the frequency function to understand its period, then create an alternative ts with an 11-year cycle as frequency (how do you define its frequency?) and plot a decomposition and stl.

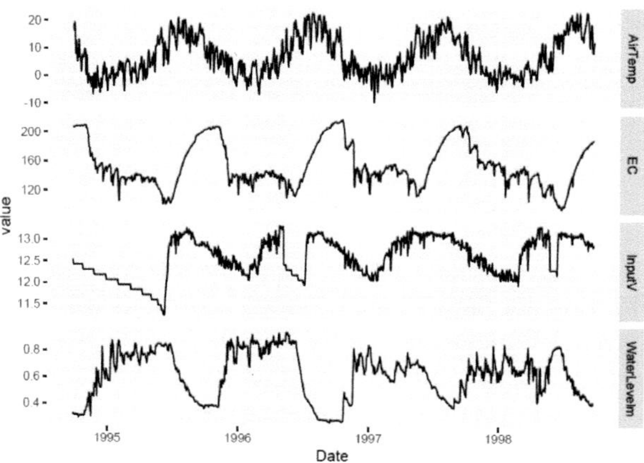

FIGURE 13.30 Facet graph of Marble Mountains resurgence data (goal)

Part V

Communication and References

14

Communication with Shiny

Communication of research is central to environmental data science, and while this can use many venues such as professional meetings and publications, outreach on the internet is especially important. This chapter will delve into probably the best way to build a web site for communicating our research: **Shiny**, used for building interactive web apps, using interactive controls to allow the user to manipulate any of the R analyses and graphics you've learned about. But as we'll see right away, these interactive controls can also be used in R Markdown.

Four ways to use Shiny:

- **Shiny Web App**: As a web application hosted on a server. This is probably the best way to communicate your work, and includes all of Shiny interactive methods. But to get it on the web, you'll need to get it hosted. One option for this is to use https://www.shinyapps.io which will host 5 apps with 25 hours usage/month for free, or you can pay for more access. We'll look at some complete Shiny Web Apps later in this chapter, but also see https://shiny.rstudio.com and https://shiny.rstudio.com/tutorial to learn more.
- **Shiny app run locally**: The same Shiny web app but run locally on your own computer in RStudio. When you're building your web app, this is where you'll start anyway, prepping your app to run well before publishing it to the web.
- **Shiny Document**: As an R Markdown document with the shiny runtime. It creates an HTML document, using interactive Shiny components. We're going to start with this method, since it's the easiest introduction to the input widgets and output options. To learn more, see Chapter 19 in R Markdown, the Definitive Guide[1] (Xie, Allaire, and Grolemund (2019)) and https://www.rstudio.com/wp-content/uploads/2015/02/rmarkdown-cheatsheet.pdf.
- **Shiny Presentation**: Also using the R Markdown system, creates an IOSlides presentation which uses interactive Shiny components.

14.1 Shiny Document

We'll start with creating a **Shiny Document** as a way to introduce Shiny interactive components, but it's also not a bad option if you just want to create an interactive environment for your own work, and it's also pretty easy to share with others, like we've already seen with R Markdown. But since Shiny requires a host to run the R code on, we can't really create it as interactive objects in this book without accessing a hosted Shiny app (which

[1] http://rmarkdown.rstudio.com/authoring_shiny.html

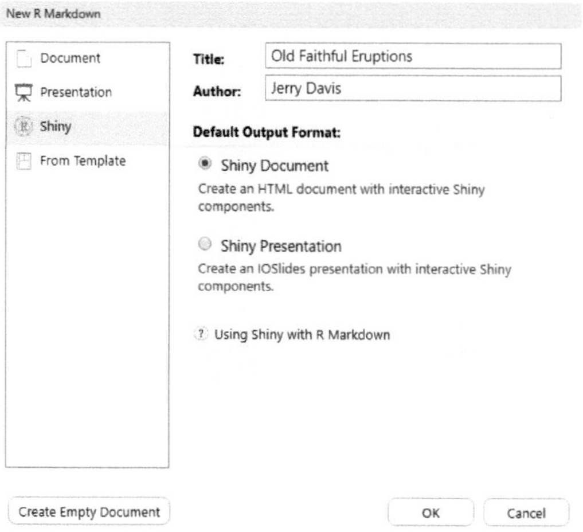

FIGURE 14.1 New Shiny Document dialog

you can do, see Xie (2021)), but we'll include snips below to show you what it looks like. After we've learned about these interactive components, we can also use them in a Shiny Web App, but there will be other structural elements we'll need to add.

To create a Shiny Document, use File>New File>R Markdown... to initiate the process, where we'll specify that we want to create a **Shiny** document and give it a title. For this first one, we'll just create its default document that accesses the **faithful** (Old Faithful eruptions) built-in data, so we'll give it that name (Figure 14.1).

When we OK this, we'll see the R Markdown document opened in the RStudio script editor (Figure 14.2).

The key to this document working in the **shiny** runtime environment is to make sure to specify this in the YAML header. At minimum, this header needs to include two settings:

```
---
output: html_document
runtime: shiny
```

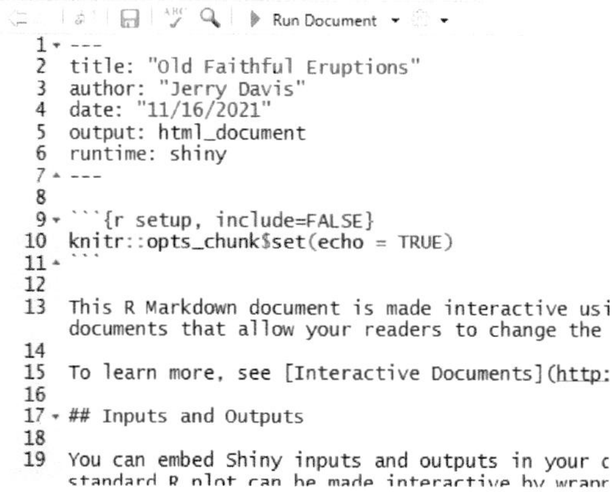

FIGURE 14.2 Shiny Document Editor

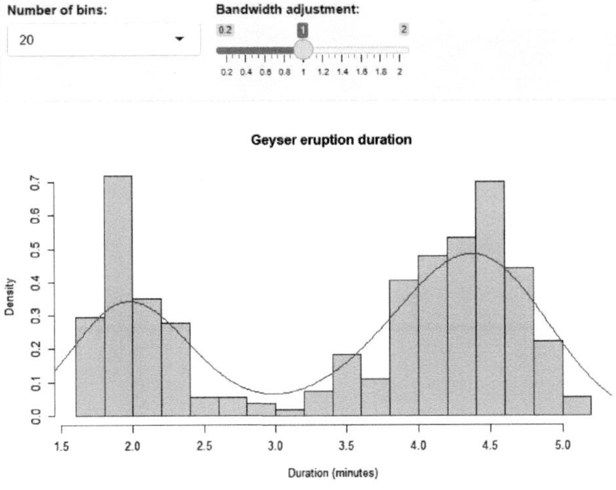

FIGURE 14.3 Old Faithful geyser eruptions Shiny interface

Let's start by going ahead and running it with the `Run Document` button, which due to the shiny runtime option replaces the `Knit` button we'd normally see in R Markdown (Figure 14.3).

So now we have a simple example to explore how it works. We'll start by exploring its components, so keep this document open to try things out.

14.1.1 Input and output objects in the Old Faithful Eruptions document

In either markdown or app mode, we can create a variety of Shiny input widgets and output objects:

- Inputs are the interactive controls (widgets) that let the user change the resulting output.
- Outputs are the graphs, maps, or tables, and are automatically updated whenever inputs change.

The code for the Old Faithful Eruptions document includes a couple of widget-controlled input settings to be used in a plot produced by `renderPlot`

- a `selectInput` that is used to set the `breaks` parameter for the `hist` function as a number of bins
- a `sliderInput` to set the `adjust` (bandwidth adjustment) parameter for the density plot

Note how the input variables are accessed by the output function as `input$n_breaks` and `input$bw_adjust`:

```
inputPanel(
  selectInput("n_breaks", label = "Number of bins:",
```

```
            choices = c(10, 20, 35, 50), selected = 20),

  sliderInput(”bw_adjust”, label = ”Bandwidth adjustment:”,
              min = 0.2, max = 2, value = 1, step = 0.2)
)
```

This is then followed by the plot object:

```
renderPlot({
  hist(faithful$eruptions, probability = TRUE, breaks = as.numeric(input$n_breaks),
       xlab = ”Duration (minutes)”, main = ”Geyser eruption duration”)

  dens <- density(faithful$eruptions, adjust = input$bw_adjust)
  lines(dens, col = ”blue”)
})
```

This is then followed by an embedded shiny app, but we won't look at it.

14.1.2 Input widgets

We'll continue to explore the various input widgets and related outputs, and stick with R Markdown. You'll still need to run the entire document, so you might want to make separate documents – just make sure they have the YAML header above. Go ahead and create a new Shiny document, the same as above, but create an empty document and then enter the YAML header and one code chunk to create a simple numericInput widget and print output (Figure 14.4).

```
numericInput(inputId=”n”, ”sample size”, value = 25)
renderPrint(print(input$n))
```

Once you save it, you'll see the Run Document button if the YAML code is right. Go ahead and run it to see the input. It's not very interesting, but you can see that the printed output changes with the input. Let's add a slider (Figure 14.5).

```
Run Document
---
output: html_document
runtime: shiny
---

```{r}
numericInput(inputId="n", "sample size", value = 25)
renderPrint(print(input$n))
```
```

FIGURE 14.4 numericInput and renderPrint code

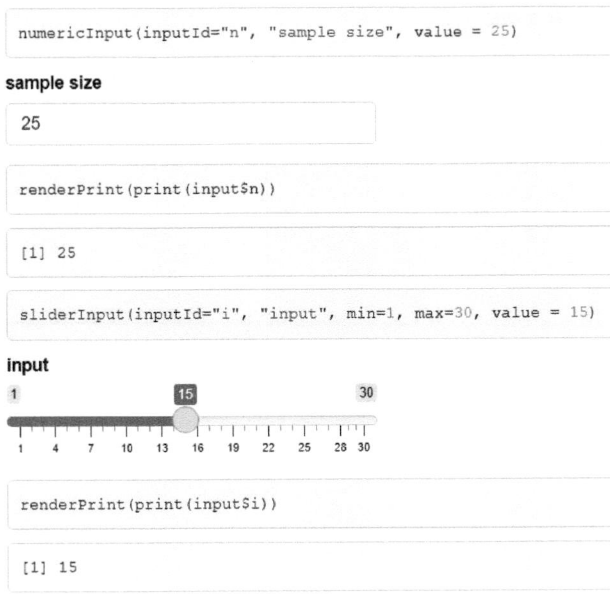

FIGURE 14.5 Numeric and slider inputs and print outputs

```
numericInput(inputId="n", "sample size", value = 25)
renderPrint(print(input$n))
sliderInput(inputId="i", "input", min=1, max=30, value = 15)
renderPrint(print(input$i))
```

We'll explore some other input widgets and outputs. You can also see these on the Shiny cheatsheet ...

https://shiny.rstudio.com/images/shiny-cheatsheet.pdf

... but just look at the part about the input and output objects. The cheatsheet is focused on creating Shiny Web Apps, so there's a lot there about building that structure, with a user interface (ui) and server. We'll get to those later.

14.1.2.1 A plot output

The above widgets went to a rendered print output, but the same simple inputs can of course be used to create a plot (Figure 14.6).

```
sliderInput(inputId="bins", "number of bins", min=1, max=30, value = 15)
renderPlot(hist(rnorm(100), breaks=input$bins))
```

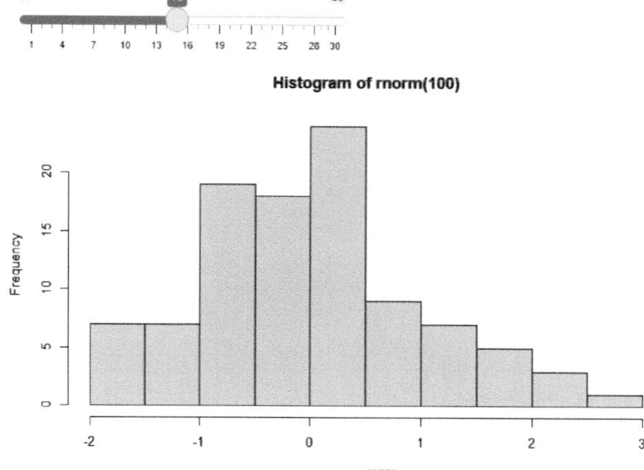

FIGURE 14.6 Plot modified by input

14.1.3 Other input widgets

There are lots of other input widgets that are pretty easy to see how they apply based on the type of control we need to set, such as radio buttons and check boxes (Figure 14.7).

- **`radioButtons()`** for choosing just one, and by default the first is chosen

```
radioButtons(inputId="which_one", label="Select:",
    choices = c("Choice 1"="Choice1","Choice 2"="Choice2","Choice 3"="Choice3"))
renderPrint(print(input$which_one))
```

- **`checkboxGroupInput()`** for choosing multiples

```
checkboxGroupInput(inputId="which", label="Select:",
    choices = c("Choice 1"="Choice1","Choice 2"="Choice2","Choice 3"="Choice3"))
renderPrint(print(input$which))
```

- **`dateInput()`**

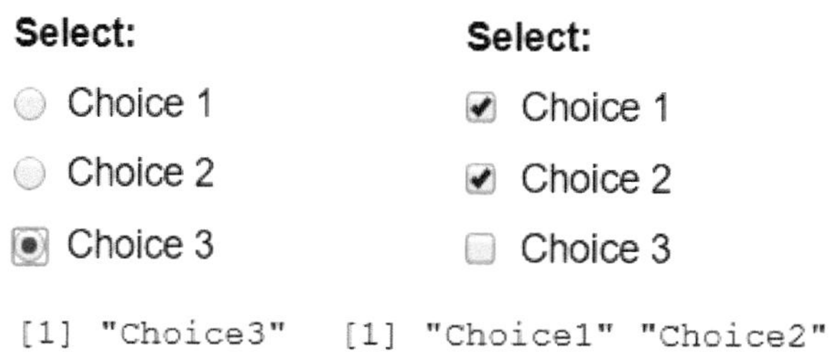

Select:
Choice 1
Choice 2
Choice 3
[1] "Choice3"

Select:
Choice 1
Choice 2
Choice 3
[1] "Choice1" "Choice2"

FIGURE 14.7 Radio buttons and check boxes

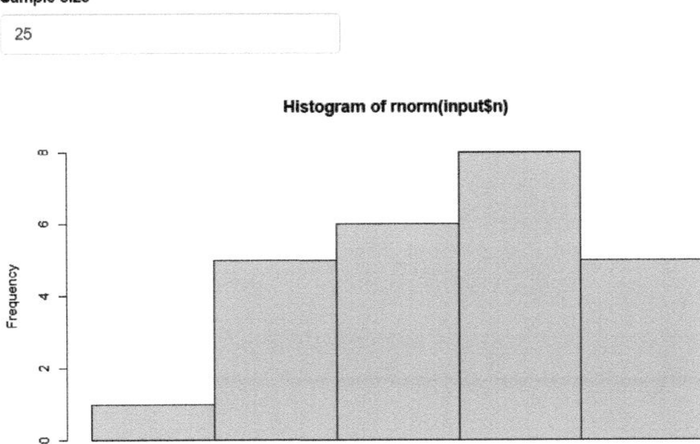

FIGURE 14.8 Simple Inline app

```
dateInput(inputId="date", label="Select date:")
renderPrint(print(input$date))
```

- **`textInput()`**

```
textInput(inputId="text", label="Enter text:")
renderPrint(print(input$text))
```

14.2 A Shiny App

The above input widgets and output objects we looked at in a Shiny document can also be built into a Shiny app, which has a place for inputs and outputs. The following simple Shiny inline app illustrates the basic elements (Figure 14.8).

- Optionally some code at the top to set things up, such as loading packages and setting up data.
- The **ui** section, which determines what the user will see, such as input widgets (like `numericInput`) and an area to plot things in, in this case `plotOutput`. In order to fit well on the page, the `fluidPage` function is used to hold it in; we could have also used `inputPanel` as we did earlier, but the result is not as well laid out.
- The **server** section, which uses two parameters (input, output) created in the user interface to:
  - provide data from the input widget (in this case `numericInput` which creates `input$n`) such as parameter settings to functions (e.g. `rnorm()` here which needs a sample size provided by `input$n`) in the rendered output (of in this case `hist`)
  - to render an output using a render_____ method such as `renderPlot` which will run everything inside it (in this case a histogram)
- Providing that `ui` and `server` to the `shinyApp` function.

```
library(shiny)
ui <- fluidPage(
    numericInput(input="n",
                 "Sample size", value=25),
    plotOutput(outputId = "hist"))
server <- function(input, output) {
    output$hist <- renderPlot({
        hist(rnorm(input$n))
    })
}
shinyApp(ui=ui,server=server)
```

This Shiny app is very short, which is useful in wrapping your head around these basic elements, which is important to understand. It can be run either as an inline app in R Markdown, in a standard R script (but all code has to be selected and then run), or as an app that can be published online.

We'll go ahead and create a new Shiny Web App to run locally (Figure 14.9). You can either create a special type of RStudio project called Shiny Web Application, in a new or existing folder, or just create a Shiny Web App script as a file, as shown here. In either case, your app file should be named `app.R`.

- Use the file menu to create a new file, specifying Shiny Web App as the type, and it'll end up being named `app.R`. (You can also name it something else, but this is the name you'll need to use to create a fully functioning Shiny Web App.)
- Go with the defaults again, and you'll end up with Shiny Web App version of the Old Faithful Eruptions document we created above, though a bit simpler, with only one input control and only a histogram.
- To run it, use the Run App button in the upper right of the code editor window.
- Have a look at what's created in the `app.R` script.

Then either create a new blank R script, or edit this one and replace the old faithful code with the above even simpler code in the figure above. Note that to get the `Run app` button in the RStudio script editor window, the code just needs to include `library(shiny)` and be a complete app, with a ui and server section, and the `shinyApp` function at the end.

```
Run App
library(shiny)
ui <- fluidPage(
  numericInput(input="n",
               "Sample size", value=25),
  plotOutput(outputId = "hist"))
server <- function(input, output) {
  output$hist <- renderPlot({
    hist(rnorm(input$n))
  })
}
shinyApp(ui=ui,server=server)
```

FIGURE 14.9 Simple inline coding

14.2.1 A brief note on reactivity

Since Shiny is an interactive environment, user operations are sensed as events, putting in motion a *reaction* of the program to that event. You can see that happening every time we adjust things in an input widget and the output changes in response. *There's a lot to reactivity*, so please review the tutorial at https://shiny.rstudio.com/tutorial to gain a better understanding than we'll be able to do here. We'll look at some examples in the longer apps described below.

14.3 Shiny App I/O Methods

One thing to note about how the Shiny web app works as opposed to just using the interactive methods in the Shiny Document we looked at earlier is the way the ui and server communicate about the plot creation: We didn't use `plotOutput` in the I/O methods; we just used the input widgets to set parameters in the `renderPlot`.

Instead, this Shiny web app has the `renderPlot` in the **server** section, and a new function `plotOutput` is used to both specify the location of the plot and set `outputId` to be used as a variable to append to `output$` in the **server** section. *This can be a little confusing*, so spend some time seeing how this works in the simple apps we just looked at, then start exploring other ones. The layout of the ui can make this even more confusing, as they're set up in various ways. This isn't surprising if you've spent much time in software development – the user interface is often the biggest challenge.

Referring to the **Outputs** section of the cheat sheet, we can see that various render*() functions work with corresponding *Output() functions:

- `renderDataTable()` works with `dataTableOutput()`
- `renderImage(      )` works with `imageOutput()`
- `renderPlot(       )` works with `plotOutput()`
- `renderPrint(      )` works with `verbatimTextOutput()`
- `renderTable(      )` works with `tableOutput()`
- `renderText(       )` works with `textOutput()`
- `renderUI(         )` works with `uiOutput()` and `htmlOutput()`

And there are others. For instance, as we'll see in the next document:

- `renderLeaflet(  )` works with `leafletOutput()`

14.3.1 Data tables

Use `renderDataTable()` and `dataTableOutput()` (Figure 14.10)

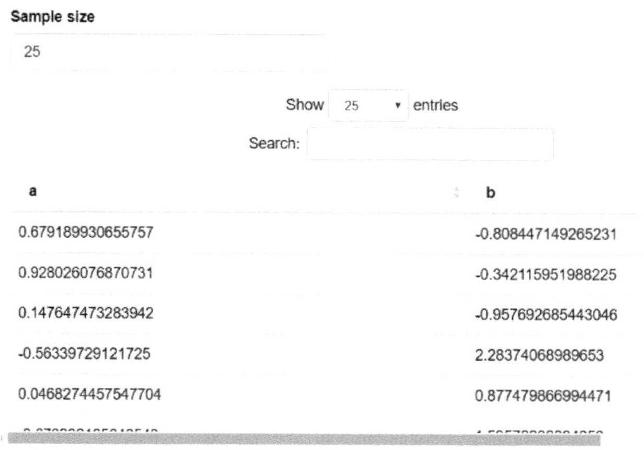

Sample size

25

Show 25 entries

Search:

a	b
0.679189930655757	-0.808447149265231
0.928026076870731	-0.342115951988225
0.147647473283942	-0.957692685443046
-0.56339729121725	2.28374068989653
0.0468274457547704	0.877479866994471

FIGURE 14.10 Rendered data table

```
library(shiny); library(tidyverse)
ui <- fluidPage(
    numericInput(input=”n”,
                 ”Sample size”, value=25),
    dataTableOutput(outputId = ”varTable”))
server <- function(input, output) {
    output$varTable <- renderDataTable(tibble(a=rnorm(input$n), b=rnorm(input$n)))}
shinyApp(ui=ui,server=server)
```

14.3.2 Text as character: `renderPrint()` and `verbatimTextOutput()`

These functions produce a mono-spaced-font output like a character variable.

```
library(shiny); library(tidyverse)
ui <- fluidPage(
    textInput(inputId=”monotxt”, label=”Enter text:”, value=”Some text”),
    verbatimTextOutput(outputId = ”txt”))
server <- function(input, output) {
    output$txt <- renderPrint(print(input$monotxt))}
shinyApp(ui=ui,server=server)
```

14.3.3 Formatted text

Use `renderText()` and `textOutput()` (Figure 14.11)

```
library(shiny); library(tidyverse)
ui <- fluidPage(
```

Enter text:

Some text

Some text

FIGURE 14.11 Text entry and rendered text

```
    textInput(inputId="text", label="Enter text:", value="Some text"),
    textOutput(outputId = "prn"))
server <- function(input, output) {
    output$prn <- renderText(print(input$text))}
shinyApp(ui=ui,server=server)
```

14.3.4 Plots

Use `renderPlot()` and `plotOutput()` (Figure 14.12)

```
library(shiny); library(tidyverse)
ui <- fluidPage(
    sliderInput(inputId="i", "input", min=1, max=30, value = 15),
    textInput(inputId="title", label="Enter title:", value="Tukey boxplot"),
    plotOutput(outputId = "Tukey"))
server <- function(input, output) {
    output$Tukey <- renderPlot(boxplot(rnorm(input$i),main=input$title))}
shinyApp(ui=ui,server=server)
```

Enter title:

Tukey boxplot

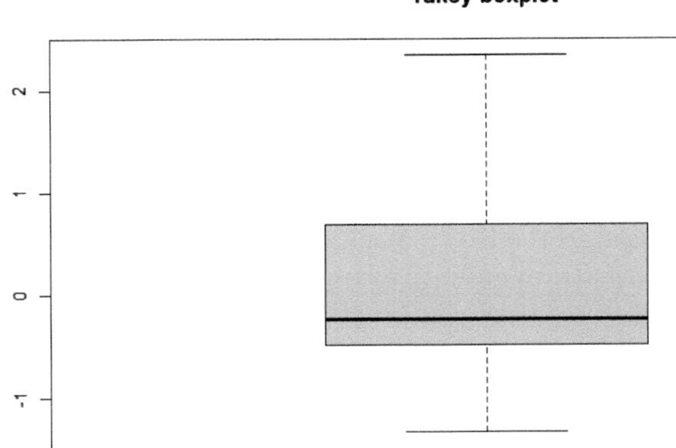

FIGURE 14.12 Rendered box plot

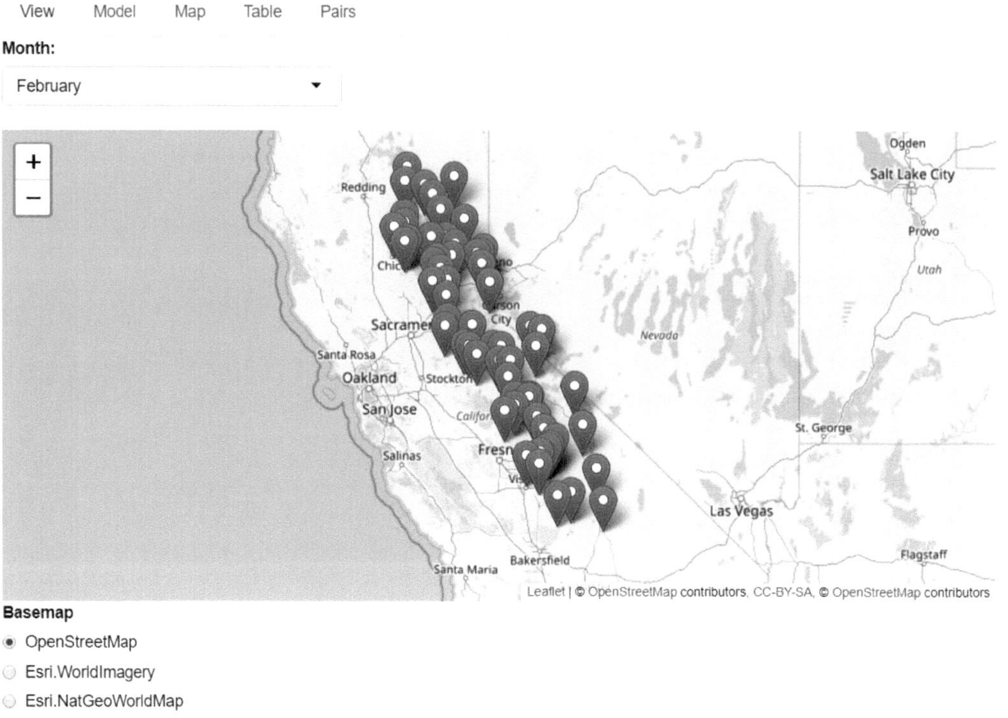

FIGURE 14.13 Shiny app of Sierra climate data, with multiple tabs available

14.4 Shiny App in a Package

A Shiny app project folder can also be stored in the `extdata` folder of a data package. Here's a `sierra` app accessed this way, which you can run from RStudio, either pasted into the console or saved in a script and run by marking the text and running. The `Run app` button won't appear even from a saved script, because what you're editing is not a full Shiny app script; it just calls one.

```
shiny::shinyAppDir(ex("sierra"))
```

Note the use of `system.file` to provide the app folder location to `shinyAppDir`. You can find this `sierra` folder in the igisci extdata folder, so as long as you have the data package installed, it should run. And the `igisci` package has a function `sierra()` that simply runs the above so that's all you have to enter if you have the library in effect with `library(igisci)`, and you'll see it appear in a new window (Figure 14.13).

We'll look at this app in detail next, so you should have it open in another window, using the above method, or simply by running:

```
igisc::sierra()
```

14.5 Components of a Shiny App (sierra)

Lets break down the components of the sierra Shiny app to see how a tabsetPanel Shiny app works. A review of the cheat sheet will show you that this is just one of a variety of Shiny app ui structures. We'll also see how reactive elements work in the server section.

14.5.1 Initial data setup

Most Shiny apps are going to need some initial code that sets up the data. I sometimes use the source method to call this code. We won't look at this now, but it's provided here to be able to identify where data in the main Shiny app code goes.

```
library(shiny); library(sf); library(leaflet); library(rgdal)
library(tidyverse); library(terra)
library(igisci)

sierraAllMonths <- read_csv(ex("sierra/Sierra2LassenData.csv")) %>%
  filter(MLY_PRCP_N >= 0) %>%
  filter(MLY_TAVG_N >= -100) %>%
  rename(PRECIPITATION = MLY_PRCP_N, TEMPERATURE = MLY_TAVG_N) %>%
  mutate(STATION = str_sub(STATION_NA, end=str_length(STATION_NA)-6))
sierraJan <- sierraAllMonths %>% # to create an initial model and var name symbols
  sample_n(0) %>%                # just gets variable names
  dplyr::select(LATITUDE, LONGITUDE, ELEVATION, TEMPERATURE, PRECIPITATION)
sierraVars <- sierraJan %>%       # Builds list of variables for map
  mutate(RESIDUAL = numeric(), PREDICTION = numeric()) %>%
  dplyr::select(ELEVATION, TEMPERATURE, PRECIPITATION, RESIDUAL, PREDICTION)

# Create basemap, using the weather station points to set the bounding dimensions

co <- CA_counties
ct <- st_read(ex("sierra/CA_places.shp"))
ct$AREANAME_pad <- paste0(str_replace_all(ct$AREANAME, '[A-Za-z]',' '), ct$AREANAME)
hillsh <- rast(ex("CA/ca_hillsh_WGS84.tif"))
hillshptsT <- as.points(hillsh)
hillshpts <- st_as_sf(hillshptsT)
CAbasemap <- ggplot() +
  geom_sf(data = hillshpts, aes(col=ca_hillsh_WGS84)) + guides(color = F) +
  geom_sf(data = co, fill = NA) +
  scale_color_gradient(low = "#606060", high = "#FFFFFF") +
  labs(x='',y='')
spdftemp <- st_as_sf(sierraAllMonths, coords = c("LONGITUDE","LATITUDE"), crs=4326)
bounds <- st_bbox(spdftemp)
sierrabasemap <- CAbasemap +
  geom_sf(data=ct) +
```

```
  geom_sf_text(mapping = aes(label=AREANAME_pad), data=ct, size = 3,
               nudge_x = 0.1, nudge_y = 0.1) +
  coord_sf(xlim = c(bounds[1], bounds[3]), ylim = c(bounds[2],bounds[4]))

# Function used by pairs plot:
panel.cor <- function(x, y, digits = 2, prefix = "", cex.cor, ...)
{
  usr <- par("usr"); on.exit(par(usr))
  par(usr = c(0, 1, 0, 1))
  r <- abs(cor(x, y))
  txt <- format(c(r, 0.123456789), digits = digits)[1]
  txt <- paste0(prefix, txt)
  if(missing(cex.cor)) cex.cor <- 0.8/strwidth(txt)
  text(0.5, 0.5, txt, cex = cex.cor * r)
}
```

14.5.2 The ui section, with a tabsetPanel structure

In this section, we're setting up the `tabsetPanel` structure, allowing the user to select whether to look at the various outputs:

- **View**: A leaflet map of the Sierra stations, and an ability to choose which month to process, and radio buttons to choose a basemap. The month choice structure took a bit to figure out; note the named indices.
- **Model**: Scatter plot and trend line of a linear model, with the ability to change the x and y variables.
- **Map**: Map of the variables in the data, or the regression or residuals. Input to choose which to display.
- **Table**: Table of the data for the chosen month. No inputs.
- **Pairs**: A pairs plot of the data, for the chosen month. No inputs.

```
ui <- fluidPage(title = "Sierra Climate",
          tabsetPanel(
            tabPanel(title = "View",
                     selectInput("month", "Month:",
                                 c("January"=1, "February"=2, "March"=3,
                                   "April"=4,    "May"=5,      "June"=6,
                                   "July"=7,     "August"=8,   "September"=9,
                                   "October"=10,"November"=11,"December"=12)),
                     leafletOutput("view"),
                     radioButtons(inputId = "LeafletBasemap", label = "Basemap",
                                  choices = c("OpenStreetMap" = "open",
                                              "Esri.WorldImagery" = "imagery",
                                              "Esri.NatGeoWorldMap" = "natgeo"),
                                  selected = "open")),
            tabPanel(title = "Model",
                     plotOutput("scatterplot"),
```

```
                varSelectInput("xvar", "X Variable:", data=sierraJan,
                               selected="ELEVATION"),
                varSelectInput("yvar", "Y Variable:", data=sierraJan,
                               selected="TEMPERATURE"),
                verbatimTextOutput("model")),
        tabPanel(title = "Map",
                plotOutput("map"),
                varSelectInput("var", "Variable:", data=sierraVars,
                               selected="TEMPERATURE")),
        tabPanel(title = "Table",
                textOutput("eqntext"),
                tableOutput("table")),
        tabPanel(title = "Pairs",
                textOutput("monthTitle4Pairs"),
                plotOutput("pairsplot"))
          )
        )
```

14.5.3 The server section, including reactive elements

This section starts with a series of specifically *reactive* functions using the `reactive` function. These are functions that are used in the main output (or in other `reactive` functions) that change the data. Here we see the following reactive functions:

- `mod()` : reads the `input$xvar` and `input$yvar` and creates a linear model
- `sierraMonth()`: reads the `input$month` and filters the month and selects all the relevant variables
- `sierradf()`: runs `mod()` and assigns `RESIDUAL` and `PREDICTION` variables
- `sierraSp()`: responds to `sierradf()` by then creating an sf
- `eqn()`: responds to `mod()` by changing the model character string to put on a map

```
server <- function(input, output) {
  mod <- reactive({
    lm(as.formula(paste(input$yvar, '~', input$xvar)), data=sierraMonth())
  })
  sierraMonth <- reactive({
    monthsel <- 201000 + as.numeric(input$month)
    sierraAllMonths %>%
      filter(DATE == monthsel) %>%
      dplyr::select(STATION,ELEVATION,LATITUDE,LONGITUDE,TEMPERATURE,PRECIPITATION)
  })
  sierradf <- reactive({
    sierraMonth() %>%
      mutate(RESIDUAL = resid(mod()), PREDICTION = predict(mod()))
  })
  sierraSp <- reactive(st_as_sf(sierradf(),
                             coords=c("LONGITUDE","LATITUDE"),crs=4326))
```

```
eqn <- reactive({
  cc = mod()$coefficient
  paste(input$yvar, " =", paste(round(cc[1],2), "+",
                              paste(round(cc[-1], digits=3),
                              sep="*", collapse=" + ",
                              paste(input$xvar))))
})
```

Then the outputs, each with a `render____` function:

- `output$view`: a Leaflet map for the View tab
- `output$map`: a ggplot map for the Map tab
- `output$scatterplot`: a scatter plot and trend line for the Model tab
- `output$model`: a summary of the lm
- `output$eqntext`: information on the model for the Table tab
- `output$monthTitle4Model`: a title for the Model tab
- `output$monthTitle4Pairs`: a title for the Pairs tab
- `output$table`: the table for the Table tab
- `output$pairsplot`: the pairs plot for the Pairs tab

```
output$view <- renderLeaflet({
  providerTiles <- providers$OpenStreetMap
  if(input$LeafletBasemap=="imagery") {
    providerTiles <- providers$Esri.WorldImagery}
  if(input$LeafletBasemap=="natgeo") {
    providerTiles <- providers$Esri.NatGeoWorldMap}
  leaflet(data = sierradf()) %>%
    addTiles() %>%
    addProviderTiles(providerTiles) %>%
    addMarkers(~LONGITUDE, ~LATITUDE,
      popup = ~str_c(ELEVATION,"m ", month.name[as.numeric(input$month)], ": ",
                  TEMPERATURE, "°C ", PRECIPITATION, "mm"),
      label = ~STATION)
})
output$map <- renderPlot({
  subTitle <- ""
  if((input$var == "RESIDUAL")|(input$var == "PREDICTION")){
    subTitle <- eqn()}
  v <- get(paste(input$var), pos=sierradf())  # just to be able to use the vector
  sierrabasemap +
    geom_sf(mapping = aes(color = !!input$var), data=sierraSp(), size=4) +
    coord_sf(xlim = c(bounds[1], bounds[3]), ylim = c(bounds[2],bounds[4]))  +
    scale_color_gradient2(low="blue", mid="ivory2", high="darkred",
                        midpoint=mean(v)) +
    labs(title=paste(month.name[as.numeric(input$month)], input$var),
        subtitle=subTitle) + theme(legend.position = c(0.8, 0.85))
})
output$scatterplot <- renderPlot({
```

```
    ggplot(data = sierradf()) +
      geom_point(mapping = aes(x = !!input$xvar, y = !!input$yvar)) +
      geom_smooth(mapping = aes(x = !!input$xvar, y = !!input$yvar), method="lm") +
      labs(title=month.name[as.numeric(input$month)])
  })
  output$model <- renderPrint({
    print(eqn())
    summary(mod())
  })
  output$eqntext <- renderText(paste(month.name[as.numeric(input$month)],
               "data. Residual and Prediction based on linear model: ", eqn()))
  output$monthTitle4Model <- renderText(month.name[as.numeric(input$month)])
  output$monthTitle4Pairs <- renderText(month.name[as.numeric(input$month)])
  output$table <- renderTable(sierradf())
  output$pairsplot <- renderPlot({
    sierradf() %>%
      dplyr::select(LATITUDE,LATITUDE,LONGITUDE,
                    ELEVATION,TEMPERATURE,PRECIPITATION) %>%
      pairs(upper.panel = panel.cor)
  })}
```

14.5.4 Calling shinyApp with the ui and server function results

And finally calling both ui and server functions with shinyApp:

```
shinyApp(ui = ui, server = server)
```

14.6 A MODIS Fire App with Web Scraping and `observe` with `leafletProxy`

We'll now look at another app, one that uses web scraping and an `observe` function that together with `leafletProxy` allows the map to maintain its scaling when we change the data.

The MODIS satellite sensor includes a fire detection layer that the USFS hosts in a way that can be accessed by (some pretty primitive) web scraping, which you can see in the code. This R Markdown document is made interactive using Shiny, and employs an Inline Application with the complete Shiny app code included (Figure 14.14). To learn more about the MODIS product from NASA, see https://modis.gsfc.nasa.gov/data/dataprod/mod14.php

The MODISfire app.R script is also in the data package, in the `MODISfire` folder, and can be run with `shiny::shinyAppDir(ex("MODISfire"))`, which simply runs the following installed as a function with no input parameters, creating a separate app window.

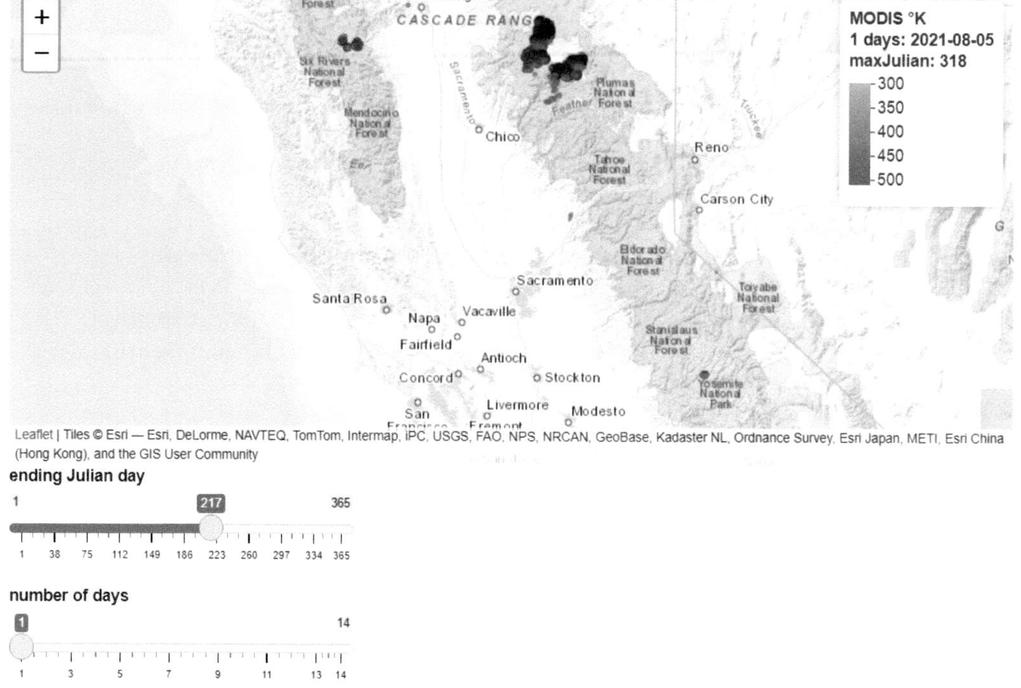

ending Julian day

1 217 365

1 38 75 112 149 186 223 260 297 334 365

number of days

1 14

1 3 5 7 9 11 13 14

year

2009 2021

2009 2011 2013 2015 2017 2019 2021

Jerry Davis, SFSU Institute for Geographic Information Science. Data source: USDA Forest Service https://fsapps.nwcg.gov/afm/gisdata.php

FIGURE 14.14 MODIS fire detection Shiny app

Note that it takes a while for the map to appear because the data is being downloaded and processed. Every time you change to a new year, there is also a delay as new data is downloaded and processed.

The code shown can also be used to create a Shiny web app by saving it as `app.R` in an RStudio Shiny app project (or just copying the `app.R` file from the `MODISfire` folder in `extdata`). It will create a `"MODISdata"` folder to hold data it downloads, to keep the location with Shiny Rmd files uncluttered with the downloaded shapefile data.

14.6.1 Setup code

Note that we'll create a folder to hold data we download, and make it the working directory.

```
library(sf); library(leaflet); library(tidyverse); library(lubridate)
library(shiny); library(here)
```

```
dataPath <- paste(here::here(),"/MODISdata",sep="")
if (!file.exists(dataPath)){dir.create(dataPath)}
setwd(dataPath)
pal <- colorNumeric(c("orange", "firebrick4"),domain=300:550)
```

14.6.2 ui

In terms of complexity of inputs, this app is pretty straightforward – the fluidPage is pretty standard for the ui section.

```
ui <- fluidPage(
    titlePanel("MODIS fire detections from temperature anomalies"),
    leafletOutput("view"),
    sliderInput(inputId = "end_jday",
                label = "ending Julian day",
                value = yday(now()), min=1, max=365, step=1),
    sliderInput(inputId = "numdays",
                label = "number of days",
                value = 1, min=1, max=14, step=1),
    sliderInput(inputId = "year",
                label = "year",
                value = year(now()),min=2009,max=year(now()),step=1,sep=""),
    helpText(paste("Jerry Davis, SFSU IGISc",
    "Data source: USDA Forest Service https://fsapps.nwcg.gov/afm/gisdata.php"))
)
```

14.6.3 Using `observe` and `leafletProxy` to allow changing the date while retaining the map zoom

In the server section, we mostly just have the leaflet map, but we need to build in code to do the web scraping and allow changing the date while retaining the zoom.

Web scraping was used to download data for a given year from the USFS. Hopefully their file naming convention stays the way it is; you can see how the shapefile name string is built with wildcard dots for characters.

`observe` and `leafletProxy`: Note that in the `renderLeaflet` section, only the year is reactive so this runs only initially or when the year changes. The `observe` function doesn't read the data anew, but just changes some parameters about the map. The trick to getting this to work was then to use `leafletProxy` to make changes to the map without recreating it. Before I figured this out, the map would start over at its beginning point and you couldn't then change the date for an area you just zoomed to. This took a while to figure out...

```
server <- function(input, output, session) {
  output$view <- renderLeaflet({ # Only year is reactive, so runs w/year change
    yrst <- as.character(input$year)
    txt <- read_file(str_c("https://fsapps.nwcg.gov/afm/data/fireptdata/modisfire_",
                           yrst,"_conus.htm"))
    shpPath <- str_extract(txt,
                           paste0("https://fsapps.nwcg.gov/afm/data/fireptdata/",
                                  "modis_fire_.........conus_shapefile.zip"))
    shpZip <- str_extract(shpPath, "modis_fire_.........conus_shapefile.zip")
    MODISfile <- str_c(dataPath,"/",str_extract(shpZip,
      "modis_fire_.........conus"),".shp")
    if(yrst == as.character(year(now())) | !file.exists(MODISfile)) {
        shpZipPath <- str_c(dataPath, "/",shpZip)
        download.file(shpPath, shpZipPath)
        unzip(shpZipPath, exdir=dataPath) }
    fires <<- st_read(MODISfile)
    leaflet() %>%
        addProviderTiles(providers$Esri.WorldTopoMap) %>%
        fitBounds(-123,37,-120,39)
    })
    observe({                    # Allows the map to retain its location and zoom
        numdays <- input$numdays; end_jday <- input$end_jday
        fireFilt <- filter(fires,between(JULIAN, end_jday - numdays, end_jday))
        yrst <- as.character(input$year)
        dat <- as.Date(end_jday-1, origin=str_c(yrst,"-01-01"))  # Julian day fix
        leafletProxy("view", data = fireFilt) %>%
            clearMarkers() %>%
            addCircleMarkers(
                radius = ~(TEMP-250)/50,  # scales 300-500 from 1:5
                color = ~pal(TEMP),
                stroke = FALSE, fillOpacity = 0.8) %>%
            clearControls() %>%   # clears the legend
            addLegend("topright", pal=pal, values=~TEMP, opacity=0.6,
                title=str_c("MODIS °K</br>",numdays, " days: ",
                           dat,"</br>maxJulian: ",as.character(max(fires$JULIAN))))
    })
}
shinyApp(ui = ui, server = server, options = list(width = "100%", height = 800))
```

14.7 Learn More about Shiny Apps

You should explore these resources to learn more about creating useful Shiny web apps:

- https://shiny.rstudio.com/images/shiny-cheatsheet.pdf
- The Shiny tutorial at https://shiny.rstudio.com/tutorial/, worth spending some time

on to learn more about getting Shiny apps working. There's a lot to understand about reactivity for instance.

To learn more about interactive documents in R Markdown, see Interactive Documents[2].

14.8 Exercises: Shiny

Exercise 14.1. Get either the sierra or MODIS fire Shiny app working, using the code provided.

Exercise 14.2. Build a Shiny document or app using your own favorite code developed earlier in this book.

[2]http://rmarkdown.rstudio.com/authoring_shiny.html

References

Applied California Current Ecosystem Studies. n.d. https://pointblue.org.

Ballard, Grant, Annie E Schmidt, Viola Toniolo, Sam Veloz, Dennis Jongsomjit, Kevin R Arrigo, and David G Ainley. 2019. "Fine-Scale Oceanographic Features Characterizing Successful Adélie Penguin Foraging in the SW Ross Sea." *Marine Ecology Progress Series*. https://doi.org/10.3354/meps12801.

Berry, Brian, and Duane Marble. 1968. *Spatial Analysis: A Reader in Statistical Geography*. Prentice-Hall.

Blackburn, Darren A, Andrew J Oliphant, and Jerry D Davis. 2021. "Carbon and Water Exchanges in a Mountain Meadow Ecosystem, Sierra Nevada, California." *Wetlands* 41 (3): 1–17. https://doi.org/10.1007/s13157-021-01437-2.

Brown, Christopher. n.d. *"R Accessors Explained"*. https://www.r-bloggers.com/2009/10/r-accessors-explained/.

Calculate Distance, Bearing and More Between Latitutde/Longitude Points". n.d. Movable Type Ltd. https://www.movable-type.co.uk/scripts/latlong.html.

Clover Valley Ranch Restoration, the Sierra Fund. n.d. https://sierrafund.org/clover-valley-ranch/.

Cohen, Jacob. 1960. "A Coefficient of Agreement for Nominal Scales." *Educational and Psychological Measurement* 20. https://doi.org/10.1177/001316446002000104.

Copernicus Open Access Hub. n.d. European Space Agency - ESA. https://scihub.copernicus.eu/.

Davis, JD, P Amato, and R Kiefer. 2001. "Soil Carbon Dioxide in a Summer-Dry Subalpine Karst, Marble Mountains, California, USA." *Zeitschrift Für Geomorphologie N.F.* 45 (3): 385–400. https://www.researchgate.net/publication/258333952_Soil_carbon_dioxide_in_a_summer-dry_subalpine_karst_Marble_Mountains_California_USA.

Davis, JD, L Blesius, M Slocombe, S Maher, M Vasey, P Christian, and P Lynch. 2020. "Unpiloted Aerial System (UAS)-Supported Biogeomorphic Analysis of Restored Sierra Nevada Montane Meadows." *Remote Sensing* 12. https://www.mdpi.com/2072-4292/12/11/1828.

Davis, JD, and GA Davis. 2001. "A Microcontroller-Based Data-Logger Design for Seasonal Hydrochemical Studies." *Earth Surface Processes and Landforms* 26 (10): 1151–59. https://doi.org/10.1002/esp.262.

Davis, Jerry. n.d. *San Pedro Creek Watershed Virtual Fieldtrip: Story Map*. https://storymaps.arcgis.com/stories/62705877a9f64ac5956a64230430c248.

Davis, Jerry D, and George A Brook. 1993. "Geomorphology and Hydrology of Upper Sinking Cove, Cumberland Plateau, Tennessee." *Earth Surface Processes and Landforms* 18 (4): 339–62. https://doi.org/10.1002/esp.3290180404.

Davis, Jerry, and Leonhard Blesius. 2015. "A Hybrid Physical and Maximum-Entropy Landslide Susceptibility Model." *Entropy* 17 (6): 4271–92. https://www.mdpi.com/1099-4300/17/6/4271.

Ellen, Stephen D, and Gerald F Wieczorek. 1988. *Landslides, Floods, and Marine Effects of the Storm of January 3-5, 1982, in the San Francisco Bay Region, California*. Vol. 1434. USGS. https://pubs.usgs.gov/pp/1988/1434/.

EPSG Geodetic Parameter Dataset. n.d. https://en.wikipedia.org/wiki/EPSG_Geodetic_Parameter_Dataset.

European Fluxes Database. n.d. http://www.icos-etc.eu/home.

Helsel, Dennis R., Robert M. Hirsch, Karen R. Ryberg, Stacey A. Archfield, and Edward J. Gilroy. 2020. "Statistical Methods in Water Resources." In *Hydrologic Analysis and Interpretation*. Reston, Virginia: U.S. Geological Survey. https://pubs.usgs.gov/tm/04/a03/tm4a3.pdf.

Hijmans, Robert J. n.d. *Spatial Data Science*. https://rspatial.org.

Horst, Allison Marie, Alison Presmanes Hill, and Kristen B Gorman. 2020. *Palmerpenguins: Palmer Archipelago (Antarctica) Penguin Data*. https://allisonhorst.github.io/palmerpenguins/.

"Hysteresis." n.d. https://en.wikipedia.org/wiki/Hysteresis.

Irizarry, Rafael A. 2019. *Introduction to Data Science: Data Analysis and Prediction Algorithms with r*. CRC Press. https://cran.r-project.org/package=dslabs.

Johnston, Myfanwy, and Bob Rudis. n.d. *Visualizing Fish Encounter Histories*. https://fishsciences.github.io/post/visualizing-fish-encounter-histories/.

Lovelace, Robin, Jakuv Nowosad, and Jannes Muenchow. 2019. *Geocomputation with r*. CRC Press. https://geocompr.robinlovelace.net/.

Marine Debris Program. n.d. NOAA Office of Response; Restoration. https://marinedebris.noaa.gov/.

Nowosad, Jakub. n.d. *Geostatistics in r*. https://bookdown-org.translate.goog/nowosad/geostatystyka/?_x_tr_sl=pl&_x_tr_tl=en&_x_tr_hl=pl.

Pebesma, Edzer. n.d. *Gstat: Spatial and Spatio-Temporal Geostatistical Modelling, Prediction and Simulation*. https://cran.r-project.org/web/packages/gstat/index.html.

Powell, Cynthia, Leonhard Blesius, Jerry Davis, and Falk Schuetzenmeister. 2011. "Using MODIS Snow Cover and Precipitation Data to Model Water Runoff for the Mokelumne River Basin in the Sierra Nevada, California (2000–2009)." *Global and Planetary Change* 77 (1-2): 77–84. https://doi.org/10.1016/j.gloplacha.2011.03.005.

Simple Features for r. n.d. https://r-spatial.github.io/sf/.

Sims, Stephanie. 2004. *Hillslope Sediment Source Assessment of San Pedro Creek Watershed, California*. https://geog.sfsu.edu/theses/.

Studwell, Anna, Ellen Hines, Meredith L Elliott, Julie Howar, Barbara Holzman, Nadav Nur, and Jaime Jahncke. 2017. "Modeling Nonresident Seabird Foraging Distributions to Inform Ocean Zoning in Central California." *PLoS ONE*. https://doi.org/10.1371/journal.pone.0169517.

Thiessen, A. 1911. "Precipitation Averages for Large Areas." *Monthly Weather Review* 39 (7): 1082–89.

Thompson, A, JD Davis, and AJ Oliphant. 2016. "Surface Runoff and Soil Erosion Under Eucalyptus and Oak Canopy." *Earth Surface Processes and Landforms*. https://doi.org/10.1002/esp.3881.

Tomlin, C Dana. 1990. *Geographic Information Systems and Cartographic Modeling*. Englewood Cliffs, N.J: Prentice Hall.

Tukey, John W. 1962. "The Future of Data Analysis." *The Annals of Mathematical Statistics* 33 (1): 1–67.

———. 1977. *Exploratory Data Analysis*. Reading, Mass: Addison-Wesley.

Voronoi, G. 1908. "Nouvelles Applications Des Paramètres Continus à La Théorie de Formes Quadratiques." *Journal Für Die Reine Und Angewandte Mathematik* 134: 198–287.

Wang, Earo, Dianne Cook, and Rob J Hyndman. 2020. "A New Tidy Data Structure to Support Exploration and Modeling of Temporal Data." *Journal of Computational and Graphical Statistics* 29 (3): 466–78. https://doi.org/10.1080/10618600.2019.1695624.

Wickham, Hadley, and Garrett Grolemund. 2016. *R for Data Science: Visualize, Model,*

Transform, Tidy, and Import Data. O'Reilly Media, Inc. https://www.tidyverse.org/learn/.

Xie, Yihui. 2021. *Bookdown: Authoring Books and Technical Documents with r Markdown*. Boca Raton, Florida: Chapman; Hall/CRC. https://bookdown.org/yihui/bookdown/.

Xie, Yihui, JJ Allaire, and Garrett Grolemund. 2019. *R Markdown: The Definitive Guide*. 1st ed. Boca Raton, Florida: Chapman; Hall/CRC. https://bookdown.org/yihui/rmarkdown/.

Yang, W, H Kobayashi, C Wang, J Shen M abd Chen, B Matsushita, Y Tang, Y Kim, et al. 2019. "A Semi-Analytical Snow-Free Vegetation Index for Improving Estimation of Plant Phenology in Tundra and Grassland Ecosystems." *Remote Sensing of Environment* 228: 31–44. https://doi.org/10.1016/j.rse.2019.03.028.

Index

For Product Safety Concerns and Information please contact our EU representative GPSR@taylorandfrancis.com Taylor & Francis Verlag GmbH, Kaufingerstraße 24, 80331 München, Germany